T0183375

Lecture Notes
in Business Information Processing   **389**

More information about this series at http://www.springer.com/series/7911

Witold Abramowicz · Gary Klein (Eds.)

Business
Information Systems

23rd International Conference, BIS 2020
Colorado Springs, CO, USA, June 8–10, 2020
Proceedings

 Springer

Editors
Witold Abramowicz ⓘ
Poznań University of Economics
and Business
Poznan, Poland

Gary Klein ⓘ
University of Colorado
Colorado Springs, CO, USA

ISSN 1865-1348   ISSN 1865-1356   (electronic)
Lecture Notes in Business Information Processing
ISBN 978-3-030-53336-6   ISBN 978-3-030-53337-3   (eBook)
https://doi.org/10.1007/978-3-030-53337-3

This Springer imprint is published by the registered company Springer Nature Switzerland AG
The registered company address is: Gewerbestrasse 11, 6330 Cham, Switzerland

Preface

During the 23 years of the International Conference on Business Information Systems (BIS), it has grown to be a well-renowned event of the scientific community. Every year the conference joins international researchers for scientific discussions on modelling, development, implementation, and application of business information systems based on innovative ideas and computational intelligence methods.

The 23rd edition of the BIS conference was co-organized by the BIS Steering Committee and the University of Colorado, USA. Due to the COVID-19 pandemic, the conference was organized fully online this year.

The growth of big data and ever-increasing connectivity combine to present both unique opportunity and risky exposure. Access to enormous quantities of data leads organizations to an interest in exploiting big data in a smart way. However, due to the attractiveness of accessible data stores, organizations must battle increased threats to data security. A broad approach allowing the extraction of valuable knowledge from data is required for companies to make profits, be more competitive, and survive in the even more dynamic and fast-changing environment, all while protecting the data that assures their competitive edge, contains their intellectual property, and countenances their continued existence.

Data science and data security are professions of today and the future, each requiring its unique set of knowledge and body of practice. Both rely on multiple fields of study, such as mathematics, statistics, economics, psychology, and information systems, and use various scientific and practical methods, tools, and systems. The key objective is to extract valuable information and infer knowledge from data for multiple purposes, starting from decision making, through to product development, and up to trend analysis and forecasting. Throughout the acquisition, storage, and manipulation of data, organizations must however secure their data and system assets. Security requires effective technology and practices that deter, detect, and correct security violations, and an understanding of behaviors both benevolent and malevolent.

The BIS 2020 conference fostered the multidisciplinary discussion about data science and security from both the scientific and practical sides, and its impact on current enterprises. Thus, the theme of the BIS 2020 was "Data Science and Security in Business Information Systems." Our goal was to inspire researchers to share theoretical and practical knowledge of the different aspects related to data science and security, and to help them transform their ideas into the innovations of tomorrow.

The first part of the BIS 2020 proceedings is dedicated to the topics of Data Security, Big Data and Data Science, and Artificial Intelligence. This is followed by other research directions that were discussed during the conference, including ICT Project Management, Applications, Social Media, and Smart Infrastructures. Finally, the proceedings end with two papers from the BIS 2019 conference that have not been published in the previous BIS proceedings.

The Program Committee of BIS 2020 consisted of 91 members who carefully evaluated all the submitted papers. Based on their extensive reviews, 30 papers were selected.

We would like to thank everyone who helped to build an active community around the BIS conference. First of all, we want to express our appreciation to the reviewers for taking the time and effort to provide insightful comments. We wish to thank all the authors who submitted their papers as well as all the participants of BIS 2020.

June 2020

Witold Abramowicz
Gary Klein

Organization

BIS 2019 was organized by the Steering Committee and the University of Colorado, USA.

Steering Committee

Witold Abramowicz (Co-chair)	Poznań University of Economics and Business Poznan, Poland
Gary Klein (Co-chair)	University of Colorado, Colorado Springs, CO, USA
Rainer Alt	Leipzig University, Germany
Adrian Paschke	Freie Universität Berlin, Germany
Kurt Sandkuhl	Rostock University, Germany

Program Committee Members

Dimitris Apostolou	University of Piraeus, Greece
Timothy Arndt	Cleveland State University, USA
Gustavo Arroyo-Figueroa	Instituto Nacional de Electricidad y Energias Limpias, Mexico
Sören Auer	TIB Leibniz Information Center Science & Technology and University of Hannover, Germany
Óscar Javier Ávila Cifuentes	University of Los Andes, Colombia
Eduard Babkin	LITIS Laboratory, INSA, France, and TAPRADESS Laboratory, State University – Higher School of Economics, Russia
Akhilesh Bajaj	University of Tulsa, USA
Rafael Batres	Tecnológico de Monterrey, Mexico
Morad Benyoucef	University of Ottawa, Canada
Matthias Book	University of Iceland, Iceland
Dominik Bork	University of Vienna, Austria
Nizar Bouguila	Concordia University, Canada
Alberto Cano	Virginia Commonwealth University, USA
François Charoy	Université de Lorraine, LORIA, Inria, France
Rafael Corchuelo	University of Seville, Spain
Beata Czarnacka-Chrobot	Warsaw School of Economics, Poland
Sergiu Dascalu	University of Nevada, Reno, USA
Christophe Debruyne	Trinity College Dublin, Ireland
Suzanne Embury	The University of Manchester, UK
Werner Esswein	Technische Universität Dresden, Germany
Joerg Evermann	Memorial University of Newfoundland, Canada

Agata Filipowska	Poznań University of Economics and Business, Poland
Ugo Fiore	Federico II University, Italy
Adrian Florea	University of Sibiu, Romania
Paul Fodor	Stony Brook University, USA
Philippe Fournier-Viger	Harbin Institute of Technology, China
Naoki Fukuta	Shizuoka University, Japan
Valeriy Gavrishchaka	West Virginia University, USA
Claudio Geyer	UFRGS, Brazil
Alfonso González Briones	BISITE Research Group, Spain
Jaap Gordijn	Vrije Universiteit Amsterdam, The Netherlands
Volker Gruhn	Universität Duisburg-Essen, Germany
Hele-Mai Haav	Institute of Cybernetics at Tallinn University of Technology, Estonia
Inma Hernandez	University of Seville, Spain
Constantin Houy	Institute for Information Systems at DFKI (IWi), Germany
Christian Huemer	Vienna University of Technology, Austria
Björn Johansson	Lund University, Sweden
Monika Kaczmarek	Universität Duisburg-Essen, Germany
Pawel Kalczynski	California State University, Fullerton, USA
Marite Kirikova	Riga Technical University, Latvia
Ralf Klischewski	German University in Cairo, Egypt
Ralf Knackstedt	University of Hildesheim, Germany
Jun Kong	North Dakota State University, USA
Agnes Koschmider	Karlsruhe Institute of Technology, Germany
Dalia Kriksciuniene	Vilnius University, Lithuania
Chahrazed Labba	Université de Lorraine, France
Elżbieta Lewańska	Poznań University of Economics and Business, Poland
Da-Yin Liao	National Taiwan University, Taiwan
Jun-Lin Lin	Yuan Ze University, Taiwan
Julie Yu-Chih Liu	Yuan Ze University, Taiwan
Peter Lockemann	Universität Karlsruhe, Germany
Fabrizio Maria Maggi	Institute of Computer Science, University of Tartu, Estonia
Andrea Marrella	Sapienza University of Rome, Italy
Eric T. Matson	Purdue University, USA
Raimundas Matulevicius	University of Tartu, Estonia
Andreas Oberweis	Karlsruhe Institute of Technology, Germany
Toacy Cavalcante de Oliveira	COPPE/UFRJ, Brazil
Eric Paquet	National Research Council, Canada
Placide Poba-Nzaou	ESG-UQAM, Canada
Birgit Proell	FAW, Johannes Kepler University Linz, Austria
Luise Pufahl	Hasso Plattner Institute, University of Potsdam, Germany

Elke Pulvermueller	Institute of Computer Science, University of Osnabrueck, Germany
António Rito Silva	Universidade de Lisboa, Portugal
David Romero	Tecnológico de Monterrey, Mexico
Duncan Ruiz	Pontificia Universidade Católica do Rio Grande do Sul, Brazil
Virgilijus Sakalauskas	Vilnius University, Lithuania
Demetrios Sampson	Curtin University, Australia
António Rito Silva	Universidade de Lisboa, Portugal
Elmar Sinz	University of Bamberg, Germany
Alexander Smirnov	SPIIRAS, Russia
Stefan Smolnik	University of Hagen, Germany
Davide Sottara	Mayo Clinic, USA
Dimitris Spiliotopoulos	University of Houston, USA
Milena Stróżyna	Poznań University of Economics and Business, Poland
Bob Travica	University of Manitoba, Canada
Herve Verjus	Université Savoie Mont Blanc, LISTIC, France
Krzysztof Węcel	Poznań University of Economics and Business, Poland
Hans Weigand	Tilburg University, The Netherlands
Mathias Weske	HPI, University of Potsdam, Germany
Anna Wingkvist	Linnaeus University, Sweden
Ouri Wolfson	University of Illinois, USA

Organizing Committee

Milena Stróżyna (Chair)	Poznań University of Economics and Business, Poland
Dan Lemack	University of Colorado, USA
Piotr Kałużny	Poznań University of Economics and Business, Poland
Elżbieta Lewańska	Poznań University of Economics and Business, Poland
Włodzimierz Lewoniewski	Poznań University of Economics and Business, Poland

Additional Reviewers

Christian Anschütz
Alina Bockshecker
Kuang-Ting Cheng
Jakub Ciesiółka
Julia Couto
Dominik Filipiak
Tomasz Granosik
Olivia Hornung
Rui Huang
Dominik Janssen
Piotr Kałużny
Vahid Khorasani Ghassab
Izabella Krzemińska
Meriem Laifa
Juarez Monteiro
Neeraj Parolia

Sigurour Gauti Samuelsson
Yide Shen
Hua Sun
Marcin Szmydt

Laura Angelica Tomaz da Silva
Kangning Wei
Nuha Zamzami
Grzegorz Łuczyna

Contents

Smart Infrastructures

BIS 2019

Data Security

Legal Requirement Elicitation, Analysis and Specification for a Data Transparency System

Christian Janßen$^{(\boxtimes)}$ (iD) and Jonas Kathmann (iD)

Department Very Large Business Applications, University of Oldenburg,
Ammerländer Heerstr. 114-118, 26129 Oldenburg, Germany
{christian.janssen,jonas.kathmann}@uni-oldenburg.de

Abstract. Within the growing amount of data through new applications, processes and technologies in companies, legal frameworks according to the processing of data become more important. The new General Data Protection Regulation (GDPR) especially has the intention, to strengthen the rights of Data Subjects in transparency (e.g. Art. 12) and self-control (e.g. Art 15–22). This research aims to develop non-functional-requirements (NFR) for a Data Transparency System for the category legal-contractual. Therefore, we follow the requirement engineering process according to Rupp [29]. As a general source for the development, qualitative expert interviews have been carried out. In order to extend our findings and form categories, we also did a systematic literature review and a structured text analysis of the GDPR. In total, we were able to generate 18 NFR and organized them into the categories Purpose, Obligation, Ownership, Procedures and Integrity and Transparency.

Keywords: Requirements Engineering · Non-functional-requirements · Data Transparency System · GDPR

1 Introduction

Data is rapidly becoming a universal currency of our economy, a digital good whose value does not diminish with use and whose benefits are realized only when it can flow [22]. In general, data is a driving innovation success factor in enterprises [19]. Within the growing amount of data through new applications, processes and technologies in companies, legal frameworks according to the processing of data becomes more important. The European Union (EU) decided to reform the basic data protection regulation from the 1990s and adopted the new GDPR on the 25th May 2016. The EU member states obligated themselves, after the ratification to apply the new law in May 2018. Through the GDPR, a multiplicity of changes and reforms became effective. New definitions of roles allow a clear assignment of tasks and duties. For example, a Data Subject could be an individual, who's personal data is regulated under the GDPR [1]. A Data

© Springer Nature Switzerland AG 2020
W. Abramowicz and G. Klein (Eds.): BIS 2020, LNBIP 389, pp. 3–17, 2020.
https://doi.org/10.1007/978-3-030-53337-3_1

Controller could be a person or legal entity that controls both purpose and means of personal data processing [1]. In combination, a Data Processor processes personal data on behalf of the Data Controller [1]. A new definition of personal data is given as follow: *"Personal data means any information relating to an identified or identifiable natural person (Data Subject); an identifiable natural person is one who can be identified, directly or indirectly, in particular by reference to an identifier such as a name, an identification number, location data, an online identifier or to one or more factors specific to the physical, physiological, genetic, mental, economic, cultural or social identity of that natural person"* [1]. This leads to new responsibilities for Data Controller and stronger rights for Data Subjects. For example, the principle Art. 5 GDPR - lawfulness, fairness and transparency - determines that personal data should be processes lawfully, fairly and transparent in a way which is reasonable for the data subject [1]. As a result, Art. 12 GDPR demonstrates transparent information, communication and modalities for the exercise of the rights of the Data Subject. These rights are specified in Art. 13, 14 GDPR for ex ant transparency, Art. 15 GDPR for ex post transparency and Art. 16–21 GDPR for general purposes [1]. However, it must be mentioned, that the EU regulation does not treat personal data as property so far. As a consequence, e.g. Data Ownership is a highly ongoing contentious legal debate. The emergence in the data-driven economy and the adoption of new legal frameworks has resulted in the need for solutions, which supports on the one side, Data Subjects to make self-responsible decisions to his data base and on the other side, Data Controller to comply with legal framework conditions in a transparent manner. To face these challenges, a Data Transparency System will be developed in an ongoing doctoral project to assists, the different parties that are involved in the data processing of personal data. Following the design science research process of Pfeffers et al. [26] in the doctoral project, we elaborated and published, as a first step, the problem definition and motivation and, as a second step, the goals of the system in [18]. The next step will be the design and development of a system model and the prototype implementation. Requirements are needed to describe the range of the system. Since the elicitation, analysis and specification of functional requirements (FR), and non-functional requirements (NFR) on a technical and legal level, would go beyond the scope of this paper, we decided to focus on NFR inside this work. Therefore, the following research question (RQ) will be implemented: **What are the non-functional requirements from the point of view of Data Protection, Data Ownership and Data Transparency for a Data Transparency System?** The rest of the paper is structured as follows: For a better understanding of different parts of the Data Transparency System, Sect. 2 gives a short overview of components, that could be used within a business environment. Section 3 describes the research methodology for the requirements elicitation, analysis and specification. Section 4 shows the results and findings, but also, a discussion and classification of the requirements. Finally, Sect. 5 concludes the paper.

2 Data Transparency System

The implementation of the Data Transparency System as a technical system is oriented on the design science research process according to Pfeffers et al. [26]. The Fig. 1 gives a short overview about different components of the system.

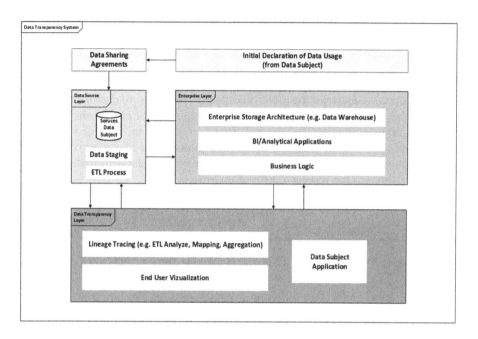

Fig. 1. Data Transparency System components

As a central element of the system, the development of monitoring components for the purpose of tracing data streams will be considered. On the one side, a Data Subject should be supported to make independent decisions regarding its data base. On the other side, Data Controller should have the ability to interact with Data Subjects and in general to comply with legal framework conditions and compliance policies. Therefore, the Data Subjects needs to give an initial declaration on data usage, which will be transferred into Data Sharing Agreements and handled together with the data in the Data Source Layer. Consent and constraints on data usage will be directly added to the data and available in the different layers of the system. Within the Enterprise Layer, components such as a Data Warehouse, Analytical Applications or general Business Logic could be used. As a cross sectional component, the Data Transparency Layer includes the Lineage Tracing (e.g. Data Lineage) and the End User Visualization (e.g. Data Maps, Table Lineage Info). Besides this, the main component in this section, the Data Subject Application joins all the information of data-related interaction processes between the involved entities and gives especially the Data Subject the possibility for self-responsible decisions making.

3   Research Methodology

The requirement elicitation, analysis and specification are embedded in the Requirements Engineering discipline. In particular, Requirement Engineering is a systematic approach to specify and manage requirements with the aim of knowing the relevant requirements, satisfying wishes and needs of stakeholders and concretizing the requirements [28]. A requirement is a documented necessary characteristic or ability of a system, that solves a problem or achieve a specific goal [16]. In other words, requirements are the basis for the creation of a system model or a system architecture. Therefore, the requirements as qualitative criteria for good requirements should be complete, atomic, technically solution-neutral, consistent, testable, necessary, traceable, realizable and clear [29]. For a better structure and categorizing, requirements can be divided into functional and non-functional aspects. A FR is a requirement, relating to a result or behaviour that will be provided by a single function of a system [27]. NFRs often described as properties or qualities that a system must provide [6]. The Fig. 2 shows the specific research process for this work.

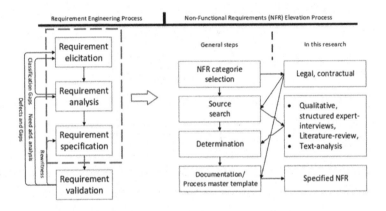

Fig. 2. Research process of the requirement elicitation, analysis and specification (in accordance with [29])

Out of the Requirement Engineering Process according to RUPP [29], the requirement elicitation, analysis and specification will be needed. For this research we selected the category legal contractual. Within the source search and determination step, we decided to collect data through qualitative structured expert interviews, a literature review and a text-analysis of the GDPR. To structure the specified NFR, we used the template process mater from Fig. 3. In their full deontic sense, the terms shall, will and should reflects degrees of accountability and responsibility. Shall requirements are mandatory in the realization. Should requirements represent the needs and wishes of different stakeholders and will requirements document future intentions.

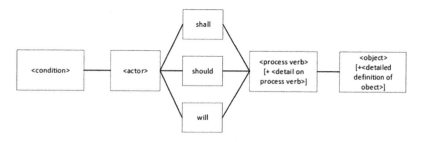

Fig. 3. Requirements process master (in accordance with [29])

The text-analysis of the GDPR with reference to transparency and self-control follows the methodical approach to Früh [10], Merten [24] and Krippendorff [21]. Besides a structured table of articles and recitals regarding to GDPR transparency and self-control as a output result (published in [18]), we were able to use additional information for the development of requirements. As a another source for the requirement extraction, a systematic structured literature review was carried out. For that reason, we used the methodology of Webster and Watson [35], Cato [5], Feak and Swales [8] and following the described process steps: 1) Definition of search strategy and items (including selection of literature-databases with focus on Computer Science), 2) Identification and selection of relevant papers (including forward search, backward search and preliminary results discussion) and 3) Analysis of relevant papers (includes specification of findings). The subsequent table gives an overview about the paper results, that was used (Table 1).

Table 1. Paper results of the literature review

Search String	Databases paper results		
	IEEE Xplore	DBLP	The collection of Computer Science Bibliographics
GDPR AND (Data Ownership OR Data Transparency)	18	11	65
Data Transparency And requirement	289	2	4
Data Ownership AND requirement	225	0	3
Data Ownership AND definition	43	1	3

Qualitative structured expert interviews were conducted as the primary source for the data collection part in this work. Because of this, the examination design is oriented to Flick [9] and Endres [7]. The method is placed in the area of qualitative research as a more structured variant as conventional interviews. The focal points of the experts are perspectives and methods of action. We decided to conduct 14 interviews to adequately cover the topic area. The Table 2 gives an overview about the acquired experts for the structured qualitative expert interviews.

Table 2. Overview of acquired experts

No.	Expert (roles)	Description
1	Data Protection Officer (company)	Two experts in this field could be interviewed. The first expert works in a medium-sized company. The second expert works for a subsidiary of a large company. Both are familiar with the legal provisions from a data protection perspective and assume the role of a Data Protection Officer.
2	Data Protection Officer (federal-state-country)	On the public side, the Data Protection Officers of the federal states of Lower Saxony and Bremen could be interviewed.
3	Senior and Junior Business Manager	As an interface between developers and users, three Junior and Senior Business Managers were interviewed. The heterogeneity of different IT areas and industries was taken into account.
4	Data Scientists	The analysis of large amounts of data requires appropriate knowledge and experience. For this reason, two Data Scientists were interviewed in order to transfer their views on the processing and analysis of personal data.
5	IT-Security Experts	Within the GDPR, IT-Security relevant measures are taken up in the course of the processing of personal data. For this purpose, an IT-Security Expert was interviewed.
6	Product Owners	A Product Owner has great knowledge in the technical area of IT products. For the interview, two product owners were interviewed who dealt with data protection aspects during the development phase of IT products.
7	Chief Executive Officer	In order to show a strategic point of view, the CEO of a medium-sized software development and IT consulting company was interviewed.
8	Project Manager	For the operative planning and execution of projects, a Project Manager of an IT consulting company with a focus on data protection, IT infrastructure and information security was interviewed.

In order to follow the structured approach, an interview guideline was developed which includes three different scenarios within the area of this research. Depending on the specific knowledge of the experts, additional questions were asked and open answers were welcomed. The period and length of the interviews was no longer than 30 min. For the evaluation and transcription of the content, we used the Tool MAXQDA. Subsequently, the generated text-basis was qualitative analyzed through the object-analysis approach according to Ginz [13].

4   Results and Discussion

In Sect. 3, we showed the data sources and research approach in order to extract our NFR for the category legal, contractual. As a result out of the expert interviews and the literature review, we have created five categories to structure our findings: Purpose, Obligation, Ownership, Procedures and Integrity and Transparency. In combination, for each category and requirement the respective expert statements, relevant articles of the GDPR and literature expressions will be given. We have also decided to divide the requirements based on the results into hypothetical (Hypo) and necessary conditions to be compliant (Comp) with the GDPR.The first category views the Purpose and is presented in Table 3.

Table 3. NFRs main category: purpose

Category	Current NFR-No.	Requirement
Purpose description for data processing	NFR1 (Comp)	For the processing of personal data, the Data Controller shall provide a precise and detailed purpose description in accordance to Art. 5 GDPR.
Consent to data processing	NFR2 (Comp)	For the transfer of personal data, a legal authorization or the consent from the Data Subject shall be given in accordance with Art. 6, 7 GDPR.

Within the GDPR context, purpose limitations with regard to data processing and the collection of personal data are pointed out. During the interviews, it became clear that purpose limitations in connection with consents are an essential component for a transparent basis of trust between Data Subjects and Data Controller. Especially the following expert statements supports this statement: *"Only the purpose, which I gave as consent, can be as suitable answer [...]"*, *"I need to know as a data subject why my data is needed and what I have to expect [...]. Therefore, the consent should be declared invalid if it is violated"*, *"The Data Subject needs to agree first on the release and transfer of personal data [...] this is alternativeness"*, *"[...] The purpose ultimately regulates the conclusion about a person. That should be regulated and protected by the GDPR. Everything that goes beyond, should not be possible"*. This could be connected with the GDPR Art. 5, 6, 7 and the recitals 32, 33, 39–50 and 171 which we extracted and analyzed during the text analysis. The second category deals with Obligations and is presented in Table 4.

The GDPR stipulates legal obligations as e.g. lawfulness basis for processing, especially for Data Controller and Data Processor. These obligations can be used to support Data Subjects rights. The interviewed experts have expressed a common opinion: *"Its all about what rights I have when I am the Data Subject [...] If this is defined properly, I have the right to request information from any Data Controller or Data Processor"*, *"As a customer of an application, data should*

Table 4. NFRs main category: obligation

Category	Current NFR-No.	Requirement
Obligation	NFR3 (Comp)	If personal data is collected from third parties, the Data Controller shall inform the Data Subject with information on the collected data in accordance with Art. 14 GDPR.
	NFR4 (Comp)	When personal data are collected directly from the Data Subject, the Data Controller shall provide the Data Subject with information on the collected data in accordance with Art. 13 GDPR.

only be collected that is really needed", "The Data Subject needs the ability to interact with e.g. the Data Controller [...]". During the text analysis, the GDPR Art. 6, 9, 13–14, 22, 46–47, 49, 89 and the recitals 60–62 were extracted and can be used to describe this more specific. The third category addresses Ownership and is presented in Table 5.

Table 5. NFRs main category: ownership

Category	Current NFR-No.	Requirement
Data Owner-ship in general	NFR5 (Hypo)	To demonstrate Data Ownership, the processing of personal data shall be based on a contract or other agreement between the Data Controller and the Data Subject in accordance with Art. 6 GDPR.

Right at the beginning, the GDPR defines what personal data is about and Data Subjects as a natural person should have control over their own personal data. As a logical response of this expression, it could be interpreted that ownership rights takes place in the new regulation. In reality, this is diluted and presented in vague terms. Questions regarding to "Who owns the Data?" or "Within the organization, how do we assign ownership?" can still not be answered through the new regulation. At least higher standards e.g. to obtain consents will be addressed more specific compared to prior legislations. We decided to develop a NFR in this area because a Data Transparency System should be able to distinguish between property arrangements. Through Data Sharing Agreements, specifications on a contractual basis can be implemented. Because of this, Data Controller and Data Processor are able to know fine-granular the actual owner of a dataset and can act within the framework of defined contractual specifications. In the interviews several statements describes this more precisely: *"We need a basic contract in which the consent and use of data is associated", "In the area of data ownership, the prerequisite is not to link everything together but to consider the purpose behind it and to coordinate this", "Due to non-existent definitions about data ownership, speculations naturally arises here [...]. This should be demonstrated and structured [...]".* In addition, also statements and text passages from GDPR Art. 6, 9–10, 23 and recitals 39–50, 171 can be useful for this purpose. The fourth category deals with Procedures and Integrity and is presented in Table 6. The term data integrity refers to the accuracy and consistency of data in information systems. The GDPR also defines a principle (Art. 5(f) GDPR) on the security side.

Table 6. NFRs main category: procedures and integrity

Category	Current NFR-No.	Requirement
Integrity of data	NFR6 (Comp)	The Data Controller shall ensure storage and processing of personal data in accordance with Art. 5, 25 and 32 GDPR.
Anonymization and pseudonymization	NFR7 (Comp)	When passing or processing personal data, the Data Controller will ensure as defined in Art. 32 GDPR appropriate procedures regarding anonymization or pseudonymization.

Data Controller and Data Processor should ensure to have appropriate security measures regarding to storage and processing of personal data. In combination, anonymization and pseudonymization can be used for guidance, Data Controller or Data Processor to masking personal data. Data that could been linked to an individual will be removed or encrypted. On the one side, by using pseudonymization, personal information can be linked back through an encryption key if needed. On the other side, anonymization procedures are permanent and cannot be reversed. During the interviews, several experts expressions supports the importance of appropriate measures: *"[...] this is related to the integrity of data. Integrity means adequate rules [...], no passing on without consent and deletion aspects [...]. These are certain points that I actually expects from legislation to prevent abuse", "If data will be transferred between Data Controller and Data Processor, the anonymization or pseudonymization should already takes place at the Data Controller [...]".* Within the GDPR, Art. 5, 25, 32, 40, 42, 89 and recitals 39, 75–79, 83 illustrates the possibility for masking personal data. Standardized procedures, especially for data processing could also useful to ensure transparency. Article 12–15 GDPR in particular calls for technical means to support the obtaining of explicit consent from Data Subjects and the provision of transparency with the respect to personal data processing and sharing [4]. In course of the interviews, different categories and areas have emerged and presented in Table 7. The first NFR section deals with transparent rights of access in the field of requests, obligations and purposes. With regard to the interviewees opinions, transparent information in a clear and understandable shape should be presented by the Data Controller. Also a detailed purpose description (comparable to Table 3) should be done: *"There should be a summary or dictionary with e.g. keywords regarding to my data. And it should give a clear overview about my consents [...]"*, *"[...] exactly what I would wish so that the trustworthiness in the processing is verifiable justified [...]"*, *"[..] full transparency what really happens to the data [...]"*. The second NFR section deals with general aspects about transparency in terms of processing and transfer. Data, that will be transferred or processed especially to third parties should always be transparent and handled in the given framework of consents. Insights from the interviews are the following: *"I want to make sure, that my data stays with me and that is not passed on to anyone without my consent"*, *"[...] I should get an exact listing of these data and who still access the data [...] and if something connects to my personal data I would like to know this"*.

In connection, the NFR sections information obligation and the enhancement of personal data with additional data have been linked to the first two sections as major concerns from the interviewees: *"I would expect a right of information, that is made transparent to me and shows which conclusions have been drawn from it"*, *"With regard to consent: As soon as it is possible to draw conclusions about an action, I must be informed"*, *"[..] a need for protection arises through the combination of data and traceability to one person. Accordingly, particular attention should be paid here if data will be enhanced through other data or non personal data"*, *"[...] but I link personal data with e.g. open data and will draw conclusions from it. In this case, transparency is needed for myself as a Data Subject"*, *"I would grant them the right to use my data to create new data. At the same time, I also want to have access to those new generated data. I have the right to know what they did with my data"*. For the transparency category in its entirety, GDPR Art. 5–6, 9–10, 12–23, 34, 46–47, 49, 89, as well as the recitals 39–50, 60–64, 171 could be used to assists the NFR and the interviewees opinions.

To join the results of the interviews and text analysis with the findings out of the literature review, the Table 8 have been created. In the further course, we will go briefly into the literature and show the reasons, why this literature was included in the creation of NFR.

1. Category Purpose
   Gantchev [11] introduces requirements regarding to the purpose and consents. Also Seinen et al. [30] and Hand [14] gives specific requirements for the lawfulness of processing requirements. In addition, consent requirements are presented in the context of control.
2. Category Obligation
   Abitebo. et al. [2] looks at regulatory frameworks that aims, to protect the rights of individuals to ensure equitable treatment of services and to bring

Table 7. NFRs main category: transparency

Category	Current NFR-No.	Requirement
Transparent rights of access	NFR8 (Comp)	In case of a Data Subject request regarding to data access, the Data Controller shall provide the Data Subject with transparent information on the processing of personal data according to Art. 15 GDPR.
	NFR9 (Comp)	Within the framework of the obligation to provide information as specified in Art. 12, 15 GDPR, the Data Controller shall provide the Data Subject with legible, clear and understandable presentations of the processing of his or her personal data.
	NFR10 (Comp)	In the case of access to personal data as defined in Art. 15 GDPR, the Data Controller shall provide the Data subject with a detailed description of the purpose for which his personal data will be processed.
Data Transparency in general	NFR11 (Comp)	When collecting personal data, the Data Controller should be able to provide at any time transparent information's on the purpose of collection to the Data Subject in accordance with Art. 5, 12, 13, 14 GDPR.
Information obligation	NFR12 (Comp)	The Data Controller shall inform as specified in Art. 13, 14 GDPR the Data Subject as soon as possible, if conclusions can be drawn from the processing of the Data Subjects personal or property data.
	NFR13 (Hypo)	As soon as conclusions can be identified during the collection of processing of personal or property data, a legal authorization from the Data Subject to process his personal data to the Data Controller is required in accordance with Art. 6 GDPR.
Data enhancement with general data	NFR14 (Comp)	Where personal data linked to other data, the Data Controller shall inform the Data Subject in accordance with Art. 13 GDPR.
	NFR15 (Comp)	If factual data is linked to other data which allows conclusions about the Data Subject, the Data Controller shall inform the Data Subject in accordance with Art. 14 GDPR.
	NFR16 (Hypo)	If personal data should linked with other data, the Data Controller shall be legally authorized by the Data Subject to processes his personal data in accordance with Art. 6 GDPR.

data transparency to data driven algorithmic processes in an early stage. In combination, Berg [3] and Wolters [36] describes requirements for property and data subject rights.

Table 8. Joint NFR results with literature review findings

Category	Requirement	Publication
Purpose	NFR1, NFR2	[11,14,30]
Obligation	NFR3, NFR4	[2,3,36]
Ownership	NFR5	[12,15,32,34]
Procedures and Integrity	NFR6, NFR7	[31,33]
Transparency	NFR8, NFR9, NFR10 NFR11, NFR12 , NFR13, NFR14 , NFR15, NFR16	[4,17,20,23,25,30]

3. Category Ownership
   For the description of Data Ownership in general, van Alstyne et al. [32] and Wallis et al. [34] could be used to describe incentive principles to promote the exchange of information and questions regarding to Data Authorship, Ownership and Responsibility. Besides this, factors such as Data Ownership are included. These can be picked up and extended through Geers [12] and Hornung et al. [15] motivations for a contractual law.
4. Category Procedures and Integrity
   Singh et al. [31] introduces a concept of decision provenance and did a tech-legal exploration into its potential for assisting accountability in algorithmic systems. Related to this, Wachter [33] try to design trustworthy and privacy enhancing systems and services. For this reason, requirements and integrity principles to inform Data Subjects have been developed.
5. Category Transparency
   For the development of requirements in the field of transparency, Meis et al. [23] developed a taxonomy of requirements and includes transparency require-ments from different sources. Murmann and Fischer-Hübner [25] did a specific view on transparency enhancing tools. In combination, Jaatinen [17], Krem-pel and Beyerer [20], Seinen et al. [30] and Bonatti et al. [4] extends this view and have been used, to develop especially the NFR transparency requirements for lawfulness of processing, personal data, consent requirements for further processing and different mechanism.

5 Conclusion

5.1 Main Contribution

Data Transparency is one of the major challenges that comes with the new GDPR. In combination, the GDPR can be also seen as a regulatory implemen-tation of subject rights that provides Data Subject a greater control over their data.

Our primary objective in this paper was to develop NFR for the category legal, contractual. Therefore we follow a research process in accordance with

Rupp [29] and used a process master for the structured documentation of the requirements. Out of the research process we did the steps requirement elicitation, analysis and specification and exploited a qualitative structured expert interview, a literature review and a text analysis of the GDPR as data sources. With regard to the research question from Sect. 1, we were able to create 16 NFR for the to be developed Data Transparency System. We divided NFR1, NFR2, NFR3, NFR4, NFR6, NFR7 in the point of view Data Protection, NFR5 in the point of view Data Ownership and NFR8, NFR9, NFR10, NFR11, NFR12, NFR13, NFR14, NFR15, NFR16 in the point of view Data Transparency.

5.2   Limitation and Future Work

Like in any research contribution, this work has also some limitations. Concerning the research process, we followed the first three steps. The last step, the requirement validation has not been included so far, because the transfer of the requirements into a prototype is still a future step. Since the defining and redefining of requirements is time-consuming, test cases should be added as part of the validation process. Users of the Data Transparency System need to know, how the system could help them to support their needs. Also we decided to elevate NFR first, because of the large thematic area.

These limitations lead also to the need of future research. FR have to be develop as a next step to describe the Data Transparency System and the functional components in more detail on a technical level. In combination, a first prototype have to be design to validate the requirements and faces the needs of the users.

References

1. Regulation (EU) 2016/679 of the European Parliament and of the Council - of 27 April 2016 - on the protection of natural persons with regard to the processing of personal data and on the free movement of such data, and repealing Directive 95/46/EC (General Data Protection Regulation) (2016)
2. Abiteboul, S., Stoyanovich, J.: Transparency, Fairness, Data Protection, Neutrality: Data Management Challenges in the Face of New Regulation (2019). arXiv:1903.03683 [c.s.]
3. Berg, C.: Privacy, property, and discovery. In: The Classical Liberal Case for Privacy in a World of Surveillance and Technological Change. PSCL, pp. 153–166. Springer,Cham (2018). https://doi.org/10.1007/978-3-319-96583-3_9
4. Bonatti, P., Kirrane, S., Polleres, A., Wenning, R.: Transparent personal data processing: the road ahead. In: Tonetta, S., Schoitsch, E., Bitsch, F. (eds.) SAFE-COMP 2017. LNCS, vol. 10489, pp. 337–349. Springer, Cham (2017). https://doi.org/10.1007/978-3-319-66284-8_28
5. Cato, P.: Einflüsse auf den Implementierungserfolg von Big Data Systemen. Verlag Dr. Kovac, Hamburg (2016)
6. Ebert, C.: Systematisches Requirements Management: Anforderungen ermitteln, spezifizieren, analysieren und verfolgen, 1. aufl edn. dpunkt-Verl, Heidelberg (2005)

7. Enderes, C.: Experteninterview. Der Leitfaden für die Bachelorarbeit (2018). https://www.bachelorprint.de/experteninterview/
8. Feak, C.B., Swales, J.M., Swales, J.M., Feak, C.B.: Telling a Research Story: Writing a Literature Review. The Michigan Series in English for Academic & Professional Purposes. University of Michigan Press, Ann Arbor Mich (2009)
9. Flick, U.: Sozialforschung: Methoden und Anwendungen: ein Überblick für die BA-Studiengänge, 3, auflage edn. Rowohlt Taschenbuch Verlag, Rororo Rowohlts Enzyklopädie (2016)
10. Früh, W.: Inhaltsanalyse: Theorie und Praxis, vol. 7. UVK, Konstanz (2011)
11. Gantchev, V.: Data protection in the age of welfare conditionality: respect for basic rights or a race to the bottom? Eur. J. Soc. Secur. **21**, 3–22 (2019)
12. Geer, D.E.: Ownership. IEEE Secur. Priv. **17**, 4 (2019)
13. Ginz, M.: Requirements Engineering I. Kapitel 4 - Anforderungsermitttlung und -analyse (2010)
14. Hand, D.J.: Aspects of data ethics in a changing world: where are we now? Big Data **6**(3), 176–190 (2018)
15. Hornung, G., Goeble, T.: "Data Ownership" im vernetzten Automobil. Computer und Recht **31**(4) (2015)
16. IEEE Standards Board: IEEE Recommended Practice for Software Requirements Specifications. Technical report, IEEE (1998)
17. Jaatinen, T.: The relationship between open data initiatives, privacy, and government transparency: a love triangle? Int. Data Priv. Law **6**(1), 28–38 (2016)
18. Janßen, C.: Towards a system for data transparency to support data subjects. In: Abramowicz, W., Corchuelo, R. (eds.) BIS 2019. LNBIP, vol. 373, pp. 613–624. Springer, Cham (2019). https://doi.org/10.1007/978-3-030-36691-9_51
19. Kirrane, S., et al.: A scalable consent, transparency and compliance architecture. In: Gangemi, A., et al. (eds.) ESWC 2018. LNCS, vol. 11155, pp. 131–136. Springer, Cham (2018). https://doi.org/10.1007/978-3-319-98192-5_25
20. Krempel, E., Beyerer, J.: The EU general data protection regulation and its effects on designing assistive environments. In: Proceedings of the 11th PErvasive Technologies Related to Assistive Environments Conference on - PETRA'18, pp. 327–330. ACM Press, Corfu, Greece (2018)
21. Krippendorff, K.: Reliability in content analysis: some common misconceptions and recommendations. Hum. Commun. Res. **30**(3), 411–433 (2004)
22. Maguire, S., Friedberg, J., Nguyen, M.-H.C., Haynes, P.: A metadata-based architecture for user-centered data accountability. Electron. Markets **25**(2), 155–160 (2015). https://doi.org/10.1007/s12525-015-0184-z
23. Meis, R., Wirtz, R., Heisel, M.: A taxonomy of requirements for the privacy goal transparency. In: Fischer-Hübner, S., Lambrinoudakis, C., Lopez, J. (eds.) TrustBus 2015. LNCS, vol. 9264, pp. 195–209. Springer, Cham (2015). https://doi.org/10.1007/978-3-319-22906-5_15
24. Merten, K.: Inhaltsanalyse: Einführung in Theorie, Methode und Praxis. Springer VS, Wiesbaden (1983). https://doi.org/10.1007/978-3-663-10353-0
25. Murmann, P., Fischer-Hübner, S.: Tools for achieving usable ex post transparency: a survey. IEEE Access **5**, 22965–22991 (2017)
26. Pfeffers, K., Gengler, T., Ross, C., Hui, W., Virtanen, V., Bragge, J.: The design science research process: a model for producing and presenting information systems. In: Proceedings of DESRIST, Claremont (2006)
27. Pohl, K.: Requirements Engineering: Grundlagen, Prinzipien, Techniken, 2, korrigierte aufl edn. dpunkt-Verl, Heidelberg (2008)

28. Pohl, K., Rupp, C.: Basiswissen Requirements Engineering: Aus- und Weiterbildung zum "Certified Professional for Requirements Engineering": Foundation Level nach IREB-Standard, 4, überarbeitete auflage edn. dpunkt.verlag, Heidelberg (2015)

29. Rupp, C.: Requirements-Engineering und -Management: aus der Praxis von klassisch bis agil, 6, aktualisierte und erweiterte, auflage edn. Hanser, München (2014)

30. Seinen, W., Walter, A., van Grondelle, S.: Compatibility as a mechanism for responsible further processing of personal data. In: Medina, M., Mitrakas, A., Rannenberg, K., Schweighofer, E., Tsouroulas, N. (eds.) APF 2018. LNCS, vol. 11079, pp. 153–171. Springer, Cham (2018). https://doi.org/10.1007/978-3-030-02547-2_9

31. Singh, J., Cobbe, J., Norval, C.: Decision provenance: harnessing data flow for accountable systems. IEEE Access **7**, 6562–6574 (2019)

32. Van Alstyne, M., Brynjolfsson, E., Madnick, S.: Why not one big database? Principles for data ownership. Decis. Support Syst. **15**(4), 267–284 (1995)

33. Wachter, S.: Ethical and normative challenges of identification in the Internet of Things. In: Living in the Internet of Things: Cybersecurity of the IoT - 2018. Institution of Engineering and Technology, London, UK (2018)

34. Wallis, J.C., Borgman, C.L.: Who is responsible for data? An exploratory study of data authorship, ownership, and responsibility. Proc. Am. Soc. Inf. Sci. Technol. **48**(1), 1–10 (2011)

35. Webster, J., Watson, R.: Analyzing the past to prepare for the future: writing a literature review. MIS Q. **36**, 13–23 (2002)

36. Wolters, P.T.J.: The Control by and Rights of the Data Subject Under the GDPR, p. 14 (2018)

The Black Mirror: What Your Mobile Phone Number Reveals About You

Nicolai Krüger[1]([⊠]), Agnis Stibe[2], and Frank Teuteberg[1]

[1] Accounting and Information Systems, University of Osnabrueck, Osnabrueck, Germany
nikrueger@uni-osnabrueck.de
[2] TRANSFORMS.ME, Paris, France

Abstract. In the present era of pervasive mobile technologies, interconnecting innovations are increasingly prevalent in our lives. In this evolutionary process, mobile and social media communication systems serve as a backbone for human interactions. When assessing privacy risks related to this, privacy scoring models (PSM) can help quantifying the personal information risks. This paper uses the mobile phone number itself as a basis for privacy scoring. We tested 1,000 random phone numbers for their matching to social media accounts. The results raise concerns how network and communication layers are predominately connected. PSMs will support future organizational sensitivity for data linkability.

Keywords: Privacy · Information privacy · Privacy scoring model · Social media privacy · Mobile phone privacy · Mobile Device Management

1 Introduction

Today, the omnipresence of smartphone use extends far beyond the professional context, as in the past. Instead, it reaches the most private areas of people's lives. With this comes the need for ethical considerations arising from the usage of such technologies and the possibility of either observing or manipulating users' behaviour. Messeging services such as WhatsApp and online social networks (OSN) are very popular and may be underestimated in terms of their sociotechnical concerns.

The public discussion regarding smartphone and social media privacy has changed radically, owing to the last US presidential elections and the scandal concerning the activities of Cambridge Analytica. Prominent security faults such as this might raise the perceived need for privacy in practice. Furthermore, recent literature has highlighted the specific risks pertaining to knowledge leakage and personal information disclosure by mobile devices [1, 2].

As responsible information systems (IS) researchers, we investigate privacy risks that are often overlooked by mobile phone users and mobile service providers. More specifically, this research carries out a study based on telephone numbers as a digital footprint. While the body of literature offers well-elaborated approaches to the measurement of privacy at the application level [3, 4], a research gap exists in terms of the combination of the mobile phone network layer and OSN. First, we hypothesise that an

© Springer Nature Switzerland AG 2020
W. Abramowicz and G. Klein (Eds.): BIS 2020, LNBIP 389, pp. 18–32, 2020.
https://doi.org/10.1007/978-3-030-53337-3_2

individual's phone number can be traced as a footprint throughout OSN. We aim to test our hypothesis using a modified privacy scoring model (PSM).

The relevance of this problem can be derived from several different angles. (i) Existing and well-established systems face new threats owing to the development of new attack models. Data science tools and practices, such as the advanced web scraping and robot process automation (RPA) employed in this study, create a new category of possible privacy attacks. This method of gathering publicly accessible information, referred to as open source intelligence (OSINT), is actively used by police and intelligence agencies [4, 5], and could potentially be misused by other authorities. (ii) De-anonymising of phone numbers is a prevalent problem in academia and business [6, 7]. (iii) Interpreting mobile phone signals has already become a market in its own right, like interpreting mobile phone data to visualize the instore movements of shoppers [8]. (iv) As [9] suggest, a clear sample in terms of technology, use cases and users is needed beyond the existing survey-driven approaches in the body of knowledge. The aim of this paper is to move a step forward in this research direction. Through addressing this dilemma concerning potential privacy issues, the objective of our paper is to enrich the understanding of mobile-network-based privacy attacks that aim to obtain the personal information of users. Thus, we formulate the following research questions:

RQ: Which requirements and implications arise for social media and mobile phone privacy from a privacy scoring model (PSM)?

To answer this question, we build upon a privacy dimension framework [10]. Furthermore, this paper enriches the existing model with the help of knowledge obtained from the body of literature on mobile phone network security, in order to suggest an applied PSM. Subsequently, we randomly select 1,000 phone numbers and test the PSM attributes and dimensions to evaluate our model with real-world data. The paper contributes to research an practice with mainly two artefacts: First, a PSM for the given context (mobile and social media privacy in combination) will be derived from the empirical data. Second, a prototype of a real-life implementation of our paper is presented.

The structure of this paper is as follows: the following section describes the relevant concepts found in the background literature. We then describe the methodology concerning the PSM employed in this paper and build a framework for a PSM that reflects the mobile phone network layer and the selected OSN attributes and dimensions. Following this, we test our model to present the findings and analysis pertaining to personal information disclosure. Lastly, we discuss our results and outline the potential contributions and limitations of our research.

2   Related Work

2.1   Privacy Scoring Models

PSMs are an appropriate method of measuring the individual privacy of a user from a user-centric point of view. A conglomeration of existing approaches for building privacy scores can be found within the PScore, published by Petkos et al. [10]. In general, three separate dominant motives for using PSMs can be found in the existing body of knowledge. Firstly, they allow for risk monitoring to increase the awareness of users about this topic. Secondly, PSMs provide recommendations for privacy settings. Lastly, they generate a privacy score that also enables a sociological study of the measured (usage) behaviour [11]. This approach was expanded by Vidyalakshmi et al. [12], as their PSM focused on a user's friends within OSN, and aimed to provide a classification of the user's friends and their trustworthiness. Numerous further examples exist in the field of OSN. Hamed and Ayed [13] introduced a privacy score for OSN and carried out an experiment with mobile and stationary users. In their study, they drew the conclusion that the permanent online connection of mobile devices creates a higher risk of tracking and potential privacy threats, as they established a correlation between the browsing time and the privacy scores of mobile users.

In contrast to the PSMs described above, Kaffel and Ayed [14] have drawn attention to web-tracking based on cookies, JavaScript and iFrames. They also propose a specific privacy scoring for this problem. In [13], the researchers' proposed PSM differentiated between mobile and desktop tracking. They subsequently arrived at the conclusion that mobiles are at a higher risk of privacy threats due to their permanent connection to the web. Researchers have also addressed privacy concerns pertaining to the mobile phone network [15, 16].

To the best of our knowledge, no existing PSM includes the mobile phone network layer, as introduced by us in the following section. Our PSM also differentiates from existing models by building a chain of information: instead of measuring single data points and quantifying them, our model builds upon the revealed information of a user and uses these data for further investigations.

2.2 Social Media and Mobile Phone Privacy

The new General Data Protection Regulation (GDPR) of the European Union (EU) has been in force since May 2018, and proclaims privacy to be a human right: "The protection of natural persons in relation to the processing of personal data is a fundamental right" [17]. In future, IS research and curricula might be pressed even more than before to fulfil the task of establishing an educational and scientific body of knowledge on privacy-enhancing technologies (PETs) and privacy by design.

However, [18] addressed the need for upcoming mobile technologies to answer arising privacy concerns. Their study presented several different classes of PETs: identity management, anonymous communication, anonymous access to services, privacy-preserving authorisation and data management. Fundamental research artefacts applying these findings can be found in the recent literature. [12] conceptualises privacy as a service. Several examples can be found in the IS literature that reflect the privacy aspects of OSN and internet communication in general. For instance, [19] presented an approach for de-anonymising users based on pattern recognition within domain name system (DNS) traffic. [20] generated a model of cultural differences in self-disclosure technologies within instant messaging services (IMS). Furthermore, [21] examined an OSN study using Facebook in Turkey. Particularly in the current political situation, privacy concerns are paramount. The researchers identified several privacy threats, and suggested the inclusion of privacy sensitivity within educational programs. Beside political reasons, the complexity of OSN is a reason for continuous research in that field, as recently shown in [22]. The authors suggest OSN users to make active use of privacy settings, adding less people to their personal network and sharing fewer private data.

Even in the early stages of mobile communication, privacy was a significant matter of concern. At the end of the last century, Kesdogan and Fouletier [23] argued that the privacy of users will be threatened owing to the decreasing size of network cells. In addition, [24] identified the home location register (HLR) as a potential bottleneck for the growing mobile communication sector. Different techniques for mobile phone positioning are available through HLR lookups, network triangulation and silent SMS [25, 26], which can be misused by attackers to achieve information disclosure or to carry out criminal activities in the physical world, such as attempting burglaries as soon as a resident leaves for a holiday abroad. [16] analysed possible solutions for overcoming such privacy threats, although none of the suggested safeguards could be initiated by users; most of these referred to the GSM infrastructure and need to be implemented by the service providers.

3   Methodology

3.1   Towards a PSM for Mobile Phone Number Privacy

By studying existing PSMs, as presented in Sect. 2.1, we learned that the flexibility and rigour of the PScore framework proposed in [10] provides a solid approach for our research purposes. Each privacy dimension needs operationalisation by the researcher, which is well described by the framework. In order to apply the general PScore framework (Fig. 1) to mobile communication systems, the development of a matrix of privacy dimensions, attributes and values within the given domain is the primary step. These dimensions represent the organisational structure of the private or sensitive information of a user. Each dimension has a number of attributes as a sub-categorisation, and each attribute contains a set of values. As stated by Petkos et al. [10], this list of domain-specific values is an iterative set that may evolve in the course of future research; it represents the status quo in terms of the current state of knowledge, the body of literature, and technical possibilities. This point will be considered further in the section describing the limitations of our study.

The PScore framework provides a horizontal structure of dimensions and a vertical structure of scores at each node of the scheme, which are weighted on the basis of their confidence, sensitivity, viability, source of confidence (declared/inferred), support and level of control. We excluded the parameter of sensitivity mentioned in the original framework, and set the parameter of confidence to a constant (=1), since [10] employed a user survey to calculate these parameters. In our setting, we test the PSM using 1,000 phone numbers, meaning that we cannot efficiently build upon user involvement at this scale. Based on the information provided in the previous sections and some test iterations with the data available, we selected (i) HLR, (ii) WhatsApp and (iii) Facebook as the overall dimensions for our scraping activities. Next, we present the privacy dimensions (Table 1) and the parameter setting (Table 2).

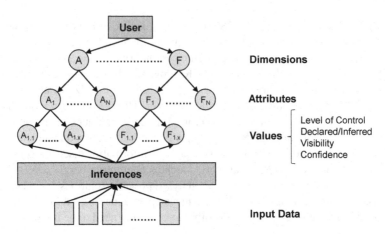

Fig. 1. Privacy scoring framework, adapted from Petkos et al. (2015)

Table 1. Overview of privacy dimensions

	Dimension	Attributes	Values	Range	Level of control	Declared/ inferred	Visibility	Confi- dence
A	HLR	Number	Original provider	{lookup successful = 1; no data = 0}	0	1	1	1
			Ported provider	{lookup successful = 1; no data = 0}	0	1	1	1
		User	Subscriber status	{unknown = 0; all other = 1}	0.25	1	1	1
			Roaming	{lookup successful = 1; no data = 0}	0	1	1	1
			HLR status	{lookup successful = 1; no data = 0}	0	1	1	1
B	Whats-App	User	WA_ Account	{lookup successful = 1; no data = 0}	0	1	1	1
		Profile	Last_Seen	{OCR successful = 1; no data = 0}	1	1	1	1
			Status	{OCR successful = 1; no data = 0}	1	1	1	1
			Profile_Pic	{data accessible = 1; no data = 0}	1	1	1	1
C	Face-book	User/Name	Account	{lookup successful = 1; no data = 0}	1	1	1	1
		Profile	URL	{lookup successful = 1; no data = 0}	0.5	1	1	1
			Timeline	{data accessible = 1; no data = 0}	1	0.5	1	0.5

Table 2. Parameter settings of the PSM

Parameter	Value	Description
Level of control	0	No influence
	0.25	Control is not impossible, but in contradiction to use case
	0.5	User might influence variable (up to the user)
	0.75	Control is possible and not in contradiction to use case (but a potential privacy issue)
	1	User has full control (but a clear threat)
Declared/inferred	0	Inferred
	1	Declared
Visibility	0	Private
	1	Public
Confidence	0	Unconfident
	1	Confident

3.2 Sample Number Creation

Generating random phone numbers, accessing a public HLR provider to verify them, and then using social media APIs to extract profiles is a common method of investigation, as these profiles are immediately converted into accumulated statistics. To test the adapted model, this paper follows the approach taken by [27], employing HLR lookups and data matching to build user profiles. Using this method, we test our PSM under realistic conditions. As a first step in data collection, a valid list of phone numbers is required. We created lists of potential phone numbers based on the service providers list from the federal network agency Germany. As the federal network agency only imposes the rule of a maximum of nine digits in a phone number, we created a list with potential phone numbers to crosscheck their validity, as a first step towards the PSM employing a HLR lookup [28]. We excluded all non-functional numbers in this step to give a final list of 1,000 existing numbers. After this initial preparatory step, the following section will address the primary data-mining part of our project. Using our list of 1,000 validated phone numbers, we wanted to find out which sensitive information we could.

3.3 Using OSINT to Identify Privacy Issues

We executed a thorough HLR lookup for the entire list of 1,000 phone numbers to fill our PSM data flow. We then wanted to investigate the WhatsApp data. At first, we conducted experiments with yosum, an unofficial API for WhatsApp (as no official API existed during the data gathering phase), but we were blocked by WhatsApp due to the massive number of contacts added within a short time frame. However, we were allowed by WhatsApp to add all of the numbers (enriched with pseudonyms) as a CSV file to a dummy phone. We also discovered WhatsApp Web to be an appropriate gateway to overcome the issue of the absence of an API.

We implemented the scraping activities with robot process automation (RPA) and Python scripts, which allowed us to download the current profile picture of each user, his/her status, and the last seen timestamp. It also allowed us to interpret the data using optical character recognition (OCR). In order to find out more personal details about the targets, we wanted to include at least one social network in our research. We discovered that several OSNs utilise phone numbers not only for the purpose of two-factor authentication but also as an identification criterion, for instance, to recover a Twitter or Facebook account in case of forgotten passwords. However, at the time when the project data were collected, parsing a specific phone number for a Facebook search was technically allowed, although this was stopped after the Cambridge Analytica leak. We implemented a Python script that processed all the generated phone numbers and searched for them on Facebook. When successful, our script saved a Facebook hyperlink to our database. Because of using RPA technology, we could avoid accessing Facebook only via official APIs. It was then possible for us to open the unique Facebook hyperlink and scrape the profile information. With respect to the users' privacy and research ethics, we did not mine the user data itself, but only classifier, e.g. if we were able to access sensitive information. Doing so, we computed the individual privacy score based on the dimensions and attributes described above.

4 Data Analysis and Results

The following sections present the results of the four steps of data collection and a combined analysis of the privacy scoring. We present the data in chronological order as generated during the research and aggregated in our PSM.

In total, we initially generated more than 3,000 phone numbers, from which we were able to verify 1,030 active and connected devices via HLR lookups (Fig. 2). This allowed us to further trace activities. The conversion rate of almost 30% indicated that our phone number generation procedure was adequate. In comparison with other research articles, this seems to be a good basis for creating a list of numbers for telephone surveys [7]. Furthermore, our script generated a wide spread over the most common mobile network providers.

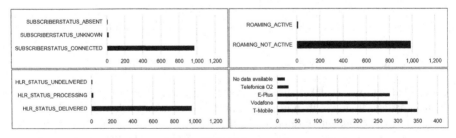

Fig. 2. HLR lookup results

The majority of cases were connected to the GSM network, delivered a valid HLR status signal, and were registered in their home country. However, we identified 11 users in roaming mode. Although we do not publish personal status information in this section, we provide a classification of the status messages (Table 3) that were found.

Table 3. WhatsApp status analysis

Status classification	Total amount	Total percentage	Percentage of WhatsApp users
Not available: The status of the WhatsApp contact was invisible to unauthorised contacts	168	16.31%	33.47%
Empty: Ability to access the status was granted, but there was no input by the user	23	2.23%	4.58%
Standard status: The user selected one of the standard WhatsApp status messages such as "Hello there"	130	12.62%	25.90%
Individual status: The user input an individual status message	181	17.57%	36.06%
Σ	502	48.76%	100%

Sharing the last online timestamp with any contact is currently a pre-selected functionality of WhatsApp. In other words, about 40% users deactivated this feature. As mentioned above, we only captured the last-seen variable once and not as a time series, which would easily have been possible. Online-experiments[1] have shown that very accurate sleep/wake profiles of users can be generated in this way. In total, our approach revealed the targeted WhatsApp data in 20.50% of the total number bucket.

In a similar way to WhatsApp, we scraped the Facebook data by looking up personal profiles based on their phone numbers. To protect the privacy of the randomly chosen users, we first verified whether a Facebook account matched a phone number. In case of success, we handed that parameter over to a target list and only stored a categorisation of the information accessible (0 = nothing found, 1 = person identified and some information accessible, 2 = person identified and all information accessible). Full data access was on this OSN possible in 16 cases (1,6%). Assuming that these profiles are no fake but real profiles, this means a full deanonymization plus access to personal information.

The personal data itself, such as home town and so forth, was not downloaded but was available in several cases (Table 4). Once again, we observed a correlation between the WhatsApp and Facebook profiles, as we found zero cases where a Facebook profile was present but no WhatsApp profile was found.

By default, the link between the cellphone numbers and personal profiles of users is activated after users have entered their personal mobile numbers on Facebook. Thus, 72

---

[1] See https://www.onlinestatusmonitor.com, last accessed 2019/12/01.

Table 4. Facebook profile analysis

Status classification	Total amount	Total percentage	Percentage of WhatsApp users
1 = Person identified and some information accessible	55	5.34%	76.39%
2 = Person identified and all information accessible	17	1.65%	23.61%
$\sum$	72	6.99%	100%

users were confirmed as having a linked Facebook account. There are likely to be more users without a linked phone number or with higher data privacy settings.

Finally, we computed our PSM based on the results shown above and categorised the PSM scores in groups of zero and one PSM points (Fig. 3). We also weighted some PSM attributes higher than others in order to increase the mean of the PSM score: complete de-anonymisation of a phone number should have a high impact on the PSM score, even when other attributes are more secure. Thus, we chose factor three for the real name; highly sensitive information (gender and age) and sensitive information with a strong privacy impact (timeline data and profile picture) were weighted with factor two; and roaming information was weighted with factor 1.5. This was also done to give better coverage of the PSM groups with respect to a normal distribution.

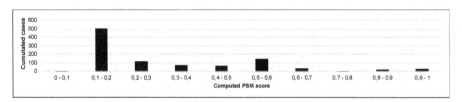

Fig. 3. Categorised PSM results

5   Discussion

Figure 4 shows grouped, weighted and visualised PSM model for further analysis. Three dimensions and six attributes were defined and tested with 1,000 users. This paper addresses a new aspect of data linkability in the privacy domain of IS research and thus provides several theoretical and practical contributions.

5.1   Theoretical Contribution

Research by [9] encouraged IS privacy researchers to conduct sample-driven, well-contextualised studies. Our research contributes to IS knowledge in exactly this field:

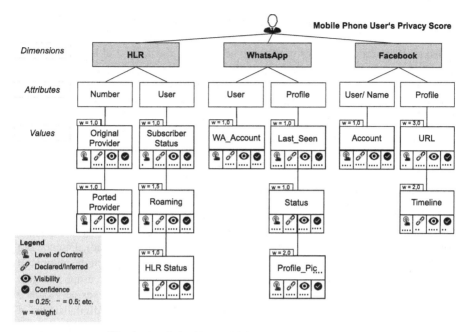

Fig. 4. Applied PSM model for mobile phone privacy

we recommend a PSM for mobile phone users as an indication of the perceived privacy of users and their behaviour in terms of a careful choice of security settings. Previous research has shown that awareness about privacy concerns when using a mobile social network does not influence users' behavioural intentions [29]. Our research builds upon this, as our hypothesis involved the underlying technical layer. We verified our hypothesis that a phone number represents an easily traceable footprint in both the GSM mobile phone network and the online network (OSN in particular) and should definitely be considered a personal privacy risk. Our research also included the body of literature with respect to privacy from an IS perspective, mobile phone privacy and PSM. This is IS research, since IS artefacts and their effects on users are studied, and particularly their privacy concerns. With respect to PSM as a quantification of users' privacy behaviour, we built our PSM based on an existing PSM framework [10] and selected dimensions, attributes, and values for the given domain. Due to the nature of this method, it can be assumed to be rigorous and to generate a reproducible result; we followed the framework of [10] and carefully documented all of the research steps for our part. The addition of an applied PSM to the existing body of knowledge is, in our view, an important step within the IS discipline.

The lack of literature addressing privacy issues at the GSM layer (and also in relation to the online communication layer) has been presented in the literature review section of this paper. To fill this gap, this study answers the research question by presenting an approach to user tracing across different communication layers and services. Our findings can be associated with privacy by design and privacy-enhancing technologies (PET) within IS research: storing unencrypted mobile phone numbers of the users of a system could be problematic with respect to the GDPR, as this paper verifies our hypothesis of linkability. From a broader perspective, this paper aims to contribute to the overall IS privacy and security research activities in the context of big data. In our view, IS plays a major role in ethical and moral discussions about the usage of OSINT and other applied big data disciplines, particularly following the Cambridge Analytica scandal. With its interdisciplinary approach, IS should communicate within and beyond the research community about privacy issues and approaches for self-protection. As shown above, the model presented works well with real-life OSN data, and we hope to deliver new insights to researchers in the field of mobile communication privacy.

5.2 Practical Contribution

Our findings have important practical implications. Firstly, it is useful for the protection of minors. Children and young people use smartphones intensively, sharing their mobile phone number for many purposes (such as activating an online profile). Parents and schools need to educate children, and to let them know that using their phone number in a mobile app or service can be used to track their behaviour and online usage patterns. This educational aspect also serves to impede child pornography and sexual violence against children and young people [29]. We do not aim to stop the digitalisation of schools; however, for good reasons related to data protection, some federal states of Germany prohibit the use of WhatsApp as an official communication tool between children and teachers [30]. Thus, tools with better privacy options and GDPR-compliant services should be preferred. At the very least, parents should carefully monitor the privacy settings of both the messenger services and OSNs of their children and educate them about potential risks.

Secondly, our findings have shown that simply requesting data from the HLR can be an anchor point for further attacks, and particularly those that focus on political VIPs, enterprises or infrastructural organisations. We present a practical, useable and verified approach in Fig. 4 to test a user's mobile phone number for privacy issues. Pentesters or IT security managers can follow our PSM data flow to create a user-centric visualisation (comparable to a data-driven mindmap) in Maltego Teeth, which is a widely used pentesting tool for Linux. We will cover that important aspect in Fig. 5, considering also the further point.

Thirdly, IT security managers should (where it matters) blacklist at least those messaging apps where no restrictive privacy settings can be applied. A good practice here is to use Threema, in which communication partners can add each other without uploading their contact databases to the provider [31]. Although WhatsApp updated its privacy setting options during the writing of this paper, tracking the online status of a user is still possible [2].

As a further artefact of our research, we present a prototype of a single-page application (SPA) using the RPA technology in the background.[2] IT managers in organizations will have the possibility to calculate the PSM of a mobile phone number. By this automized approach, a privacy audit of Mobile Device Management (MDM) becomes feasible. Further, individual users can also specify which privacy aspects matter for them.

---

[2] The prototype can be accessed via https://github.com/swingingcode/bis2020_blackmirror.

Fig. 5. Prototype of an RPA-implementation as single-page web application frontend

Thus, finetuning of the personal PSM configuration will be possible. As we developed the privacy checker on an open-source base, it will be easily applicable, for instance to other OSNs or company-specific aspects (e.g. LinkedIn visibility).

6 Conclusion

Privacy and security concerns are becoming extremely important, as human nature faces challenges in keeping pace with the ongoing technological revolution that we are all currently experiencing. Although some people are more cautious, and carefully review each new technology that they use, many others, especially younger people, grow up with rather careless attitudes towards the potential privacy risks that new mobile services can generate [32]. As previous research has shown, the behavioral goal of users in OSN, which is sharing information with as many as people as possible, contradicts with strict and limiting privacy settings [22]. The present study provides concrete evidence for this vulnerability and calls for closer attention from all the stakeholders in this context, especially scholars, professionals, and users.

Today, the omnipresence of mobile phones in all corporate, institutional and personal infrastructures offers major opportunities for seamless communication and comfortable user experiences. Although mobile phone privacy was already an important research field before the most recent leaks regarding Facebook, through the subsequent massive media coverage, a global wave of awareness of profiling activities based on online services has been created. We investigated users' behaviour in terms of their security settings and the role of phone numbers in this context, grounded on the hypothesis that mobile phone numbers could be used as a footprint and a link between the GSM network and an OSN. Based on the literature concerning privacy scores, we extended existing approaches to measure privacy issues by quantifying data points in the HLR and selected online networks. We verified that a simple phone number is a sufficient starting point for gathering information about a victim's mobile phone status, OSN usage, and in some cases even for gathering the real name and further sensitive information such as gender, age, and so on. Such privacy attitudes can be transformed using novel design methods in human-technology interaction [33].

Our study has certain limitations. Firstly, the phone numbers used in our study represent only phone numbers from Germany. Secondly, a distinction between corporate and private phone users was outside of the scope of this work. Thirdly, we did not actively include users within the research process, as we wanted a large sample of test data. Thus, it was up to the research team to adjust the parameters within the PSM framework.

Future work could build upon our PSM to include further OSNs and other data sources. IoT devices connected via 5G networks will play a major role in future privacy issues, for instance privacy threats due to connected cars or smart devices. Dedicated search engines for this purpose already exist and will underline the linkability of mobile phone numbers and organizational data more.[3] Future IS research could make use of our GSM-based PSM and enhancing IoT privacy.

References

1. Agudelo-Serna, C.A., Ahmad, A., Bosua, R., et al.: Strategies to mitigate knowledge leakage risk caused by the use of mobile devices: a preliminary Study. In: ICIS 2017: Transforming Society with Digital Innovation, pp. 1–19. Seoul (2017)
2. Holland, M.: WhatsApp ermöglicht weiterhin Überwachung beliebiger Nutzer. https://www.heise.de/newsticker/meldung/WhatsApp-ermoeglicht-weiterhin-Ueberwachung-beliebiger-Nutzer-3857506.html. Accessed 07 Dec 2019
3. Dwyer, C., Hiltz, S.R., Passerini, K.: Trust and privacy concern within social networking sites: a comparison of Facebook and MySpace. Am. Conf. Inf. Syst. **123**, 339–350 (2007)
4. Bundesnachrichtendienst: OSINT. http://www.bnd.bund.de/DE/Auftrag/Informations-gew innung/OSINT/-osint_node.html. Accessed 07 Dec 2019
5. Europol: Cyber Intelligence. https://www.europol.europa.eu/activities-services/services-sup port/intelligence-analysis/cyber-intelligence. Accessed 07 Dec 2019
6. Meffert, H., Burmann, C., Kirchgeorg, M.: Marketing. Grundlagen Marktorientierter Unternehmensführung. Springer, Wiesbaden (2012)
7. Kunz, T., Fuchs, M.: Pre-call validation of RDD cell phone numbers. A field experiment. In: JSM Proceedings (2011)
8. T-Systems International GmbH: Den Überblick über die wichtigen Orte haben. https://www.t-systems.com/blob/384750/cdfde6863685cf0f08068a53e9e18a84/DL_Flyer_Motionlogic.pdf. Accessed 07 Dec 2019
9. Bélanger, F., Crossler, R.E.: Privacy in the digital age: a review of information privacy research in information systems. MIS Q. **35**(4), 1017–1041 (2011)
10. Petkos, G., Papadopoulos, S., Kompatsiaris, Y.: PScore: a framework for enhancing privacy awareness in online social networks. In: 10th International Conference on Availability, Reliability and Security, pp. 592–600 (2015)
11. Liu, K., Terzi, E.: A framework for computing the privacy scores of users in online social networks. ACM Trans. Knowl. Discov. Data **5**(1), 1–30 (2010)

---

[3] An example of such a search engine is http://shodan.io, last accessed 2019/12/07.

12. Vidyalakshmi, B.S., Wong, R.K., Chi, C.H.: Privacy scoring of social network users as a service. In: Proceedings - 2015 IEEE International Conference on Services Computing, SCC 2015, New York, pp. 218–225, New York (2015)
13. Hamed, A., Ayed, H.K.B.: Privacy scoring and users' awareness for Web tracking. In: 6th International Conference on Information and Communication Systems, pp. 100–105. Fort Worth (2015)
14. Kaffel, H., Ayed, B.: Privacy risk assessment for web tracking. In: 2016 IEEE Canadian Conference on Electrical and Computer Engineering (CCECE), Vancouver (2016)
15. Fife, E., Orjuela, J.: Mobile phones and user perceptions of privacy and security. In: International Conference on Mobile Business, p. 23, Delft (2012)
16. Rechert, K., Meier, K., Wehrle, D., et al.: Location privacy in mobile telephony networks – conflict of interest between safety, security and privacy. In: IEEE International Conferences on Internet of Things, and Cyber, Physical and Social Computing, pp. 508–513. Dalian (2011)
17. European Union: Regulation (EU) 2016/679 of the European Parliament and of the Council of 27 April 2016 on the protection of natural persons with regard to the processing of personal data and on the free movement of such data, and repealing Directive 95/46/EC, Official Journal of the European Union, pp. 1–88 (2018)
18. Deswarte, Y., Aguilar Melchor, C.: Current and future privacy enhancing technologies for the Internet. Ann. Telecommun. **61**, 399–417 (2006)
19. Herrmann, D.: Privacy issues in the domain name system and techniques for self-defense. IT-Inf. Technol. **57**(6), 388–393 (2015)
20. Lowry, P.B., Cao, J., Everard, A.: Privacy concerns versus desire for interpersonal awareness in driving the use of self-disclosure technologies: the case of instant messaging in two cultures. J. Manag. Inf. Syst. **27**(4), 163–200 (2011)
21. Külcü, Ö., Henkoglu, T.: Privacy in social networks: an analysis of Facebook. Int. J. Inf. Manag. **34**, 761–769 (2014)
22. Lankton, N.K., McKnight, D.H., Tripp, J.F.: Understanding the antecedents and outcomes of Facebook privacy behaviors: an integrated model. IEEE Trans. Eng. Manag. 1–15 (2019)
23. Kesdogan, D., Fouletier, X.: Power control in cellular radio systems. In: IEEE Wireless Communication System Symposium, pp. 35–40, London (1995)
24. Palat, S.K., Andresen, S.: User profiles and their replication for reduction of HLR accesses and signalling load. In: Proceedings of ICUPC 5th International Conference on Universal Personal Communications, IEEE, pp. 865–869, Cambridge (1996)
25. Arapinis, M., Ilaria Mancini, L., Ritter, E., et al.: Analysis of privacy in mobile telephony systems verifying interoperability requirements in pervasive systems. Int. J. Inf. Secur. **16**(5), 491–523 (2016)
26. Eren, E., Detken, K.-O.: Mobile Security. Risiken mobiler Kommunikation und Lösungen zur mobilen Sicherheit. Carl Hanser Verlag, München, Wien (2006)
27. Costin, A., Isacenkova, J., Balduzzi, M., et al.: The role of phone numbers in understanding cyber-crime schemes. In: 11th Annual Conference on Privacy, Security and Trust, pp. 213–222, Tarragona (2013)
28. Bundesnetzagentur: Numbering plan for the numbering space for public telecommunications. https://www.bundesnetzagentur.de/SharedDocs/Downloads/EN/Areas/Telecommunications/Companies/NumberManagement/numbering_space/NP_numbering_space_2016.pdf?blob=publicationFile&v=1. Accessed 07 Dec 2019
29. Abramova, O., Krasnova, H., Tan, C.-W.: How much will you pay? Understanding the value of information cues in the sharing economy. In: ECIS 2017 Proceedings, pp. 1011–1028, Guimarães (2017)
30. Baden-Württemberg, Kultusministerium.: Kommunikationsplattformen am Beispiel WhatsApp. https://it.kultus-bw.de/,Lde_DE/Startseite/IT-Sicherheit/Kommunikations-plattformen?QUERYSTRING=whatsapp. Accessed 07 Dec 2019

31. Karaboga, M., Masur, P., Matzner, T., et al.: Selbstdatenschutz. In: Schütz et al. (ed.) Schriftenreihe Forum Privatheit und selbstbestimmtes Leben in der digitalen Welt, Fraunhofer-Institut für System- und Innovationsforschung ISI, Karlsruhe (2014)
32. Frischmann, B., Selinger, E.: Re-Engineering Humanity. Cambridge University Press, Cambridge (2018)
33. Stibe, A., Cugelman, B.: Social influence scale for technology design and transformation. In: Lamas, D., Loizides, F., Nacke, L., Petrie, H., Winckler, M., Zaphiris, P. (eds.) INTERACT 2019. LNCS, vol. 11748, pp. 561–577. Springer, Cham (2019). https://doi.org/10.1007/978-3-030-29387-1_33

Big Data and Data Science

REM Sleep Stage Detection
of Parkinson's Disease Patients with RBD

Pinar Bisgin[1] , Salima Houta[1], Anja Burmann[1(✉)], and Tim Lenfers[2,3]

[1] Fraunhofer Institute for Software and Systems Engineering, HealthCare,
Dortmund, Germany
{pinar.bisgin,anja.burmann}@isst.fraunhofer.dev
[2] Department of Medical Informatics, Heilbronn University, Heilbronn, Germany
[3] Ruprecht-Karls University Heidelberg, Heidelberg, Germany

Abstract. REM sleep behavior disorder (RBD) is commonly associated with Parkinson's disease. In order to find adequate therapy for affected persons and to seek suitable early Parkinson Patterns, the investigation of this phenomenon is highly relevant. The analysis of sleep is currently done by manual analysis of polysomnography (PSG), which leads to divergent scoring results by different experts. Automated sleep stage detection can help deliver accurate, reproducible scoring results. In this paper, we evaluate different machine learning models from the PSG signals for automatic sleep stage detection. The highest accuracy of 87.57% was achieved by using the Random Forest algorithm.

Keywords: Parkinson's disease · Neurodegenerative disease · PCompanion · Sleep stage · PSG · RBD · Classification

1 Introduction

Parkinson's disease is a chronic progressive disease and the most common neurodegenerative movement disorder. Due to the aging population, the prevalence has risen rapidly in recent years with currently about 400.000 patients in Germany. The suffering of those affected is immense and although much has already been done in therapy, new options for early detection and treatment are urgently needed. In Parkinson's disease, nerve cells in the substantia nigra die off continuously - something that cannot yet be stopped by medication. These cells produce the messenger substance dopamine. A dopamine deficiency leads to the typical motor symptoms, which, however, become objectively apparent to the patient and doctor when more than 50% of the nerve cells have already been lost. The neurodegenerative process already has slowly spread over the entire brain, potentially over decades, when the typical motor symptoms such as the small-step gait, the increased muscle tone (muscle stiffness, rigor) and the general movement slowdown (bradykinesia) become visible. The tremor also leads to speech disorders, reduced facial expressions and, as the disease progresses, to

ⓒ Springer Nature Switzerland AG 2020
W. Abramowicz and G. Klein (Eds.): BIS 2020, LNBIP 389, pp. 35–45, 2020.
https://doi.org/10.1007/978-3-030-53337-3_3

sudden changes in motor symptoms during the course of the day (motor fluctuations), including sudden freezing of movements. In addition to these motoric symptoms, a number of so-called non-motoric symptoms occur, some of which lead to considerably restrictive functional disorders of the vegetative system (gastrointestinal passage, cardiovascular regulation) and the psychic-cognitive system (sleep disorders, depression, anxiety, psychoses, hallucinations, concentration and drive disorders, dementia) and even other sensory organs such as a disturbed sense of smell or pain perception, and impaired vision [1,2]. All these symptoms severely restrict everyday activity and quality of life.

Sleep activity during the REM phase is crucial evidence that neurodegenerative disease such as Parkinson's disease can occur [3,4]. Therefore, an automated sleep phase detection is subject of investigation in the PCompanion project. In order to achieve the correct diagnosis and treatment of Parkinson's disease based on the biological records, an accurate sleep assessment is considered a critical part of the process. To date, conventional visual scoring is still the most acceptable approach, although it includes visual data interpretation of these signals [5]. Qualitative scoring is subject to some pitfalls, including the experience of experts, which may lead to deviating scoring results from deviating experts [6,7]. In addition, visual inspection is a time-consuming process for full-day EEG labeling. Therefore, automatic assessment is seen as an efficient approach [8,9].

In this paper, we evaluate the performance of different machine learning models proposed in the literature for automatic sleep scoring. For the diagnosis of sleep issues, all night polysomnographic (PSG) recordings, including electroencephalogram (EEG), electrooculogram (EOG) and electromyogram (EMG), are taken from RBD patients and the recordings are scored by a well-trained expert according to the Rechtschaffen & Kales (R&K) rules presented in 1968 [10]. According to the R&K rules, each epoch (i.e., 30 s of data) is classified into one of the sleep stages, including wakefulness (Wake), non-rapid eye movement (stages 1−4, from light to deep sleep) and rapid eye movement (REM). Stages 3 and 4 were frequently combined as the slow-wave sleep stage. Instead of the R&K rules, the scoring rules developed by the American Academy of Sleep (AASM) became the clinical standards [11,12], recently. 19 polysomnography data sets of RBD patients of the University Hospital Aachen are used. Furthermore, we demonstrate the advantages of using multimodal sensor data by integrating EOG and EMG to complement EEG usage. As RBD disease has low prevalence and patients sleep is plagued by arousals [13], sleep classification is more challenging than in healthy volunteers.

The paper is organized in the following manner: Sect. 2 presents a brief literature review about Parkinson's disease and the detection of stages during sleep. Section 3 describes the project PCompanion including the components of the technical solution. Section 4 describes briefly the data set of the SOMNOscreen^{TM} employed in our research. This section presents information regarding the methods used in this study, as well as information related to the performance evaluation criteria employed in the study. Section 5 provides the assessment procedures used and the experimental results obtained.

Finally, Sect. 6 describes the conclusion derived from the study and some thoughts with regard to future work.

2 Related Work

The manual analysis of polysomnography to classify sleep stages is time-consuming and subject to large interindividual fluctuations. Therefore, automated sleep phase classification has been researched for 50 years. In the following, previous work on sleep phase classification will be presented.

Boostani et al. (2017) [14] compared different characteristics that describe different procedures of feature selection and models of machine learning from a large number of publications of sleep phase classification with single-channel EEG-signals. The polysomnographic datasets "Sleep-EDF" and "CAP sleep" from Physionet were used and each dataset contained the same number of healthy and diseased subjects. For the classification, the models k-NN, SVM, Random Forest, LDA, Neural networks, Gaussian Mixture Models, Hidden Markov Models, Clustering of Machine-Learning have been compared. There has been studied a 5-Class-Problem. The authors achieved an accuracy of 88.66% on healthy subjects and 66.96% on RBD patients by using the Random Forest algorithm. The different results can be explained by the more frequent position changes and movements of RBD patients. The EEG signals thus comprise more movement artifacts with a high amplitude compared to healthy volunteers. As a result, misclassifications often occurred in the sleep phases REM and N1.

In the article by Khalighi et al. (2013) [15], different methods for sleep phase classification were applied to a self-generated data set of 42 patients. Six EEG channels, two EOG channels and one EMG channel of PSG were used for analysis. SVM, LDA, Naïve Bayes and AdaBoost were used for classification. The authors achieved an accuracy of 88.87% in distinguishing between the awake and sleep phases (2-class-problems) and 81.74% in classifying all sleep phases (5-class-problems). By adding the EOG and EMG channels, an accuracy of 86.99% was achieved in the classification of the REM sleep phase.

Zhu et al. (2014) [16] sleep phase classification based on graphical domain characteristics on a single-channel EEG. Sleep phase recognition was performed with 8 subjects from the "Sleep-EDF". 4 out of the 8 volunteers were healthy, the others had slight problems falling asleep. The authors divided the sleep class classification into different class problems. They achieved an accuracy of 92.6% in distinguishing between awake phase, NREM sleep and REM sleep with an SVM classifier.

Previous publications have achieved good results in sleep stage recognition in healthy people. Nevertheless, Parkinson's patients with a different sleep pattern have not yet been considered, which means that the recognition of sleep stages is not possible with the previous work. In the context of this work, the recognition of the REM sleep phase is of particular importance in order to recognize patterns of neurodegenerative disease and thus to obtain early recognition of Parkinson's disease, e.g. Thus, a 2-class problem is considered in this paper. A distinction is

made only between REM sleep and not REM sleep (NREM). The sleep stages
N1, N2, and N3 are combined with the awake state to form NREM. A data
processing pipeline is being developed that can achieve sleep stage recognition
with the help of machine learning algorithms.

3   Project PCompanion

The aim of the joint development is the ParkinsonCompanion - the first mobile,
patient-oriented screening, and monitoring system that can be used in the early
diagnosis of Parkinson's disease. It is also the first patient-oriented system for
the screening of Rapid Eye Movement (REM) sleep and vegetative disorders,
which also has an adaptive, interactive interface and allows user type modeling.
In order to predict the course of Parkinson's disease or to differentiate between
severe and atypical forms, Parkinson's Companion uses predictive analytical
methods such as data mining, complex event processing and machine learning.
This integrates the analysis of vegetative functions, sleep, movement and cognitive functions. Users receive direct feedback/interpretation of observations, test
results and trends in the course of the disease. Through integration into the
telemedical video monitoring system (MVB Parkinson GmbH), which has been
established for Parkinson's patients for more than 10 years, patients can also
use a 24-h telephone service. With the help of a mobile patient-oriented monitoring system, the signals from the sensor system and the recorded data, such
as context information from the app on conspicuous Parkinson's patterns, can
provide scientists and doctors with much more reliable data for a better diagnosis, since the frequency and severity of the seizures can be better recorded. The
anonymization and cross-patient aggregation of the data also enables clinical
research, for example on the drug that most effectively reduces symptoms or on
various context parameters that trigger Parkinson's-associated patterns.

The consortium of the project PCompanion coordinated by the neurolo-
gist of the University Hospital Aachen consists of five institutions and two
associated partners in Germany: Department of Neurology at the University
Hospital Aachen, Fraunhofer Institute for Software and Systems Technology
ISST, Department and Institute of Ergonomics of University Aachen (RWTH),
Medizinische Videobeobachtung in Koblenz, SOMNOmedics GmbH Rander-
sacker, Deutsche Parkinson Hilfe e.V. [German Parkinson Help], and practicing
doctors.

4   Methods

4.1   Sensor Platform

The sensor system is based on a PSG which was developed by SOMNOmedics.
SOMNOscreen^{TM} plus is a portable PSG system with 33 channels, upgradeable
to up to 58 channels, which you can handle flexibly in diagnostics. No matter

whether you evaluate according to the R&K or AASM standard or with individual specifications for studies, the user decides what to focus on and accordingly the measurement configuration. The SOMNOscreen^{TM} plus can be upgraded at any time and thus offers the possibility to perform polygraphy screenings, outpatient and inpatient PSGs as well as neurological applications with complete EEG 32 derivation in one device - all this with maximum patient comfort (see Fig. 1).

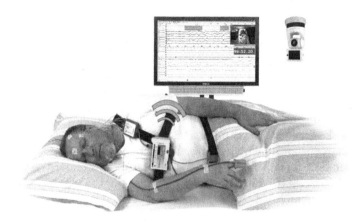

Fig. 1. The SOMNOscreen^{TM}.

4.2 Study

For sleep recognition, a data set of the University Hospital Aachen with manually annotated sleep phases is used. The data set contains the PSGs of 19 male patients with an average age of 65 ± 14 years (range 36–89 years). The total time of all sleep recordings is 137 h. The EEG, EOG and EMG channels are relevant for sleep phase classification according the AASM. All 19 patients suffer from RBD. The data is extracted via the SOMNOmedics software "Domino" (version 3.0.0).

4.3 Sleep Stage Classification

For data analysis python^{TM}, version 3.7.3., was used. Python is a widely-used programming language in machine learning and has a large community that maintains a large number of libraries. For sleep phase classification, the PSG channels are reduced to the AASM channels: EEG C4A1, EEG C3A1, EOG left A2, EOG right A2 and EMG chin. Furthermore, the manually annotated sleep phases of the somnologist of the University Hospital Aachen were extracted to develop a supervised algorithm (see Fig. 2).

Fig. 2. Machine-learning pipeline.

Preprocessing: To reduce noise and unwanted interferences in the signal, the signals were filtered with a Notch filter with a cut-off frequency of 50 Hz and a Butterworth 6th order Bandpass filter. The Butterworth filter had a lower limit at 0.5 Hz and an upper limit at 45 Hz. In the next step, the filtered signals were divided into the typical 30-second epochs according to R&K and AASM.

Feature Extraction and Feature Selection: Features were extracted from raw sleep data using a window size of 30 s without overlapping between consecutive windows. Features can be computed from the segmented data, usually, contain time and frequency domain features. The time-domain features are including [14]: (1) arithmetic mean, (2) median, (3) variance, (4) standard deviation, (5) skewness and curtosis (6) 25th and 75th Percentile and (7) Hjorth Parameters. The frequency-domain features are: (1) power spectral density, (2) relative power spectral density, (3) spectral entropy, (4) Petrosian fractal entropy estimation. After that, the generated features were standardized with z-score function [17]. Features were selected by use of the Mutual Information (MI) [18], Pearson Correlation Coefficient (PC), SVM recursive feature elimation (SVM-RFE) and minimum Redundancy Maximum Relevance (mRMR) [19].

Validation: K-Fold-cross validation was performed for the validation data set by calculating the agreement for each subject through training the system using data from the rest of the subjects.

Classification: For our intended goal of embedded system classification, we focused on classifiers that could be implemented in a computationally efficient manner. There are a large set of classifiers. Our choice was to classify the data with the k-Nearest-Neighbor (k-NN), Logistic Regression (LR), Support Vector Machines (SVM), Naïve Bayes, Decision Tree (DT) and Random Forest (RF).

The k-NN is an instance-based classifier based on majority voting of its neighbors: The k-NN algorithm finds a close group of k objects in the training dataset with the target object in the training data and predicts the class of closest objects to the target object [20]. The logistic regression model arises from the desire to model the posterior probabilities of the K classes via linear functions in x, while at the same time ensuring that they sum to one and remain in $[0, 1]$ [21]. Support Vector Machines (SVM) is a method from the area of functions for learning linear models. For linearly separable data it can be shown that a maximum margin hyperplane is optimal with respect to generalization properties (i.e. ability to classify unseen data.) The maximum margin hyperplane separates the two classes so that the distance from the hyperplane to the closest example(s) is maximal [22]. For the classification of sleep stages, we used linear kernels. Naïve Bayes (NB) is a simple probabilistic method based on Bayes' theorem. For an unknown example from the test data set, the class with the highest probability (given the features) is returned. NB is based on the basic assumption that the individual features are independent of each other when the class is given. When used on continuous features, the assumption is also made that the features are normally distributed [23]. A decision tree classifier is a straight-line approach in which a class is found out by successive rules. Decision trees are ordered. A pattern is classified by starting at the root node with the properties of the pattern that can be queried. The branches correspond to different possible values. In the next step, the following is calculated according to the answer of a branch which is considered the new root node of the subtree. Each connection must be able to divide the pattern uniquely so that only one branch is selected. The steps are repeated until a leaf node is reached [21]. The random forest classifier consists of a combination of tree classifiers where each classifier is generated using a random vector sampled independently from the input vector, and each tree casts a unit vote for the most popular class to classify an input vector [24].

Evaluation Metrics: The performance metric that is generally most widely used is Accuracy (ACC) [25]. It describes the proportion of correct classifications in all classifications made.

$$ACC = \frac{TP + TN}{TP + TN + FP + FN} \tag{1}$$

where TP, TN, FP, and FN denote true positives, true negatives, false positives, false negatives, respectively [26]. For a given sleep stage A, TP denotes the number of A windows that were correctly classified as A, FP the number of non-A windows that were falsely classified as A and FN the number of A windows that were falsely classified as non-A.

Computation Time and Memory Requirements: The processing of sleep data places high demands on the hardware, as the data set is large due to the high sampling rates and long recording times. The algorithm is particularly demanding on the CPU. The preprocessing of the data required a lot of RAM.

5 Results

The experimental results of the different algorithms with different feature selection methods are summarized in Table 1. With Random Forest and the SVM-RFE an accuracy of 87.57% is obtained.

The classifiers depend to varying degrees on the feature selection procedures and evaluation procedures used. The Random Forest classifier delivers the highest result with an accuracy of 87.84%, and is therefore best suited for

Table 1. Accuracy of the Machine-Learning algorithms.

	mRMR	SVM-RFE	PC	MI
LR	87.36%	87.40%	87.50%	87.02%
k-NN	85.59%	85.35%	86.09%	85.28%
NB	87.36%	36.56%	38.54%	36.82%
SVM	87.36%	87.36%	87.36%	87.36%
DT	77.62%	78.89%	78.97%	80.17%
RF	85.10%	87.57%	87.84%	87.00%

the detection of the REM sleep phase. The Naive Bayes classifier is not suitable for the recognition of sleep stages since the probability of classifying the correct sleep phase is lower than with a coin toss. The only exception is the use of the mRMR feature selection with the NB classifier.

A challenge in the pre-processing of the data was the required resources due to the large data set. The records consist of a total of 137 h of recording time. For each patient, 5 channels were used for processing, most of which had a sampling rate of 256 Hz. Then processing the data, special attention had to be paid to the performance of the data processing chain, because a large amount of data requires a lot of computing power on the one hand, and processing takes a lot of time on the other. Due to the unequal distribution of the NREM and REM sleep phases, because the REM phase rarely occurs in relation to NREM, a clear identification of the sleep stages is not possible.

The methods shown have already achieved promising results in the REM sleep stage classification. This is the first step towards the early detection of Parkinson's disease in the PCompanion project. The next step would be to extend the classifiers by an automated differentiation between healthy REM sleep phases and pathological REM sleep phases of RBD patients. This prediction can serve the project in the early detection of Parkinson's disease.

The computing time is very long due to the long signal recording during one night. Therefore the algorithm needs a high computing time to process the files window by window. The more signals are viewed from the PSG, the more time is needed. However, a general trend cannot be observed, the calculation time depends on the specific sensor set and its optimized properties.

6 Conclusion and Future Work

The results shown within this paper are promising regarding REM sleep phase detection. Across all considered classifiers a high accuracy was achieved, with one exception to be mentioned. Nonetheless, the following aspects must be taken into account, when interpreting these promising results. Due to the large data set of overall 137 h recording time a focus needed to be set on performance within the machine-learning pipeline. That resulted in omitting the leave-one cross-validation as an evaluation procedure as a result of excessive computing time.

Furthermore, all of the evaluated data sets were gathered from RBD patients within the clinical context. Consequently, the REM sleep phase detection on sleep data from healthy subjects could lead to diverging results. Another bias could result from the limited number of medical professionals that carried out the manual labeling of the profiles of the sleep data sets.

With the algorithm it is possible to detect the REM phases also in RBD patients. Thus, the results can help to support the diagnosis of Parkinson's disease. Initially, the algorithm can be used to assist in the recognition of the sleep phases. In combination with the activity detection in the REM phase, which is characteristic for RBD, it can finally be diagnosed whether Parkinson's disease is involved. Thus, in the next step, a second algorithm will be finally developed to determine the activity in the REM phase.

Future work on this topic should address the above-mentioned aspects: the data processing pipeline should be optimized in order to process large sleep data sets more efficiently. Consequently, further features, classifiers and evaluation procedures should be taken into account. The application of neural networks e.g. could be worth considering [27], as well as the weighted training of classification in order to counteract the unequal share of REM sleep and non-REM sleep in the overall sleep duration.

The next step would be to extend the classifiers by an automated distinction between the REM sleep stage of healthy subjects and the REM sleep stage of RBD patients. This differentiation can potentially contribute to the early recognition of Parkinson's disease.

Acknowledgement. The authors acknowledge the public funding by the Federal Ministry of Education and Research of Germany in the framework of PCompanion (project number V5IKM011).

References

1. Prashanth, R., Roy, S.D., Mandal, P.K., Ghosh, S.: High-accuracy detection of early Parkinson's disease through multimodal features and machine learning. Int. J. Med. Inform. **90**, 13–21 (2016)
2. Doppler, K., et al.: Dermal phosphor-alpha-synuclein deposits confirm REM sleep behaviour disorder as prodromal Parkinson's disease. Acta Neuropathol. **133**(4), 535–545 (2017)
3. Postuma, R., et al.: Quantifying the risk of neurodegenerative disease in idiopathic REM sleep behavior disorder. Neurology **72**(15), 1296–1300 (2009)
4. Iranzo, A., et al.: Rapid-eye-movement sleep behaviour disorder as an early marker for a neurodegenerative disorder: a descriptive study. Lancet Neurol. **5**(7), 572–577 (2006)
5. Younes, M., Thompson, W., Leslie, C., Equan, T., Giannouli, E.: Utility of technologist editing of polysomnography scoring performed by a validated automatic system. Ann. Am. Thorac. Soc. **12**(8), 1206–1218 (2015)
6. Malhotra, A., et al.: Performance of an automated polysomnography scoring system versus computer-assisted manual scoring. Sleep **36**(4), 573–582 (2013)

7. Collop, N.A.: Coring variability between polysomnography technologists in different sleep laboratories. Sleep Med. **3**(1), 43–50 (2002)
8. Ferri, R., et al.: A new quantitative automatic method for the measurement of non-rapid eye movement sleep electroencephalographic amplitude variability. J. Sleep Res. **21**, 212–220 (2012)
9. Chiu, C.C., Hai, B.H., Yeh, S.J.: Recognition of sleep stage based on a combined neural network and fuzzy system using wavelet transform features. Biomed. Eng.: Appl. Basis Commun. **26**(2), 1450021–1450029 (2014)
10. Rechtschaffen, A., Kales, A. (eds.): A manual of standardized terminology, techniques and scoring system for sleep stages of human subjects, no. 204. National Institutes of Health Publications, U.S. Government Printing Office (1968)
11. Iber, C., Ancoli-Israel, S., Chesson, A., Quan, S.F.: The AASM Manual for the Scoring of Sleep and Associated Events, 1st edn. American Academy of Sleep Medicine, Westchester (2007)
12. Moser, D., et al.: Sleep classification according to AASM and Rechtschaffen & Kales: effects on sleep scoring parameters. Sleep **32**(2), 139–49 (2009)
13. Boeve, B.F., et al.: Pathophysiology of REM sleep behaviour disorder and relevance to neurodegenerative disease. Brain **130**(11), 2770–2788 (2007)
14. Boostani, R., Karimzadeh, F., Nami, M.: A comparative review on sleep stage classification methods in patients and healthy individuals. Comput. Methods Programs Biomed. **140**, 77–91 (2017)
15. Khalighi, S., Sousa, T., Pires, G., Nunes, U.: Automatic sleep staging: a computer assisted approach for optimal combination of features and polysomnographic channels. Expert Syst. Appl. **40**(17), 7046–7059 (2013)
16. Zhu, G., Li, Y., Wen, P.: Analysis and classification of sleep stages based on difference visibility graphs from a single-channel EEG signal. IEEE J. Biomed. Health Inform. **18**(6), 1813–1821 (2014)
17. Mohamad, I.B., Usman, D.: Standardization and its effects on K-means clustering algorithm. Res. J. Appl. Sci. Eng. Technol. **6**(17), 3299–3303 (2013)
18. Ross, B.C.: Mutual information between discrete and continuous data sets. PLoS ONE **9**(2), e87357 (2014)
19. Yun, C., Shin, D., Jo, H., Yang, J., Kim, S.: An experimental study on feature subset selection methods. In: 7th IEEE International Conference on Computer and Information Technology (CIT 2007), pp. 77–82. IEEE (2007)
20. Agrawal, R., Ram, B.: A modified k-nearest neighbor algorithm to handle uncertain data. In: 2015 5th International Conference on IT Convergence and Security (ICITCS), pp. 1–4. IEEE (2015)
21. Friedman, J., Hastie, T., Tibshirani, R.: The Elements of Statistical Learning. Springer Series in Statistics, vol. 1, no. 10. Springer, New York (2001)
22. Cristianini, N., Shawe-Taylor, J., et al.: An Introduction to Support Vector Machines and Other Kernel-based Learning Methods. Cambridge University Press, Cambridge (2000)
23. Witten, I.H., Frank, E., Hall, M.A., Pal, C.J.: Data Mining: Practical Machine Learning Tools and Techniques, 3rd edn. Morgan Kaufmann, Burlington (2017)
24. Breiman, L.: Random forests - random features technical report 576, Statistical Department, UC Berkeley, USA (1999)
25. Kumar, M., Sheshadri, H.: On the classification of imbalanced datasets. Int. J. Comput. Appl. **44**(8), 1–7 (2012)

26. Kirchner, J., Faghih-Naini, S., Bisgin, P., Fischer, G.: Sensor selection for classi-fication of physical activity in long-term wearable devices. In: IEEE Sensors, pp. 1–4 (2018)
27. Zhang, J., Yao, R., Ge, W., Gao, J.: Orthogonal convolutional neural networks for automatic sleep stage classification based on single-channel EEG. Comput. Methods Programs Biomed. **183**, 105089 (2020)

Towards an Automatized Way for Modeling Big Data System Architectures

Matthias Volk[✉], Daniel Staegemann[✉], Felix Prothmann[✉],
and Klaus Turowski[✉]

Otto-von-Guericke-University Magdeburg, Universitaetsplatz 2, 39106 Magdeburg, Germany
{matthias.volk,daniel.staegemann,felix.prothmann,
klaus.turowski}@ovgu.de

Abstract. Although the term of big data and related technologies received lots of attention in recent years, many projects are less successful than anticipated. One of the most crucial steps in the planning of a system includes the modeling of the underlying architecture. However, as of now, no standardized approach exists that facilitates the modeling of big data system architectures (BDSA). In this research, a systematic approach is presented that delivers a foundation towards a standard for the modeling of BDSA. Further, a prototype is introduced that automatizes the creation of those models reducing the required effort and simultaneously increasing the maintainability.

Keywords: Big data · System architecture · Modeling · Deployment diagram · Prototype · Literature review · Design science research

1 Introduction

Over that last decades, information technologies (IT) turned into an essential part of our society. Hand in hand with their increasing maturity and capability, also the effort needed for their provisioning increased. This includes the construction and deployment of hardware and software architectures, as well as their appropriate interconnection and integration. Altogether, this can be vaguely summarized under the term system architecture. The International Organization for Standardization (ISO), defines a system architecture as "fundamental concepts or properties of a system in its environment embodied in its elements, relationships, and in the principles of its design and evolution" [1]. In the context of this, not only the system itself but also the integration of stakeholders, hardware- and software components, as well as other systems needs to be considered [1, 2], resulting in a highly complex scenario. Hence, the planning and construction represents a sophisticated endeavor, at which assisting techniques may appear as a promising aid. In context of this, modeling constitutes a well-established approach for illustrating complex correlations by providing common specifications and notations for all relevant components of the envisioned architecture. In general, this discipline "has become a common practice in modern software engineering" [3]. Hence, the application of standardized modeling

© Springer Nature Switzerland AG 2020
W. Abramowicz and G. Klein (Eds.): BIS 2020, LNBIP 389, pp. 46–60, 2020.
https://doi.org/10.1007/978-3-030-53337-3_4

techniques seems to be an auspicious way to prevent misunderstandings, misinterpretations and information gaps [3, 4]. However, as it turns out, with the ongoing evolution of new information and communication technologies, not only the system architectures but also the modeling of integrative scenarios is becoming increasingly complex. This applies especially in times, at which novel technologies are combined and used to overcome the storing, processing and management of multi massive data. In course of this, *big data systems* proofed to be as a propitious solution. According to [5], big data "consists of extensive datasets primarily in the characteristics of volume, velocity, variety, and/or variability that require a scalable architecture for efficient storage, manipulation, and analysis". The consideration of those specifics complicates the general planning, creation, and integration of big data system architectures, as shown by many authors, such as [6]. In addition to the handling of massive amounts of data (volume), a multitude of different sources, data types (variety) and the required speed for the processing, also systems have to be robust and flexible for occurring changes (variability) [7].

Although approaches for the provisioning of concrete implementation details in the form of documentations or reference architectures exist, these often lack in the general way of presentation. Today, in most of the cases, self-defined depictions are chosen, sometimes composed out of different elements from multiple Unified Modeling Language (UML) diagram types [8, 9]. This situation is reinforced when it comes to the modeling of massive and complex systems, such as in the case of a big data system, which often comprises a multitude of different technologies. Not only the time spent for the correct identification of needed elements but also the alignment of those and the later maintenance can be a sophisticated and time-consuming task. Due to the aforementioned reasons, big data projects are not always as successful as intended. To overcome this problem the following research question (RQ) shall be answered in the course of this work:

"How can the comprehensiveness and systematics of modeling of big data system architectures (BDSA) be improved by using information technologies?".

Since, to our knowledge, no comprehensive solution exists, and there are also no sufficient preliminary works to build upon, it is necessary to approach this problem in a stepwise and systematic way. First, to obtain an understanding of the most relevant elements and information of a BDSA, existing models need to be identified and basic requirements derived. To further reduce the effort spent on the construction and maintenance of complex BDSA models, a suitable way to harness IT for the automatization of the modeling step will be examined. Hence, the following sub-research question (SRQ) will be answered during the course of this work:

SRQ1: *"How were BDSAs modeled so far?"*
SRQ2: *"Which information are needed for the modeling of a BDSA?"*
SRQ3: *"Are existing system architecture modeling approaches applicable for BDSA?"*
SRQ4: *"How can the modeling of BDSA be realized in an automatized way?"*

By the end, all made observations are set in relation to existing approaches for the modeling of system architectures, to identify a suitable solution for the modeling of BDSA. Eventually, all findings are intended to be used for an initial prototype that facilitates the automatized modeling of BDSA architecture only on the base of the required information.

1.1 Methodology

In order to find a suitable answer to the previously formulated research question, the design science methodology according to Hevner et al. [10] is followed. Due to the complexity of the planned undertaking, additionally, the recommended workflow according to Peffers et al. was used [11] which proposes six consecutive steps. After the problem was identified, (1) and the objectives of a solution defined (2), the design and development are taking place (3). Within the subsequent steps, the artifact is demonstrated (4), evaluated (5) and presented (6). Especially the transition from the second to the third steps requires an in-depth understanding regarding the particular domain and the required theory. Due to the existing lack of available contributions in this field, a thorough literature review was conducted, that attempts to stepwise approach the aforementioned research problem, following the guidelines provided by [12, 13]. Based on the used methodology, the contribution at hand is structured as follows. While the first section provides a thorough description of the identified problem, the main research question and the intended objectives of this research, in section two, the review protocol of the conducted literature review is outlined. Apart from the sole purpose of the identification of the related work, the results shall be used to answer the previously derived sub-research questions. All findings will be presented and discussed in the third section. Here, an approach towards the actual *standardization* is introduced and a prototype for the (automatized) modeling presented. Additionally, an evaluation is conducted, at which existing BDSA from research papers are *remodeled*. Eventually, the paper ends with a conclusion.

2 Literature Review

For the identification of a suitable approach, to model big data system architectures in a systematic way, first, an overview of the current state of the art is needed. This shall not only give an idea about possible approaches but also attempts to find an answer to the previously formulated SRQs. Consequently, to provide a better overview and enhance the comprehensibility as well as the reproducibility of the findings, the subsequent review protocol describes the performed steps of the conducted literature review in detail. Particular, a keyword-based search in combination with the forward-backward search, according to Webster & Watson, is utilized [12, 13].

2.1 Review Protocol

As a starting point for the collection of the needed material, six literature databases were queried, namely AIS Electronic Library (AISEL), Science Direct (SD), ACM Digital Library (ACM), Institute of Electrical and Electronics Engineers Xplore Digital Library (IEEE), Scopus und Wiley Online Library (Wiley). While most of these databases are operated by publishers, others, such as Scopus, serve more as a kind of meta-database.

Although the used query engines exhibit certain differences, in all of them, a keyword-based search was conducted, applying terms, such as "big data, system, architecture, modeling and model". For the reduction and refinement of the initially found out 5027 papers, a two-stepped refinement procedure was chosen, at which various criteria were checked, as they are described in Table 1. Whenever one of the inclusion criteria was not applicable or one of the exclusion criteria fulfilled, the paper was rejected from the search process.

Table 1. Inclusion and exclusion criteria

Inclusion criteria	Exclusion criteria
The paper must be published either in conference proceedings, a book, or in a journal	The papers predominantly deal with algorithms or frameworks to be implemented
The paper tackles at least one of the research questions	The paper is not written in English or German
The paper was published between 2013–2019	

While some of those criteria were checked during the query itself, like the publishing data, most of them were observed during the further course of the literature review. Within the first phase, only the title, keywords and abstract were examined and all duplicates originating from meta-databases removed. The remaining papers were afterward qualitatively analyzed in detail, constituting the second phase of the refinement. Eventually, 60 relevant papers were identified, representing 1.2% of the initial search set. An overview of the results is depicted in Table 2. Apart from the sole number of papers remaining after each phase, additionally, the concrete paper is referenced. As mentioned before, as a third step the forward & backward search [12] was subsequently conducted. This resulted in another nine contributions, focusing especially on the realization of reference architectures. Those can be typically used to support the engineering procedure in terms of an extensive and complex combination of tools and technologies for a targeted purpose. After all, each found out paper was qualitatively analyzed.

2.2 Findings

During the qualitative analysis of the found out papers, it was noticed from the very beginning that currently, no standardized approach exists to model BDSA. This may imply one the hand that a specific modeling approach is not required due to the reason that these do not differ very much from *traditional* IT architectures, or on the hand that authors are insecure about the modeling, due to the absence of a common approach. Especially the latter became apparent during the examination of the results. While in 11 cases architectures were depicted through the (partial) use of well-known diagrams from the UML, in the remaining 43 cases self-created notations were detected. The distribution with regard to the used modeling is depicted in Table 3. Frequently used approaches, originating from the UML, comprise diagrams such as components [8,

Table 2. Results of the literature review

Database	Initial query	Phase 1	Phase 2	Paper
AIS electronic library	98	16	9	[14–22]
Science direct	574	33	12	[9, 23–33]
ACM digital library	539	23	10	[34–43]
IEEE digital library	1460	54	21	[44–64]
Scopus	2242	19	7	[39, 65–68, 70, 71]
Wiley	114	4	1	[72]
Forward & backward search	–	–	–	[8, 73–80]

9, 36, 64], deployment [8, 65] or package diagrams [8, 36]. However, modifications on notation elements and their interconnection were also partially observed in here. One particular example is the case of [8], in which a reference architecture for the realization of predictive analytics is introduced. Apart from the sole use of one specific diagram, multiple approaches are combined here, structuring the architecture on multiple levels. In [9] another reference architecture for BDSA is presented. On the base of 15 requirements, 21 architectures were compared to each other and the results transferred into one comprehensive reference architecture. In general, only a few contributions were found that either implicitly or explicitly deal with the formulation of requirements on BDSA [9, 16, 26, 71]. At this point, the ability to store, manage and process *big data* was always from mandatory interest.

Table 3. Depiction of the modeled BDSA

Elements	No.	Paper
Component diagram	5	[8, 9, 36, 64]
Deployment diagram	2	[8, 65]
Package diagram	2	[8, 36]
Flow diagram	2	[24, 54, 80]
Self-created model	43	[14, 16, 19–23, 25–31, 33, 38, 42, 44, 46–58, 60–63, 67, 69–71, 74, 75, 77, 78, 80]

With regard to the self-created models, it was noticed that these differ greatly from each other in terms of both, the information density of the BDSA and the used notation elements for the description. In particular, some of the models focus only on the general overview, describing various components [38, 46, 67], while others deal with detailed subsystems and the relationship between each of the components of the BDSA, such as in [21, 23, 26, 42, 44, 57, 61, 62]. Additionally, in multiple publications only self-defined elements have been found [14, 20–25, 27–31, 44, 46, 48, 52, 53, 56, 61, 67, 74, 80]. Those were used and modified from existing types of modeling languages or entirely new created. In [80], for instance, a system architecture for remote sensing of satellites was developed and presented. The very complex depiction groups the relevant components and their interaction into different levels, which are logically structured from bottom-up, starting with the acquisition of the remote sensing data. Although later on a flow chart is used for the general workflow, the idea remains unclear at the beginning. This is especially due to the reason that no notation elements are described and no known elements used. Hence, interpretations concerning each of the components can be made. Similar to this, in [31] an architecture is constructed that intents to forecast social and economic changes. Although the model depicts a good and structured overview, not many details are provided. Within the depiction of the system, the authors use the data life cycle to reveal the different stages, starting from the *data receiving* to the *publishing module*. In doing so, arrows between the different components are used to show the flow of data. However, no detailed transition information are presented, such as the connection on a system-level or the implemented functions. Thus, as one may note, the modeling of big data architectures exhibits large differences, not only in terms of the level of detail. However, in most of the related papers important information, which are needed for the modeling of a BDSA, were identified. In particular, this includes information to the following general elements: subsystem, components, artifact, relation, actors and software. All of them and the related contributions are listed in Table 4 and are subsequently further described.

Table 4. Elements of a BDSA

Elements	No.	Paper
Subsystem	45	[8, 9, 16, 17, 19, 21–23, 26–33, 38, 39, 42, 44, 46, 48–52, 54–58, 61–65, 67, 69–71, 74, 75, 77, 78, 80]
Component	48	[8, 9, 14, 16, 17, 19–33, 36, 38, 44, 46, 48, 50–58, 60–62, 64, 65, 67, 69–71, 74, 75, 77, 78, 80]
Artifact	43	[8, 9, 14, 17, 19–21, 23–25, 27–31, 33, 36, 38, 40, 44, 46–53, 55, 56, 58, 60–62, 64, 67, 69–71, 74, 77, 78, 80]
Relation	43	[8, 9, 20–31, 33, 36, 38, 44, 46–51, 53, 55–58, 60–62, 64, 65, 67, 69–71, 74, 75, 77, 78, 80]
Actor	3	[20, 63, 64]
Software	30	[9, 15, 20, 21, 25, 27–29, 31, 36, 39–42, 44, 47, 49, 51, 52, 56, 61, 62, 65, 66, 68–70, 72, 78, 79]

A *Component* is "a unit of computation or a data store" [81] that encapsulates several functionalities for later reuse. Often, these are further distinguished into hardware and software components. Typically, they are either part of *subsystems* or the system itself. In turn, a *subsystem* refers to the collection of elements, components and connections to fulfill a specific purpose within a system architecture [82]. The investigation of the contributions revealed that most of these elements reoccurred as part of system architecture, although they had huge differences in terms of the used name. Further complications were ascertained if the level of detail was very low. In this case, sometimes the subsystem did not deliver any specific information and acted more like a single component. Found out types of subsystems comprise, for instance: data sources, data storage, data collection, data preparation, data integration, data preprocessing, analysis, visualization, and management. Data sources, in turn, consisting out of *components*, such as sensors, machines, databases or streams, were almost always part of the proposed model of the architecture [14, 19, 21–23, 29, 31, 44, 48, 52, 61, 73, 74, 77, 78]. The same applies to the data storage that was often described by the HDFS, data warehouses or SQL-, NoSQL- and cloud storages [19, 21, 22, 29, 31, 39, 42, 48, 52, 55, 57, 61, 63, 64, 71, 78, 80]. The used analysis subsystem, consisting out of components such as data mining, predictive analytics, stream analysis or data rating, was also of major interest [16, 19, 22, 27, 29, 31, 33, 44, 52, 62, 70, 71, 74, 77, 78, 80].

Sometimes those components revealed concrete insights regarding the interfaces, functionalities or the inner structure. At this point, executable (e.g. scripts, programs) or non-executable elements (e.g. web data, documents, videos) were often used as input and output and aligned to the term of an *artifact*. This conforms also the definition according to the UML that make use of those elements in their deployment diagrams. In their documentation, an artifact "represents some (usually reifiable) item of information that is used or produced by a software development process or by operation of a system" [82].

Relations are used in almost all of the found out papers. In most of the cases, these were depicted either using undirected lines or arrows. Noticeable was the vague use of most of the connections. Sometimes legends or needed information were provided, but frequently not. At this point, the corresponding text sections had to be read or interpretations made. A closer look at this notation element revealed that the connection was often used for different purposes. For instance, in [20, 22, 23, 61, 75] it was used for the exchange of messages, such as inquires for reports, uploads, and others. Whereas in [8, 25] deployments, as they are known from deployment diagrams, were implicitly depicted by this symbol. Around 73% of the use cases dealt in general with the communication between different elements [8, 9, 16, 20, 22–24, 27, 28, 30, 31, 33, 36, 38, 44, 46–48, 50, 51, 53, 55–58, 61–63, 69, 71, 77, 78, 80]. In most of the cases, the data was from major interest. For instance, the general use of the data as described in [8, 20, 62], whereas the logical and processing flow delineated in [21, 23–26, 29–31, 44, 55, 58, 60, 64, 70, 75]. The regulation of the data access is depicted in [21, 24, 26, 44, 57, 64], either implicitly through the communication between different components or directly in terms of the data source.

An *actor* in this context describes a person that interacts with the intended BDSA. Although these were addressed only in a few papers directly [20, 63, 64], they represent an important piece of information that especially reveals in which way the system as a whole, single components or the software is going to be used. The *software* element directly refers to specific implementation details regarding the used software products within the architecture. While some of the papers contained detailed information, in either the depicted architecture or the related paragraphs, others just described general functionalities without naming concrete tools. Although a multitude of modeling approaches for a BDSA were identified, no uniform applicable method was found within the contributions. Only a few made use of established and well-accepted diagrams of the UML to display the BDSA in parts (cf. Table 3). However, even in those contributions, missing parts, incorrect use of notation elements and other obstacles were identified. It can be assumed that this originates partially out of the targeted audience, which are in here rather researchers than system architects. In any case, again, it highlights the absence of a unified approach and the necessity to find a suitable way for the depiction of BDSA as a whole.

3   A Systematic Modeling Approach for BDSA

By definition, diagram types such as the deployment diagram already offer a good way to present system architectures [82], even if they do not necessarily deal with big data-related elements. According to the Object Management Group (OMG), which is in charge of the UML, a deployment diagram "specifies constructs that can be used to define the execution architecture of systems and the assignment of software artifacts to system elements" [82]. Compared to the component diagrams, they can also be used to give general insights, and thus, to provide a conceptual overview of the architecture rather than a concrete functional implementation. In cases, such as the previously identified ones, the adoption and modification of deployment diagrams appears to be a good choice towards the standardization of a BDSA modeling approach. In order to perform manipulations for the realization of "more elaborate deployment models" [82], profiles and meta-models can be used. By considering the aforementioned results from the literature review, several changes were made to the conventional deployment diagrams.

The element of a person, known from use-case diagrams, is integrated. This reflects the fact that often a number of stakeholders interact with the system in different ways. The same applies to the other frequently used elements such as the cloud or specific subsystems. Furthermore, already existing elements of the diagram were changed, particularly focusing on the referred profiles. Artifacts that depict input and output received the stereotypes `processed_data`, `audio`, `video`, `picture`, and `webdata`. While most of these are self-explaining, the latter includes data especially originating from the web, such as social media. Sensors, predominately used in the IoT domain and smart cities, are defined through the node stereotype `sensor`. If complex networks, that handle a number of different components and artifacts, need to be displayed, the stereotype of the `network` for a node can be used. Since the processing speed of the data and its creation plays a major role, the connection of the elements needs to be adjusted. This can be done through additional connection information `data_stream`.

In order to simplify the modeling process itself, an automatized procedure was chosen, that creates a BDSA model out of an existing JSON configuration file. This is realized not only due to the reason that sometimes related architectures are getting very complex and errors occur or elements are missing, but also for future improvements and required changes of the model itself. In doing so, the entire layout of the model will be built automatically. An excerpt from this JSON file is depicted on the left-hand side of Fig. 1. To ensure that the model can be successfully created, a potential user is prompted to follow the predefined structure of the input file. By using a self-developed validator, the structure is checked before the actual modeling is taking place. This includes, for instance, the verification of missing elements, parameters or connections and the identification of circulations, multiple occurrences of an element and syntax errors. In general, the structure of the JSON file is comprised of the elements `settings`, `artifacts`, `nodes`, `components`, `actors`, `subsystems` and `connections`. Within the first area, basic characteristics for the desired model can be defined, such as the align-ment of the elements. For a more appealing visualization, the tag `leftToRight` was defined to either build the model in a horizontal or vertical direction. This shall ensure that even though numerous elements are included, still, the readability is achieved. All other elements are used as described before. `Actors` represent stakeholders that are directly interacting with the system. Those are symbolized by a person. Relevant sys-tem elements are described through `nodes`, as either a separate entity or children of a `subsystem`. If concrete information about the functionalities and defined interfaces need to be displayed, `components` can be used. The definition of each of those, except the `settings` and `connections`, needs to be performed by a unique ID, as well as an expressive name. The technical implementation of the prototype is realized by a client-server architecture using Java as a runtime environment, Vaadin as a web interface and PlantUML as the modeling tool. After a thorough comparison of various model-ing tools, such as Modelio, Papyrus, BOUML, Eclipse UML2 and PlantUML, only the latter turned out to be a promising solution for the pursued task. Criteria, used for the comparison, were, for instance, the support of existing UML diagrams, supported pro-gramming languages, diagram layouting, frequent updates, exemplary code fragments and comprehensive documentation. In Fig. 1, an easy to use prototype is depicted, which includes the aforementioned changes on the existing diagrams, the JSON input field, and the field for the compiled model.

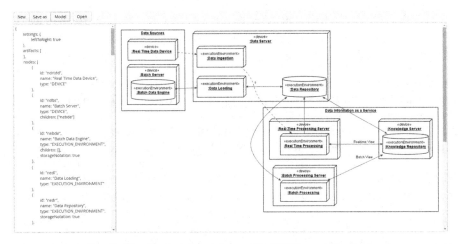

Fig. 1. The developed prototype of the BDSA modeling tool

In accordance with the used DSR methodology [10, 11] the proposed artifact was continuously evaluated throughout all stages. The starting point is formed by architec-tures, which were previously identified during the literature review and mostly introduce self-defined notation elements. In particular, the following contributions were exemplary selected for the evaluation [8, 9, 21, 31, 48, 56, 80]. Each of them was reconstructed by following the proposed modeling approach. First, for every single architecture, the needed input information were extracted out of the existing model and the corresponding text section. Secondly, the configuration was manually created and afterward imported into the developed prototype. Due to the huge number of elements and size of the final model, only one of those architectures [21] is depicted on the right-hand side of Fig. 1. Although it was once more noticed that most of the architectures did not explicit any concrete insights about the connected parts and specific tools to be used, it was possible to (re-) model each of those without discarding system architecture-relevant parts. Hence, the functionality of the developed modeling approach, as well as the constructed prototype, was successfully evaluated.

3.1 Discussion

The demonstration and evaluation of the developed artifact revealed that BDSA can be modeled in a comprehensive and systematic way, independent from the level of detail. Although, seven exemplarily chosen architecture were successfully remodeled by using the modified modeling language and the developed prototype, further evaluations are in a larger scale needed. Apart from the pure consideration of additional architectures, this includes furthermore the implementation within a real-world context and the analysis of the appearance by comparing the models of the developed solution with the original ones. In doing so, additional tweaks, changes and extensions that are beneficial for the current solution might be identified. Besides the extension of the developed solution in terms of new specifications, also the realization in terms of a particular domain-specific language (DSL) is imaginable. For the initial prototype, most of the alignments, syntax, semantics, interpretations and validations were self-programmed. In the future, when the solution is getting more and more complex, even more, elaborated concepts, that offer the automatic provision of relevant code pieces, could be added. DSLs typically serve as kind of a *programming language* for a specific domain targeting only the needed functionalities and, thus, will highly reduce the cost for development and maintenance [83]. Beyond that, an integration within a concrete big data planning tool, such as proposed in the work by Volk et al. [84], appears to be sensible. In their work, they introduced an end-to-end decision support system, which shall help decision-makers with the thorough planning and realization of big data projects. In this context, they also described an automatized way for the modeling of BDSA, facilitating the visualization of the given recommendation of suitable tools and their connection.

4 Conclusion

In recent years, big data coined to be one of the key technologies for the realization of data-intensive projects, independent from the scope of application. Although it received lots of attention and research from both sides, academia as well as the industry, today lots of confusion still exists when it comes to the actual engineering of relevant architectures. One step towards the expedient planning and realization of big data endeavors is based on the grasp of the elements and their composition in BDSA. Currently, no standard or universally followed best practice for the construction of relevant models exists. Due to this, in this paper, the question, how the comprehensiveness and systematics of modeling a BDSA can be improved by using IT, was discussed. By applying the DSR methodology and conducting a structured literature review, the current state of the art is identified and a possible solution proposed. In particular, an enhanced version of a deployment diagram, suitable for the modeling of BDSA, is developed. Moreover, a prototype was developed that automatizes the IT-driven creation of those models to reduce the required effort and to increase the maintainability. This research provides researches as well as system architects a solution to cover specific information, required for the general system understanding, in an appealing way. Hence, planning and communication between different stakeholders can be greatly improved.

References

1. ISO/IEC/IEEE 42010:2011(E): Systems and software engineering—architecture description. IEEE Computer Society
2. Golden, B.: A Unified Formalism for Complex Systems Architecture (2013)
3. Chaudron, M.R.V., Heijstek, W., Nugroho, A.: How effective is UML modeling? Softw. Syst. Model. **11**, 571–580 (2012)
4. Desic, S., Gvozdanovic, D., Kusek, M., Huljenic, D.: Advantages of UML-based object-oriented system development. In: MIPRO Meeting (2011)
5. NIST Big Data Interoperability Framework, vol. 1, definitions, version 2. National Institute of Standards and Technology, Gaithersburg, MD (2018)
6. Volk, M., Bosse, S., Bischoff, D., Turowski, K.: Decision-support for selecting big data reference architectures. In: Abramowicz, W. (ed.) 22nd International Conference, BIS (Business Information Systems) 2019, pp. 3–17 (2019)
7. Staegemann, D., Volk, M., Nahhas, A., Abdallah, M., Turowski, K.: Exploring the specificities and challenges of testing big data systems. In: 15th International Conference on Signal Image Technology and Internet based Systems, SITIS, Italy (2019)
8. Geerdink, B.: A reference architecture for big data solutions introducing a model to perform predictive analytics using big data technology. In: ICITST (International Conference for Internet Technology and Secured Transactions), vol. 8, pp. 71–76 (2013)
9. Hruschka, S., Herrero, V., Romero, O., Abelló, A., Franch, X., Vansummeren, S., Valerio, D.: A software reference architecture for semantic-aware big data systems. Inf. Softw. Technol. **90**, 75–92 (2017)
10. Hevner, A.R., March, S.T., Park, J., Ram, S.: Design science in information systems research. MIS Q. **28**, 75–105 (2004)
11. Peffers, K., Rothenberger, M., Tuunanen, T., Vaezi, R.: Design science research evaluation. In: DESRIST (International Conference on Design Science Research in Information Systems), pp. 398–410 (2012)
12. Webster, J., Watson, R.T.: Guest editorial: analyzing the past to prepare for the future: writing a literature review. MIS Q. **26**, xiii–xxiii (11 p.) (2002)
13. Levy, Y., Ellis, T.J.: A systems approach to conduct an effective literature review in support of information systems research. Inform. Sci. J. **9**, 181–212 (2006)
14. Tan, C., Sun, L., Liu, K.: Big data architecture for pervasive healthcare: a literature review. In: ECIS (European Conference on Information Systems), vol. 23 (2015)
15. Gölzer, P., Cato, P., Amberg, M.: Data processing requirements of industry 4.0 - use cases for big data applications. In: ECIS (Conference: European Conference on Information Systems), vol. 23 (2015)
16. Burmeister, F., Drews, P., Schirmer, I.: Towards an extended enterprise architecture meta-model for big data - a literature-based approach. In: AMCIS (Americas Conference on Information Systems), vol. 24 (2018)

17. Goes, P.B.: Big data - analytics engine for digital transformation: where is IS? In: AMCIS (Americas Conference on Information Systems) (2015)
18. Chen, H.-M., Kazman, R., Garbajosa, J., Gonzalez, E.: Big data value engineering for business model innovation. In: HICSS (Hawaii International Conference on System Sciences), vol. 50, pp. 5921–5930 (2017)
19. Passlick, J., Lebek, B., Breitner, M.H.: A self-service supporting business intelligence and big data analytics architecture. In: Wirtschaftsinformatik 2017, pp. 1126–1140 (2017)
20. Schwarz, C., Schwarz, A., Black, W.C.: Examining the impact of multicollinearity in discovering higher-order factor models. CAIS **34**(1), 62 (2014)
21. Le Dinh, T., Phan, T.-C., Bui, T.: Towards an Architecture for big data-driven knowledge management systems. In: SIGODIS (Intelligence And Intelligent Systems), pp. 1–10 (2016)
22. Alshboul, Y., Nepali, R., Wang, Y.: Big data lifecycle: threats and security model. In: SIGSEC (Information Systems Security, Assurance and Privacy) (2015)
23. Persico, V., Pescapé, A., Picariello, A., Sperlí, G.: Benchmarking big data architectures for social networks data processing using public cloud platforms. Future Gener. Comput. Syst. **89**, 98–109 (2018)
24. Yuan, J., Chen, M., Jiang, T., Li, T.: Complete tolerance relation based parallel filling for incomplete energy big data. Knowl.-Based Syst. **132**, 215–225 (2017)
25. Song, J., Guo, C., Wang, Z., Zhang, Y., Yu, G., Pierson, J.-M.: HaoLap: a Hadoop based OLAP system for big data. J. Syst. Softw. **102**, 167–181 (2015)
26. Hsu, H.-H., Chang, C.-Y., Hsu, C.-H. (eds.): Big Data Analytics for Sensor-Network Collected Intelligence. Academic Press, London (2017)
27. Campos, J., Sharma, P., Gabiria, U.G., Jantunen, E., Baglee, D.: A big data analytical architecture for the asset management. CIRP **64**, 369–374 (2017)
28. Ahmad, A., Babar, M., Din, S., Khalid, S., Ullah, M.M., Paul, A., Goutham Reddy, A., Min-Allah, N.: Socio-cyber network: The potential of cyber-physical system to define human behaviors using big data analytics. Future Gener. Comput. Syst. **92**, 868–878 (2019)
29. Ahmad, A., Khan, M., Paul, A., Din, S., Rathore, M.M., Jeon, G., Choi, G.S.: Toward modeling and optimization of features selection in big data based social Internet of Things. Future Gener. Comput. Syst. **82**, 715–726 (2018)
30. Babar, M., Rahman, A., Arif, F., Jeon, G.: Energy-harvesting based on internet of things and big data analytics for smart health monitoring. Sustain. Comput.: Inf. Syst. **20**, 155–164 (2018)
31. Blazquez, D., Domenech, J.: Big data sources and methods for social and economic analyses. Technol. Forecast. Soc. Change **130**, 99–113 (2018)
32. Mistrík, I. (ed.): Software Architecture for Big Data and the Cloud. MK an imprint of Elsevier, Cambridge (2017)
33. Spangenberg, N., Wilke, M., Franczyk, B.: A big data architecture for intra-surgical remaining time predictions. Procedia Comput. Sci. **113**, 310–317 (2017)
34. Chen, H.-M., Kazman, R., Garbajosa, J., Gonzalez, E.: Toward big data value engineering for innovation. In: BIGDSE (International Workshop on Big Data Software Engineering), vol. 2, pp. 44–50 (2016)
35. Emmanuel, I., Stanier, C.: Defining big data. In: BDCA (International Conference on Big Data and Advanced Wireless Technologies), pp. 1–6 (2016)
36. Guerriero, M., Tajfar, S., Tamburri, D.A., Di Nitto, E.: Towards a model-driven design tool for big data architectures. In: BIGDSE (International Workshop on Big Data Software Engineering), vol. 2, pp. 37–43 (2016)
37. Khan, N., Alsaqer, M., Shah, H., Badsha, G., Ahmad Abbasi, A., Salehian, S.: The 10 Vs, issues and challenges of big data. In: ICBDE (International Conference on Big Data and Education), pp. 52–56 (2018)

38. Klein, J., Buglak, R., Blockow, D., Wuttke, T., Cooper, B.: A reference architecture for big data systems in the national security domain. In: BIGDSE (International Workshop on Big Data Software Engineering), vol. 2, pp. 51–57 (2016)
39. Nielsen, F.Å.: A new ANEW: evaluation of a word list for sentiment analysis in microblogs. In: MSM (Workshop on 'Making Sense of Microposts'), pp. 47–51 (2011)
40. Ptiček, M., Vrdoljak, B.: Big data and new data warehousing approaches. In: ICCBDC (International Conference on Cloud and Big Data Computing), pp. 6–10 (2017)
41. Sebaa, A., Nouicer, A., Chikh, F., Tari, A.: Big data technologies to improve medical data warehousing. In: BDCA (international Conference on Big Data, Cloud and Applications), vol. 2, pp. 1–5 (2017)
42. Seref, B., Bostanci, E.: Opportunities, threats and future directions in big data for medical wearables. In: BDAW (International Conference on Big Data and Advanced Wireless Technologies), pp. 1–5 (2016)
43. Zafar, M.N., Azam, F., Rehman, S., Anwar, M.W.: A systematic review of big data analytics using model driven engineering. In: ICCBDC (International Conference on Cloud and Big Data Computing), pp. 1–5 (2017)
44. Sang, G.M., Xu, L., Vrieze, P.D.: A reference architecture for big data systems. In: SKIMA (International Conference on Software, Knowledge, Information Management and Applications), vol. 10 (2016)
45. Chen, H.-M., Kazman, R., Haziyev, S.: Agile big data analytics development: an architecture-centric approach. In: HICSS (Hawaii International Conference on System Sciences), vol. 49, pp. 5378–5387 (2016)
46. Darwish, T.S.J., Abu Bakar, K.: Fog based intelligent transportation big data analytics in the Internet of vehicles environment: motivations, architecture, challenges, and critical issues. IEEE Access 6, 15679–15701 (2018)
47. Boci, E., Thistlethwaite, S.: A novel big data architecture in support of ADS-B data analytic. In: ICNS (Integrated Communication, Navigation and Surveillance Conference), pp. C1-1–C1-8 (2015)
48. Gohar, M., Hassan, S.A., Khan, M., Guizani, N., Ahmed, A., Rahman, A.U.: A big data analytics architecture for the Internet of small Things. IEEE Mag. 56, 128–133 (2018)
49. Haroun, A., Mostefaoui, A., Dessables, F.: A big data architecture for automotive applications: PSA group deployment experience. In: CCGRID (International Symposium on Cluster, Cloud and Grid Computing), vol. 17, pp. 921–928 (2017)
50. Twardowski, B., Ryzko, D.: Multi-agent architecture for real-time Big Data processing. In: IEEE/WIC/ACM International Joint Conferences on Web Intelligence (WI) and Intelligent Agent Technologies (IAT), vol. 3, pp. 333–337 (2014)
51. Kiran, M., Murphy, P., Monga, I., Dugan, J., Singh Baveja, S.: Lambda architecture for cost-effective batch and speed big data processing. In: Big Data (IEEE International Conference on Big Data), pp. 2785–2792 (2015)
52. Pavlikov, R., Beisembekova, R.: Architecture and security tools in distributed information systems with Big Data. In: AICT (International Conference on Application of Information and Communication Technologies), vol. 10 (2016)
53. Koley, S., Nandy, S., Dutta, P., Dhar, S., Sur, T.: Big data architecture with mobile cloud in CDroid operating system for storing huge data. In: CAST, pp. 12–17 (2016)
54. Din, S., Ghayvat, H., Paul, A., Ahmad, A., Rathore, M.M., Shafi, I.: An architecture to analyze big data in the Internet of Things. In: ICST (International Conference on Sensing Technology), vol. 9, pp. 677–682 (2015)
55. Wang, H., Wang, Q., Liu, P., Sun, L.: Big data and intelligent agent based smart grid architecture. In: ICA (IEEE International Conference on Agents), pp. 106–107 (2017)
56. Kashlev, A., Lu, S.: A System architecture for running big data workflows in the cloud. In: IEEE International Conference on Services Computing, pp. 51–58 (2014)

57. Agrawal, R., Imran, A., Seay, C., Walker, J.: A layer based architecture for provenance in big data. In: Big Data (International Conference on Big Data), pp. 29–31 (2014)
58. Liu, D.: Big data analytics architecture for internet-of-vehicles based on the spark. In: ICITBS (International Conference on Intelligent Transportation, Big Data and Smart City), pp. 13–16 (2018)
59. Martinez-Mosquera, D., Lujan-Mora, S., Recalde, H.: Conceptual modeling of big data extract processes with UML. In: INCISCOS (International Conference on Information Systems and Computer Science), pp. 207–211 (2017)
60. Munar, A., Chiner, E., Sales, I.: A big data financial information management architecture for global banking. In: FiCloud (International Conference on Future Internet of Things and Cloud), vol. 2, pp. 385–388 (2014)
61. Costa, C., Santos, M.Y.: BASIS: a big data architecture for smart cities. In: SAI (SAI Computing Conference), pp. 1247–1256 (2016)
62. Siriweera, T.H.A.S., Paik, I., Kumara, B.T.G.S., Koswatta, K.R.C.: Intelligent big data analysis architecture based on automatic service composition. In: IEEE International Congress on Big Data, pp. 276–280 (2015)
63. Sergeevich, K.A., Ovseevna, A.M., Petrovich, S.I.: Web-application for real-time big data visualization of complex physical experiments. In: SIBCON (2015)
64. Viana, P., Sato, L.: A proposal for a reference architecture for long-term archiving, preservation, and retrieval of big data. In: International Conference on Trust, Security and Privacy in Computing and Communications), vol. 13, pp. 622–629 (2014)
65. Canito, A., Fernandes, M., Conceição, L., Praça, I., Marreiros, G.: A big data platform for industrial enterprise asset value enablers. In: DCAI (International Conference on Distributed Computing and Artificial Intelligence), vol. 15, pp. 145–154 (2018)
66. Koren, O., Binyaminov, M., Perel, N.: The impact of distributed data in big data platforms on organizations. In: FTC (Proceedings of the Future Technologies Conference), pp. 1024–1036 (2018)
67. Lu, Y., Xu, X.: Cloud-based manufacturing equipment and big data analytics to enable on-demand manufacturing services. Robot. Comput.-Integr. Manuf. **57**, 92–102 (2019)
68. Shakhovska, N., Duda, O., Matsiuk, O., Bolyubash, Y., Vovnyanka, R.: Analysis of the activity of territorial communities using information technology of big data based on the entity-characteristic mode. In: CSIT (International Conference on Computer Science and Information Technologies), pp. 155–170 (2018)
69. Narain Singh, K., Kumar Behera, R., Kumar Mantri, J.: Big data ecosystem: review on architectural evolution. In: IEMIS (Emerging Technologies in Data Mining and Information Security), vol. 2, pp. 335–345 (2018)
70. Singh, P.K., Verma, R.K., Krishna Prasad, P.E.S.N.: IoT-based smartbots for smart city using MCC and big data. In: SIST (Smart Intelligent Computing and Applications), pp. 525–534 (2018)
71. Woo, J., Shin, S.-J., Seo, W., Meilanitasari, P.: Developing a big data analytics platform for manufacturing systems: architecture, method, and implementation. Int. J. Adv. Manuf. Technol. **99**, 2193–2217 (2018)
72. Billot, R., Bothorel, C., Lenca, P.: Introduction to Big Data and Its Applications in Insurance, Chap. 1, pp. 1–25 (2018)
73. Assunção, M.D., Calheiros, R.N., Bianchi, S., Netto, M.A.S., Buyya, R.: Big Data computing and clouds: Trends and future directions. J. Parallel Distrib. Comput. **79–80**, 3–15 (2015)
74. Borodo, S.M., Shamsuddin, S.M., Hasan, S.: Big data platforms and techniques. IJEECS **1**, 191–200 (2016)
75. Chen, H.-M., Kazman, R., Haziyev, S.: Agile big data analytics for web-based systems: an architecture-centric approach. IEEE Trans. Big Data **2**, 234–248 (2016)

76. Chen, H.-M., Kazman, R., Haziyev, S.: Strategic prototyping for developing big data systems. IEEE Softw. **33**, 36–43 (2016)
77. Demchenko, Y., Ngo, C., Membrey, P.: Architecture Framework and Components for the Big Data Ecosystem (2013)
78. Pääkkönen, P., Pakkala, D.: Reference architecture and classification of technologies, products and services for big data systems. Big Data Res. **2**, 166–186 (2015)
79. Philip Chen, C.L., Zhang, C.-Y.: Data-intensive applications, challenges, techniques and technologies: a survey on big data. Inf. Sci. **275**, 314–347 (2014)
80. Ullah Rathore, M.M., Paul, A., Ahmad, A., Chen, B.-W., Huang, B., Ji, W.: Real-time big data analytical architecture for remote sensing application. IEEE J. Sel. Top. Appl. Earth Obs. Remote Sens. **8**, 4610–4621 (2015)
81. Medvidovic, N., Taylor, R.N.: A classification and comparison framework for software architecture description languages. IIEEE Trans. Softw. Eng. **26**, 70–93 (2000)
82. OMG: Unified Modeling Language, v 2.5.1, pp. 1–796 (2017)
83. Bettini, L.: Implementing domain-specific languages with Xtext and Xtend. Learn how to implement a DSL with Xtext and Xtend using easy-to-understand examples and best practices. Packt Publishing (2016)
84. Volk, M., Staegemann, D., Pohl, M., Turowski, K.: Challenging big data engineering: positioning of current and future development. In: Proceedings of the 4th International, pp. 351–358 (2019)

Empowering Domain Experts to Preprocess Massive Distributed Datasets

Michael Behringer$^{(\boxtimes)}$, Pascal Hirmer, Manuel Fritz, and Bernhard Mitschang

Institute for Parallel and Distributed Systems, University of Stuttgart,
70569 Stuttgart, Germany
`michael.behringer@ipvs.uni-stuttgart.de`

Abstract. In recent years, the amount of data is growing extensively. In companies, spreadsheets are one common approach to conduct data processing and statistical analysis. However, especially when working with massive amounts of data, spreadsheet applications have their limitations. To cope with this issue, we introduce a human-in-the-loop approach for scalable data preprocessing using sampling. In contrast to state-of-the-art approaches, we also consider conflict resolution and recommendations based on data not contained in the sample itself. We implemented a fully functional prototype and conducted a user study with 12 participants. We show that our approach delivers a significantly higher error correction than comparable approaches which only consider the sample dataset.

Keywords: Data cleaning · Human-in-the-loop · Interactive data preprocessing

1 Introduction

Today, the amount of available data is growing very fast and more and more business processes and decisions are based on analysis of data. According to Gartner [9], spreadsheets are still one of the most common applications for analyzing data and a fundamental tool for budgeting, planning and forecasting in large enterprises. This is not surprising as spreadsheets are well-established, intuitive to use, and extensible by macros. However, the issue with spreadsheets is that they are very error-prone and up to 94% contain errors [9], which can lead to costly impacts. Some examples are provided by Gandel [6], e.g., hidden cells which are not visible in the spreadsheet anymore reappearing during export and leading to wrong assumptions. Even though spreadsheets are used for today's data volumes, they *"were never intended to be used as a data store for millions - even billions - of cells"* [9]. As a result, especially when working with large amounts of data, i.e., tables with thousands of rows, spreadsheet applications have their limitations. For huge datasets, data visualization, processing, and analysis take a lot of time or cannot be handled at all. Consequently, many applications provide a sample of the entire dataset in order to be able to work

© Springer Nature Switzerland AG 2020
W. Abramowicz and G. Klein (Eds.): BIS 2020, LNBIP 389, pp. 61–75, 2020.
https://doi.org/10.1007/978-3-030-53337-3_5

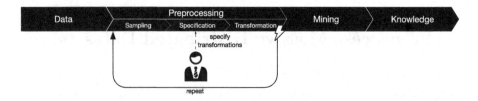

Fig. 1. Data analysis pipeline extended for interactive preprocessing.

with a small subset of the entire dataset. This is most commonly called *specification phase*. Subsequently, the changes made in the sample data is then applied to the entire dataset, which is called *execution phase*. This is depicted in Fig. 1. Even though this approach of separating phases is feasible in some scenarios, there are severe issues that arise when relying on sampled data. On the one hand, conflicts between the sample data and the entire dataset are very difficult to solve without the insights of the user. Typical conflicts are, e.g., type mismatch or feature characteristics not covered in the sample. On the other hand, most applications that provide sampling do not offer any means for interactive scalability, which is especially important in the era of big data when employing analysts' domain knowledge during the preprocessing steps. Instead, these tools connect to Hadoop-based ecosystems only for executing the specified transformations. More precisely, transformations are specified on a sample generated once and applied to the entire dataset in a subsequent step. Thus, it is difficult to adapt to occurring conflicts in records not covered by the sample.

There are three approaches to cope with this challenge. First, another sample is requested by the user. This approach has two major drawbacks: (a) the domain expert needs to draw a lot of samples until all conflicts are represented at least once, and (b) a large overhead regarding processing time. Second, the user decides to discard all records which are responsible for conflicts. However, uncleaned data can have unpredictable impacts on the quality of the subsequent analysis steps. Both of these approaches become particularly infeasible when facing big data. With each (failed) iteration, the preprocessing becomes more time consuming, the domain expert loses motivation, and impacts of an increasing number of discarded records are unpredictable. Therefore, the third approach is to notify domain experts about occurring conflicts immediately and to empower them to resolve them. This approach is the most promising one because the user stays in the loop during data preprocessing.

To realize the third approach, in this paper, we introduce an approach for scalable data preprocessing using sampling while keeping the user in the loop. In contrast to state-of-the-art approaches, we also consider conflict resolution and recommendations based on data not contained in the sample itself. Our approach consists of an architecture for scalable data processing, including a Hadoop-based data storage, as well as concepts for human interaction during data preprocessing. We reach this by establishing a fine-grained iteration between the specification and execution phases. To preserve generality, our framework is capable of using any sample algorithms as well as any automatic error recognition/data cleaning methods. The framework is validated through a case study in the business domain as well as a user study to evaluate user acceptance.

The remainder of this paper is structured as follows: In Sect. 2, we give an exemplary scenario to introduce the target users as well as problems and benefits in more detail. In Sect. 3, related work is described. In Sect. 4, we introduce our main contribution: an architecture for data preprocessing which enables to keep the human in the loop. For evaluation purposes, a user study is described which shows the benefits in Sect. 5. Finally, Sect. 6 summarizes the paper and describe future work.

2 Motivating Scenario

In this section, we present a scenario, in which we introduce the target user and point out emerging issues regarding interactive data preprocessing in the context of large-scale datasets. In order to conduct a reliable analysis and use the gained insights, it is essential to obtain an understanding of the data. During this phase, datasets are examined with only a small amount of knowledge about their correlation or context known a priori. Consequently, this explorative process is strongly characterized by trial and error. According to studies [18], more than 80% of the participants consider it as especially important to prepare data in self-service. Furthermore, there is a consensus that domain knowledge is indispensable for a reliable analysis, which is of particular interest during the preprocessing step.

Albeit, a domain expert who performs an explorative analysis is often not a trained data scientist. Rather, this individual is, according to Gartner [14], a *"person who creates or generates models that leverage predictive or prescriptive analytics but whose primary job function is outside of the field of statistics and analytics."* and is named citizen data scientist (CDS).

Figure 1 depicts the typical phases of an analysis process, such as the KDD process [5] or CRISP-DM [17], and extends them by required steps of interactive data preprocessing. Here, the amount of data is first reduced to a manageable level for the user, thus preventing the information overload problem (sampling). The citizen data scientist subsequently specifies various transformations to preprocess the data (specification), such as combining columns, filling empty values, or similar actions. Finally, these transformations are applied to the entire dataset (transformation). These steps follow a sequential order of execution in traditional interactive analysis tools (cf. Sect. 3).

A citizen data scientist thus specifies different transformations on a sample and, if the quality is satisfactory, they are applied to the entire dataset. Since, to the best of our knowledge, there is no sampling algorithm that can ensure that every eventuality is represented in the sample at least once, it is very likely that a potential conflict will occur in this step. An excerpt of a sample dataset is depicted in Fig. 2. In this illustrative example, only simple top-level domains are found in the sample of the dataset presented to the CDS, while a composite

before transformation

	...	domain	revenue	...
Frontend	...	google.com	9999.99	...
	...	google.fr	5734.43	...

	...	domain	revenue	...
Backend	...	yahoo.com	1234.56	...
	...	google.co.uk	7244.23	...
	...	...	...	...

Transformation:
split column "*domain*" on "."

after transformation

	...	domain_0	domain_1	revenue	...	...
Frontend	...	google	com	9999.99	...	...
	...	google	fr	5734.43	...	...

	...	domain_0	domain_1	domain_2	revenue	...
Backend	...	yahoo	com		1234.56	...
	...	google	co	uk	7244.23	...
	...	...	...	...	...	...

Fig. 2. Possible mismatch between perceived state (frontend) and not recognized state (backend) before and after applying a transformation using a sample.

top-level domain, e. g., *.co.uk*, is not contained in this sample. The CDS wants to split the column *domain* in order to group by country using top-level domains.

By doing so, the CDS observes only the expected separation of the *domain* column into two new columns, while the execution on the entire dataset leads to an unexpected and unperceivable third column. As a consequence, such a conflict is not acceptable for the citizen data scientist. Rather, this discrepancy between specification and particular instances only becomes perceptible at a later stage, i. e., either (a) in the transformation step, if a specified transformation cannot be executed, or even worse, (b) not before the mining results have been taken into account. This leads to many iterations and lowers the motivation of the user.

A suitable tool for interactive data preprocessing must, therefore, support the user at all times as follows: (a) to work interactively on a sample of manageable size, (b) to retain the familiar spreadsheet environment, (c) to execute the transformation in the background during the specification (d) to identify conflicts that occur, and (e) to inform the user about potential conflicts, thus, allowing for refinement and conflict resolution.

3   Related Work

A typical approach to integrate domain experts into the analysis process is using spreadsheet tools or spreadsheet-like interfaces. According to Mack et al. [13], however, the most common problems using spreadsheets arise due to limited scalability, e. g., a simple copy/paste operation of *"a few hundred thousand rows of data"* could last over 150 min. Consequently, scalability issues may occur for datasets with just 18,000 rows [13].

To prevent the freezing of spreadsheet applications, Bendre et al. [2] proposed an asynchronous approach to calculation while maintaining consistency. In this case, the current viewport is calculated first, cells that are not yet recalculated are blurred out, and the user is provided with the feeling of responsiveness.

Nevertheless, today's data volumes are far too large for humans to be able to keep track of. In addition, cells are only blurred out and the overall calculation time remains the same, which may block the user from proceeding. In order to reduce the amount of data to a manageable amount for domain experts, the most common approach is to use sampling algorithms [12]. These algorithms are typically categorized as probability sampling and non-probability sampling.

In case of probability sampling, each instance of the population has the equal and well-known probability of being selected for inclusion in the sample. The simplest approach in this category is Simple Random Sampling. In non-probability sampling, instances of the population are selected on the grounds of a subjective decision, i. e., some instances have no chance at all to be included in the sample, while others have an increased probability. Typical algorithms are Convenience Sampling or Quota Sampling. Depending on the algorithm, this can lead to an enormous effort and is domain-specific.

Another issue of spreadsheets is the fact that they are very error-prone. To address this issue, modern interactive tools continue to use the familiar spreadsheet interface, but with the approach of making the specification of transformations less error-prone. This could be achieved by program synthesis using natural language [7] or by the use of Programming-by-Demonstration [3]. Examples of the latter approach include *OpenRefine*[1], *Wrangler* [10] or the work done by Gulwani et al. [8].

In particular, *Wrangler* shows efficient functionalities to integrate the user in data preprocessing. On the one hand, there is an intuitive specification of the transformations via a user interface, which allows the user to proceed step-by-step without formulas. On the other hand, each transformation is presented in a traceable transformation history, which allows the user to adjust the executed transformations at any time, e. g., calculating the median instead of the average. Once the specification is completed, the transformations can be exported as a script, i. e., to apply the changes made on a sample to a large-scale dataset.

---

[1] OpenRefine: https://openrefine.org.

These fundamental procedures can be found in various so-called ETL tools. Well-known representatives include KNIME[2], RapidMiner[3] or Trifacta Wrangler[4], which emerged from the research project *Wrangler* described above. All of these allow either the graphical definition of a sequence of transformations via pipes-and-filters pattern or support this kind of transformation history and graphical specification of transformations. Even though it is nowadays possible to connect these tools directly to a Hadoop-based ecosystem, however, the phases (transformation specification on a sample and subsequent execution) are still strictly separated from each other. An occurring conflict that was not covered in the sample, such as a wrongly coded date of birth, is only detected during execution, or even worse not before conducting subsequent analysis. Hence, the domain expert must revise the transformations and restart the execution from scratch again. In the context of today's data volumes in the era of big data, the concept of specification and subsequent transformation is, therefore, no longer sufficient due to the enormous amount of time required and the resulting expenses. Other popular data analysis tools, like Tableau, usually request already cleaned data and are, therefore, not suitable for data preprocessing tasks.

In recent years, various papers have been published on automatic error detection and correction, e. g., by exploiting machine learning, knowledge bases or expert feedback. Abedjan et al. [1] compared multiple tools and found that none of the tools were even close to correct all errors. Moreover, it is important in which order different approaches are performed. *HoloClean* by Rekatsinas et al. [16] introduced a holistic framework leveraging different approaches in order to perform automatic data cleaning. In particular, semantic errors, however, are not automatically recognizable at all [11]. In some cases, this could be addressed by exploiting ontologies. However, this requires a high initial effort and is domain-specific [19]. Consequently, such approaches would limit explorative interactive analysis to those domains where such ontologies already exist [15]. As a result, on the one hand, there is currently no automatic tool available which is capable of cleaning data sources in a reliable way. On the other hand, no interactive tool is available that is capable of processing large amounts of data almost independent of the prevailing sample quality.

4   Architecture to Preprocess Massive Datasets

In the following section, we present an approach to cope with these limitations. We specify an architecture which enables an user (a) to work interactively on a sample of manageable size, (b) to retain the familiar spreadsheet environment. In contrast to prior work, we preserve interactivity also for massive datasets. Therefore, we add the requirements (c) to execute the transformation in the background during the specification, (d) to identify conflicts that occur, and (e) to inform the user about potential conflicts and to allow for refinement.

---

[2] KNIME: https://www.knime.com.

[3] RapidMiner: https://rapidminer.com.

[4] Trifacta Wrangler: https://www.trifacta.com.

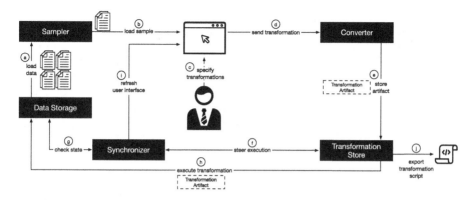

Fig. 3. Proposed architecture to empower domain experts to preprocess massive distributed datasets.

4.1 Components

Our architecture consists of multiple components which are depicted in Fig. 3 and are explained in detail below:

Data Storage. The data to be analyzed is stored in any format and in an arbitrary data storage (e. g., a relational database). For data extraction, adapters need to be provided for each specific type of database system or file format. These adapters handle data extraction and could furthermore apply basic data cleaning operations. One example of such a cleaning operation is resolving foreign keys to enable the look-and-feel of familiar spreadsheet interfaces.

Sampler. Today's amounts of data are too extensive for human beings to keep track of. Accordingly, a sampling component is necessary, which extracts a manageable amount of data from the entire dataset. It must be stated, that to the best of our knowledge, however, there is no feasible sampling algorithm that can ensure that every eventuality is represented in the sample at least once.

Converter. A common technique to combine user interaction and scalability is to convert the user's actions on a sample into a script that is executed on the entire dataset afterwards. In doing so, the converter component also uses the interactions of the domain expert to derive the underlying intention. These "actions" are applied directly to the sampled dataset and subsequently converted into a so-called transformation artifact. Note, that for each action of the user, an independent transformation artifact is created that is later integrated into the transformation script when exported (Fig. 3, j).

Transformation Artifact. The transformation artifact is one of the most important components of our architecture. The actions of the user are clearly recorded in this statement such that they can be executed elsewhere, exported, or executed again. A transformation artifact is modeled as a JSON object and contains, e. g., the operation performed, necessary parameters, and affected cells.

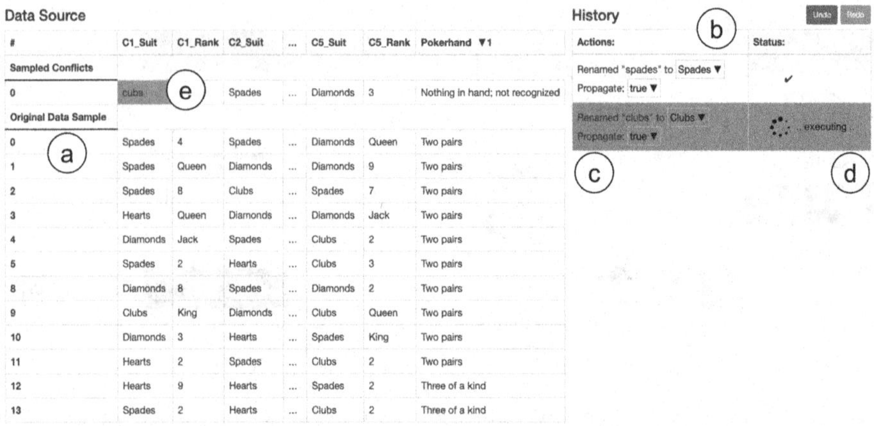

Fig. 4. User Interface with the poker hand dataset used in the conducted user study. The user interface shows a sample of the dataset (a), a transformation history which monitors all conducted steps (b), and allows the user to make modifications (c). The current execution state is shown at any time (d). Potential conflicts recognized on the compute backend are integrated to the sample keeping the user in the loop (e).

Transformation Store. Data Preprocessing is a highly iterative process in which the preliminary steps often have to be refined or modified on the basis of the knowledge gained. This leads to many different transformations which are applied, refined or discarded. Each of these transformations is stored as transformation artifact in the transformation store in order to enable their effective management. By doing so, we enable to relieve the user from repeated activities by repeating transformations that have already been carried out before.

Synchronizer. When processing large amounts of data, it can be assumed that a user generates new transformations faster than they can be processed on the entire dataset. Since these transformations could be dependent on each other, a component is required to ensure this sequence. For example, an operation cannot be applied to a new column until *combine_table* is executed on the source columns properly. Thus, a dependency graph over all transformations is generated and the Synchronizer controls the execution of the transformation artifacts on the compute cluster. The Synchronizer informs the user about the current state of the execution and any potential conflicts that may have occurred and enables the user to stay "in the loop", even when dealing with large-scale data.

User Interface. The main user interface of our prototype provides a spreadsheet-like interface (Fig. 4, a). Instead of relying on error-prone formulas, the user access is restricted to predefined operations (e. g., split columns or rename cells). Our prototype has a modular structure and can thus be easily extended with additional (predefined) operations.

Furthermore, there is a transformation history (Fig. 4, b), which allows to change parameters, affected columns or cells as well as operations at a later stage of the analysis (Fig. 4, c) which results in a subsequent re-execution of all dependent transformations with the modified parameters. The respective execution status is immediately presented to the user (Fig. 4, d).

As described before (cf. Sect. 2) various conflicts can occur during data pre-processing on a sample. This could be the case if subsequent transformations can no longer be executed as a result of parameter changes (Fig. 4, c), for instance, because the corresponding columns no longer exist or more columns are generated than perceivable in the user interface. In such cases our user interface exploits the capabilities of our architecture by informing the user immediately in two ways. First, by highlighting the transformation which could not be applied as expected (Fig. 4, b). Second, an example of the potentially conflicting records is added to the sample and highlighted as well (Fig. 4, e).

In the depicted scenario the user decided to rename all lower case *clubs* to *Clubs* to match the style of the remaining dataset. The system monitors the execution of this transformation and identifies, e. g., by exploiting the Levenshtein distance, similar instances which might indicate a typing mistake (*cubs* instead of *clubs*). Therefore, the system adds records as rows to the sample as appropriate. The user interface allows to keep the user in the feedback loop and informed about any deviations in the entire dataset and, therefore, enables to react in an appropriate way.

4.2   Walkthrough

In the following section we describe an exemplary run through our architecture (cf. Fig. 3) and show the advantages compared to conventional approaches.

We start with the extraction of the data out of the Data Storage, e. g., from a distributed file system (Fig. 3, a) and sample this data to a manageable level for the user. This subset of the dataset to be analyzed is sent to the user interface (Fig. 3, b). The user now has the opportunity to evaluate the dataset and specify transformations in a spreadsheet-like user interface (Fig. 3, c).

As soon as the first transformation is specified in the user interface, it is applied directly to the visualized dataset and simultaneously sent to the converter component (Fig. 3, d), which generates a so-called transformation artifact. This is primarily only stored in the transformation store (Fig. 3, e).

At this point, the decisive component of our architecture is utilized. The Synchronizer checks (Fig. 3, f) currently known transformation artifacts and if they are ready for execution. An executable transformation is then applied to the entire dataset (Fig. 3, h). The Synchronizer further monitors the execution (Fig. 3, g) and is therefore always up to date regarding the current execution state and any conflicts that may occur. In both scenarios, the user interface is updated immediately (Fig. 3, i). If a potential conflict occurs, the execution of further transformations is suspended (Fig. 3, f) and the user receives a visual feedback as well as an instance tuple on which the transformation led to a different result than expected. If the transformation has been successfully completed, it is also marked as completed in the user interface and the subsequent transformation is selected (Fig. 3, f) and executed (Figure 3, h). This ensures that the user is kept in the loop at all times, becomes aware of anomalies continuously, and can, therefore, process the entire dataset without further problems.

4.3 Discussion

As mentioned before, in order to create a suitable tool for interactive data pre-processing it is necessary to support users at all times as follows: (a) to work interactively on a sample of manageable size, (b) to retain the familiar spreadsheet environment, (c) execution in the background during the specification, (d) to identify potential conflicts that occur, and (e) to inform the user about potential conflicts and allowing for refinement. With our approach and architecture for scalable human in the loop data preprocessing, we fulfill these requirements. Requirement (a) is fulfilled by sampling capabilities, providing the user with suitable samples that represent the entire dataset. Based on this sample, the user can interact with the data without being overwhelmed with too many entries. Requirement (b) is fulfilled by an intuitive user interface, which provides a spreadsheet-like look-and-feel, so users do not need to gain new expertise but can rather work in a well-known environment. Additionally, this prevents from common errors that occur when using formulas in spreadsheet applications. Through constant application of the sample on the entire dataset using an efficient scalable backend, requirement (c) is fulfilled. Occurring potential conflicts are recognized on-the-fly during application and monitoring of the transformations, and the user is subsequently informed. Hence, requirements (d) and (e) are fulfilled through our architecture as well. In summary, our approach fulfills these requirements and, thus, is a suitable architecture for interactive data preprocessing.

5 Evaluation

In order to evaluate our proposed architecture (cf. Sect. 4) with regard to its suitability for everyday use and user acceptance, we implemented the architecture in a prototypical way using a Javascript-based frontend, components based on Python and Apache Spark as compute backend with an attached Hadoop Distributed File System.

Based on this prototype, we conducted a user study to explore how potential users interact with our architecture and how it compares to conventional systems. For this, we allow different levels of scalability support in our implementation: (1) no scalability, i.e., the entire dataset is provided to the user, (2) the user works on a static sample of the dataset and the created transformations are then transferred to the backend for execution, and finally (3) the presented interactive approach, in which transformations are executed during the specification phase in the backend and the user is dynamically notified about potential conflicts. We refer to these different levels of scalability support in the following as $L_{NoScalability}$, $L_{StaticSample}$ and $L_{DynamicSample}$.

Participants. We recruited 12 participants (3 female, 9 male) via mailing lists and personal contacts of one of the authors. The participants were aged between 20 and 38 years (M = 25.5, SD = 6.25). They were either computer science students, data science students or research assistants. The participants were neither rewarded for participating in this user study nor familiar with our prototype or had any knowledge of the measured variables. Furthermore, none of the participants had expert knowledge regarding data preparation.

Method. We conducted our study using a repeated measures design with the magnitude of scalability support as the only independent variable. Each participant used each of the three types of support in a counterbalanced order according to the Latin Square to reduce learning effects between the conditions.

We used objective as well as subjective measures as dependent variables. As objective measures, we used the number of corrected errors (syntactic/semantic) and the number of errors corrected per action, where an action is counted on each mouse click, e.g., sorting a column as well as opening a dialog.

Additionally, we collected quantitative subjective feedback through a 5-point Likert scale about the overall satisfaction with the current level of support. Additional qualitative feedback was collected by a semi-structured interview after all conditions. Each experiment took roughly 60 min per participant and was conducted in a quiet environment.

Apparatus. We deployed a prototypical implementation of our architecture on an IBM Pure Flex cluster managed by OpenStack with 11 compute nodes (360 Intel Xeon E5 CPUs, 2.816 TB RAM). In addition, we deployed our developed frontend prototype on a virtual machine powered by Ubuntu 18.04.4 LTS (2 VCPUs, 16 GB RAM). Since we used a virtual machine, each participant could keep his or her familiar mouse/keyboard combination. For each condition and participant, we measured the number of corrected errors, the number of actions performed as well as the time necessary per action.

For our user study, we used the poker hand dataset from the UCI Machine Learning Library [4] as ground truth. Since our evaluation requires a dataset that can be processed under all three conditions, we decided to use this rather small dataset. Note that the conditions $L_{StaticSample}$ and $L_{DynamicSample}$ would be feasible on much larger datasets thanks to the use of Apache Spark in the backend, but the $L_{NoScalability}$ condition would not. This dataset is easy to validate in terms of correctness, which is necessary to measure the number of errors unambiguously and at the same time the indispensable domain knowledge is available for all participants.

We first converted this dataset into human understandable strings according to the dataset description, e.g. the value "*6*" of the feature "*Poker Hand*" was converted into "*Full House*". Subsequently, syntactic errors (spelling mistakes) and semantic errors (empty fields) were randomly inserted into the dataset. A randomly generated fixed seed was applied to generate the sample (using Simple Random Sampling) used for $L_{StaticSample}$ and $L_{DynamicSample}$.

Procedure. After explaining the purpose of the study and the procedure, we asked the participants to execute some exemplary tasks to get familiar with an interactive data preprocessing tool. This tasks include operations on cells (renaming), on columns (sorting, combining, splitting), and rows (delete). Furthermore, we gave a short overview of other predefined operations implemented in our prototype.

Comparing the level of scalability support, the participants were asked to clean/preprocess the given dataset with each level of scalability support once in a counterbalanced order. Furthermore, we asked the participants to end one preprocessing when they are either satisfied with the reached quality or lost their motivation/confidence to solve this task. Afterwards, we asked for subjective feedback and proceeded with the next level of scalability support. After all conditions had been processed once, we conducted a semi-structured interview.

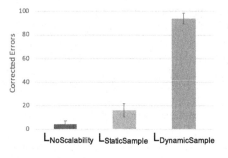
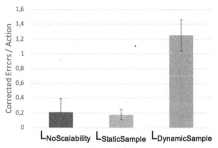

(a) Mean percentage of syntactical errors corrected during the study.

(b) Mean number of syntactical errors corrected with one action performed.

Fig. 5. Results of the conducted user study. Error bars depict the standard error.

Results. We statistically compared the conditions $L_{NoScalability}$, $L_{StaticSample}$ and $L_{DynamicSample}$ regarding the corrected errors and the number of errors corrected per action using a one-way repeated measures ANOVA.

First, we analyzed for each condition the number of syntax errors corrected. $L_{NoScalability}$ ($M = 3.75\%$, $SD = 11.70$) as well as $L_{StaticSample}$ ($M = 15.99\%$, $SD = 19.88$) were outperformed by $L_{DynamicSample}$ ($M = 94.05\%$, $SD = 16.15$). A one-way repeated measures ANOVA revealed a significant difference between our conditions ($F(2, 22) = 108.951$, $p < .0001$). A follow-up Bonferroni-corrected post-hoc test showed that $L_{DynamicSample}$ is statistically significantly better performing compared to the other conditions. The results are depicted in Fig. 5a.

Secondly, we analyzed the number of corrected semantic errors. Here, a one-way repeated measures ANOVA did not show any significant difference ($F(2, 22) = 1.313$, $p > .05$) between the different conditions.

Thirdly, we analyzed the number of errors corrected by one single action of the user. $L_{DynamicSample}$ performed best once again ($M = 1.25$, $SD = 0.74$), followed by $L_{StaticSample}$ ($M = 0.18$, $SD = 0.25$) and $L_{NoScalability}$ ($M = 0.21$, $SD = 0.63$). A one-way repeated measures ANOVA revealed a significant difference ($F(2, 22) = 13.219$, $p < .0001$) between the conditions. A follow-up Bonferroni-corrected post-hoc test shows that $L_{DynamicSample}$ is statistically significantly better performing compared to the other conditions. These findings are depicted in Fig. 5b.

Looking at the subjective ratings, the participants liked the interaction with $L_{DynamicSample}$ ($M = 3.5$, $SD = 0.80$) and $L_{StaticSample}$ ($M = 3.5$, $SD = 0.67$) more compared to $L_{NoScalability}$ ($M = 1.67$, $SD = 0.65$). These findings are depicted in Fig. 6. A Friedman ANOVA shows that these differences are statistically significant ($\chi^2(2) = 20.571$, $p < .0001$). Wilcoxon tests were used to follow-up this finding. It appeared that there is a significant difference between $L_{NoScalability}$ and $L_{StaticSample}$ ($Z = 3.175$, $p < .001$) as well as between $L_{NoScalability}$ and $L_{DynamicSample}$ ($Z = -3.175$, $p < .001$).

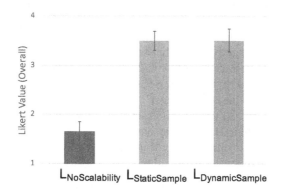

Fig. 6. The mean Likert scale values indicating how the participants liked the interaction with the different levels of scalability support. Error bars depict the standard error.

In the concluding interviews, the majority of our participants stated that they would prefer an interactive data preprocessing tool to a conventional spreadsheet application. In particular, the combination of a manageable and responsive dataset while keeping the same functionality as on the entire dataset was highly appreciated (P1, P7, P9, P10, P12).

While a few of our participants (P3, P9) were slightly confused by the asynchronously added records in $L_{DynamicSample}$, the overwhelming majority experienced these suggestions about potential conflicts as remarkably beneficial for data preprocessing (P1, P2, P4, P5, P6, P9, P10, P11).

Discussion. Our user study shows that our approach is able to outperform state-of-the-art concepts of data preprocessing (cf. Section 3) when dealing with massive datasets.

Regarding $L_{StaticSample}$ and $L_{DynamicSample}$, the measurements outperforming $L_{NoScalability}$ are as expected for large-scale data preprocessing. However, in

$L_{DynamicSample}$, significantly more errors can be found in contrast to the state of the art. In regard to semantic errors, our approach does not perform better as others, however, finding semantic errors is, in general, a very complex problem in data analysis, especially when the recognition should be automated. In addition to these findings, we show in our user study that not only more errors are found, but also that the necessary effort to correct these errors is significantly smaller.

Last but not least, our participants confirmed that our concept provides the same level of user satisfaction as state-of-the-art concepts for data preprocessing. Accordingly, based on our evaluation, we state that our approach finds more errors with less effort while maintaining equal user satisfaction.

6   Conclusion and Future Work

In this paper, we present an approach to empower domain experts to preprocess massive distributed datasets. Common approaches try to do this by using a static sample or just pretend responsiveness to the user. Instead, we achieve this by executing each transformation immediately on the compute backend while the domain expert is still specifying further transformations. Our approach employs important features, such as coping with large-scale data by employing conflict resolution capabilities. Furthermore, we include concepts for human interaction during data preprocessing steps based on literature review. Our architecture enables domain experts (a) to work interactively on a sample of manageable size, (b) to retain the familiar spreadsheet environment, (c) to execute transformations in the background during specification, (d) to identify potential conflicts that occur, and (e) to be informed about potential conflicts and allowing for continuous refinements.

For evaluation purposes, we conducted a user study with 12 participants. This study shows that our approach delivers a significantly higher error correction rate. In addition, fewer actions were required to correct an error. We also achieved a comparable user satisfaction as with state-of-the-art approaches.

In future work, we aim at even enhancing the user support by introducing further algorithms for data preprocessing. In addition, improvements in the user interface of our prototype are planned to enable better feedback to the users. These enhancements also include more control over the algorithms used for conflict detection, e. g., parameter configurations like a slider to configure the Levenshtein distance.

Acknowlegements. This research was performed in the project 'IMPORT' as part of the Software Campus program, which is funded by the German Federal Ministry of Education and Research (BMBF) under Grant No.: 01IS17051.

References

1. Abedjan, Z., et al.: Detecting data errors: where are we and what needs to be done? Proc. VLDB Endow. **9**(12), 993–1004 (2016)
2. Bendre, M., et al.: Anti-freeze for large and complex spreadsheets: asynchronous formula computation. In: Proceedings of the International Conference on Management of Data (SIGMOD) (2019)
3. Cypher, A. (ed.): Watch What I Do - Programming by Demonstration. MIT Press, Cambridge (1993)
4. Dua, D., Graff, C.: UCI machine learning repository (2017). http://archive.ics.uci.edu/ml
5. Fayyad, U.M., Piatetsky-Shapiro, G., Smyth, P.: From data mining to knowledge discovery in databases. AI Mag. **17**(3), 37 (1996)
6. Gandel, S.: Damn Excel! How the 'most important software application of all time' is ruining the world (2013). http://fortune.com/2013/04/17/damn-excel-how-the-most-important-software-application-of-all-time-is-ruining-the-world/
7. Gulwani, S., Marron, M.: NLyze: interactive programming by natural language for spreadsheet data analysis and manipulation. In: Proceedings of the International Conference on Management of Data (SIGMOD) (2014)
8. Gulwani, S., et al.: Spreadsheet data manipulation using examples. Commun. ACM **55**(8), 97–105 (2012)
9. International Business Machines Corporation: Transforming the Common Spreadsheet: A Smarter Approach to Budgeting, Planning and Forecasting, Technical report (2009)
10. Kandel, S., et al.: Wrangler: interactive visual specification of data transformation scripts. In: Proceedings of the Conference on Human Factors in Computing Systems (CHI) (2011)
11. Kemper, H.G., et al.: Datenbereitstellung und -modellierung. In: Business Intelligence - Grundlagen und praktische Anwendungen: Eine Einführung in die IT-basierte Managementunterstützung (2010)
12. Lohr, S.L.: Sampling: Design and Analysis. Brooks/Cole (2009)
13. Mack, K., et al.: Characterizing scalability issues in spreadsheet software using online forums. In: Extended Abstracts of the Conference on Human Factors in Computing Systems (CHI EA) (2018)
14. Moore, S.: Gartner says more than 40 percent of data science tasks will be automated by 2020 (2017). https://www.gartner.com/en/newsroom/press-releases/2017-01-16-gartner-says-more-than-40-percent-of-data-science-tasks-will-be-automated-by-2020
15. Reimann, P., Schwarz, H., Mitschang, B.: A pattern approach to conquer the data complexity in simulation workflow design. In: Meersman, R., et al. (eds.) OTM 2014. LNCS, vol. 8841, pp. 21–38. Springer, Heidelberg (2014). https://doi.org/10.1007/978-3-662-45563-0_2
16. Rekatsinas, T., et al.: HoloClean - holistic data repairs with probabilistic inference. Proc. VLDB Endow. **10**(11) (2017)
17. Shearer, C.: The CRISP-DM model: the new blueprint for data mining. J. Data Warehouse. **5**(4) (2000)
18. Stodder, D.: Visual Analytics for Making Smarter Decisions Faster. Technical report, TDWI, Renton, WA, USA (2015)
19. Wache, H., et al.: Ontology-based integration of information - a survey of existing approaches. In: Proceedings of the Workshop on Ontologies and Information Sharing, International Joint Conference on Artificial Intelligence (IJCAI) (2001)

Efficient Construction of Behavior Graphs
for Uncertain Event Data

Marco Pegoraro$^{(\boxtimes)}$ⓘ, Merih Seran Uysalⓘ, and Wil M. P. van der Aalstⓘ

Process and Data Science Group (PADS) Department of Computer Science,
RWTH Aachen University, Aachen, Germany
{pegoraro,uysal,wvdaalst}@pads.rwth-aachen.de
http://www.pads.rwth-aachen.de/

Abstract. The discipline of process mining deals with analyzing execution data of operational processes, extracting models from event data, checking the conformance between event data and normative models, and enhancing all aspects of processes. Recently, new techniques have been developed to analyze event data containing uncertainty; these techniques strongly rely on representing uncertain event data through graph-based models capturing uncertainty. In this paper we present a novel approach to efficiently compute a graph representation of the behavior contained in an uncertain process trace. We present our new algorithm, analyze its time complexity, and report experimental results showing order-of-magnitude performance improvements for behavior graph construction.

Keywords: Process mining · Uncertain data · Event data representation

1 Introduction

Process mining [1] is a research field that performs process analysis in a data-driven fashion. Process mining analyses are based on recordings of events and tasks within the process, stored in a number of information systems supporting business activities. These recordings are extracted and orderly collected in databases called *event logs*. Utilizing an event log as a starting point, process mining analyses can automatically extract a process model describing the behavior of the real-world process (*process discovery*) and measure deviations between execution data of the process and a normative model (*conformance checking*). Process mining is a rapidly growing field both in academia and industry. More than 25 commercial tools are available for analyzing processes. Process mining tools are used to analyze processes in tens of thousands of organizations, e.g., within Siemens, over 6000 employees use process mining to improve processes.

Commercial process mining tools are able to automatically discover and draw a process model from an event log. Most of the process discovery algorithms used

---

We thank the Alexander von Humboldt (AvH) Stiftung for supporting our research interactions.

W. Abramowicz and G. Klein (Eds.): BIS 2020, LNBIP 389, pp. 76–88, 2020.
https://doi.org/10.1007/978-3-030-53337-3_6

by these tools are based on counting the number of *directly-follows relation-ships* between activities in the process. The more often a specific activity follows another one in a process of an organization, the stronger a causality implication between the two activities is assumed to be. Directly-follows relationship are also the basis for detecting more complicated constructs in the workflow of a process, such as parallelism or interleaving of activities. These relationships are often summarized in a labeled graph called the *Directly-Follows Graph* (DFG).

Recently, a new class of event logs has gained interest: *uncertain event logs* [12]. These execution logs contain, rather than precise values, an indica-tion of the possible values acquired by event attributes. In this paper, we will consider the setting where uncertainty is expressed by either a set or an interval of possible values for an attribute, as well as the possibility of an event being recorded in the log even though it did not occur in reality. An example of an uncertain trace is shown in Table 1.

Table 1. An example of simple uncertain trace. Events e_2 and e_4 have uncertain activity labels. Event e_3 has a possible range of timestamps, rather than a precise value. Event e_5 has been recorded, but it might not have happened in reality.

Case ID	Event ID	Activity	Timestamp	Event type
945	e_1	a	05-12-2011	!
945	e_2	{b, c}	07-12-2011	!
945	e_3	d	[06-12-2011, 10-12-2011]	!
945	e_4	{a, c}	09-12-2011	!
945	e_5	e	11-12-2011	?

Existing process mining tools do not support uncertain data. Therefore, novel techniques to manage and analyze it are needed. *Uncertain Directly-Follows Graphs* (UDFGs) allow representing directly-follows relationships in an event log under conditions of uncertainty in the data. This leads to the discovery of models of uncertain logs through methods based on directly-follows relationships such as the Inductive miner [13].

An intermediate step necessary to compute UDFGs is to construct the *behav-ior graph* of the traces in the uncertain log. A behavior graph represents in a graphical manner the time and precedence relationships among certain and uncertain events in an uncertain trace. Figures 1 and 2 show, respectively, the behavior graph of the trace in Table 1 and the UDFG representing the relation-ship between activities in the same trace. Uncertain timestamps are the most critical source of uncertain behavior in a process trace: for instance, if n events have uncertain timestamps such that their order is unknown, the possible con-figurations for the control-flow of the trace are the $n!$ permutations of the events.

The construction of behavior graphs for uncertain traces is the basis of both conformance checking and process discovery on uncertain event data. It is, thus,

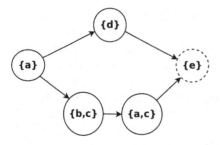

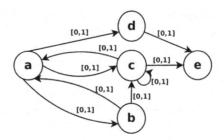

Fig. 1. The behavior graph of the uncertain trace given in Table 1. Each vertex represents an uncertain event and is labeled with the possible activity label of the event. The dashed circle represents an indeterminate event (may or may not have happened).

Fig. 2. The UDFG computed based on the behavior graph in Fig. 1. The arcs are labeled with the minimum and maximum number of directly-follows relationship observable in the corresponding trace. Here, every relationship can occur in the trace once, or not occur at all.

important to be able to build the behavior graph of any given uncertain trace in a quick and efficient manner. Constructing a behavior graph is the most computationally expensive step towards producing a process model (e.g., a Petri net using the approach in [12]). In this paper, we present a novel algorithm for behavior graph construction which runs in quadratic time complexity, therefore allowing a significant speedup for the operations of conformance checking and process discovery for uncertain event logs. We will prove the correctness of the new algorithm, as well as show the improvement in performance both theoretically, via asymptotic complexity analysis, and practically, with experiments on a number of uncertain event logs comparing computing times of the baseline method against the novel construction algorithm. The algorithms have been implemented in the context of the PROVED (*PRocess mining OVer uncErtain Data*) library[1], based on the PM4Py framework [6].

The reminder of the paper is structured as follows. Section 2 explores recent related works in the context of uncertain event data. Section 3 provides formal definitions and describes the baseline method for our research. Section 4 illustrates a novel and more efficient method to construct a behavior graph of an uncertain trace. Section 5 presents the analysis of asymptotic complexity for both the baseline and the novel method. Section 6 shows the results of experiments on both synthetic and real-life uncertain event logs comparing the efficiency of both methods to compute behavior graphs. Section 7 comments on the results of the experiments and concludes the paper.

---

[1] https://github.com/proved-py/proved-core/tree/Efficient_Construction_of_Behavio r_Graphs_for_Uncertain_Event_Data.

2   Related Work

Research concerning the topic of process mining over uncertain event data is very recent. The work that introduced the concept of uncertainty in process mining, together with a taxonomy of the various kinds of uncertainty, specifically showed that if a trace displays uncertain attributes, it contains behavior, which can be appropriately expressed through process models – namely, behavior graphs and behavior traces [12]. As opposed to classic process mining, where we have a clear cut between data and model and between the static behavior of data and the dynamic behavior of models, the distinction between data and models becomes blurry in presence of uncertainty, because of the variety in behavior that affects the data. Expressing traces through models is utilized in [12] for the calculation of upper and lower bounds for conformance scores of uncertain traces against classic reference models. A second application for behavior graphs in the domain of process mining over uncertain event data is given in [13]. Behavior graphs of uncertain traces are employed to count the number of possible directly-follows relationships between uncertain events, with the objective of automatically discovering process models from uncertain event data. The formulation used in this and previous works on uncertainty in process mining shares similarities with temporal extensions of fuzzy logic e.g. [8]; however, unlike fuzzy temporal logic, our framework is suited to compactly represent the control-flow dimension of uncertain event data as Petri nets, a graphical model capable of simulation.

Behavior graphs are Directed Acyclic Graphs (DAGs), which are commonly used throughout many fields of science to represent with a graph-like model time information, precedence relationships, partial orders, or dependencies. They are successfully employed in compiler design [2], circular dependency analysis in software [4], probabilistic graphical models [5] and dynamic graphs analytics [11].

3   Preliminaries

Let us introduce some basic notations and concepts, partially from [1]:

Definition 1 (Power Set). *The power set of a set A is the set of all possible subsets of A, and is denoted with $\mathcal{P}(A)$. $\mathcal{P}_{NE}(A)$ denotes the set of all the non-empty subsets of A: $\mathcal{P}_{NE}(A) = \mathcal{P}(A) \setminus \{\emptyset\}$.*

Definition 2 (Sequence). *Given a set X, a finite sequence over X of length n is a function $s \in X^* : \{1, \ldots, n\} \to X$, and is written as $s = \langle s_1, s_2, \ldots, s_n \rangle$. For any sequence s we define $|s| = n$, $s[i] = s_i$, $x \in s \iff x \in \{s_1, s_2, \ldots, s_n\}$ and $s \oplus s_0 = \langle s_1, s_2, \ldots, s_n, s_0 \rangle$.*

Definition 3 (Directed Graph). *A directed graph $G = (V, E)$ is a set of vertices V and a set of directed edges $E \subseteq V \times V$. We denote with $\mathcal{U}_G$ the universe of such directed graphs.*

Definition 4 (Path). *A* path *over a graph* $G = (V, E)$ *is a sequence of vertices* $p = \langle v_1, v_2, \ldots v_n \rangle$ *with* $v_1, \ldots, v_n \in V$ *and* $\forall_{1 \leq i \leq n-1}(v_i, v_{i+1}) \in E$. $P_G(v, w)$ *denotes the set of all paths connecting* v *and* w *in* G. *A vertex* $w \in V$ *is* reachable *from* $v \in V$ *if there is at least one path connecting them:* $|P_G(v, w)| > 0$.

Definition 5 (Transitive Reduction). *The* transitive reduction *of a graph* $G = (V, E)$ *is a graph* $\rho(G) = (V, E')$ *with the same reachability between vertices and a minimal number of edges.* $E' \subseteq E$ *is a smallest set of edges such that* $|P_{\rho(G)}(v, w)| > 0 \implies |P_G(v, w)| > 0$ *for any* $v, w \in V$. *The transitive reduction of a directed acyclic graph is unique.*

This paper analyzes *uncertain event logs*. These event logs contain uncertainty information explicitly associated with event data. A taxonomy of different kinds of uncertainty and uncertain event logs has been presented in [12]; we will refer to the notion of *simple uncertainty*, which includes uncertainty without probabilistic information on the control-flow perspective: activities, timestamps, and indeterminate events. Event e_4 has been recorded with two possible activity labels (a or c). This is an example of strong uncertainty on activities. Some events, e.g. e_3, do not have a precise timestamp, but have a time interval in which the event could have happened has been recorded: in some cases, this causes the loss of the precise order of events (e.g. e_3 and e_4). These are examples of strong uncertainty on timestamps. As shown by the "?" symbol, e_5 is an indeterminate event: it has been recorded, but it is not guaranteed to have happened. Conversely, the "!" symbol indicates that the event has been recorded in a correct way, i.e. it certainly occurred in reality (e.g. the event e_1).

Definition 6 (Universes). *Let* $\mathcal{U}_I$ *be the set of all the* event identifiers. *Let* $\mathcal{U}_C$ *be the set of all* case ID identifiers. *Let* $\mathcal{U}_A$ *be the set of all the* activity identifiers. *Let* $\mathcal{U}_T$ *be the totally ordered set of all the* timestamp identifiers. *Let* $\mathcal{U}_O = \{!, ?\}$, *where the "!" symbol denotes* determinate events, *and the "?" symbol denotes* indeterminate events.

Definition 7 (Simple uncertain events). $e = (e_i, A, t_{min}, t_{max}, o)$ *is a simple uncertain event, where* $e_i \in \mathcal{U}_E$ *is its event identifier,* $A \subseteq \mathcal{U}_A$ *is the set of possible activity labels for* e, t_{min} *and* t_{max} *are the lower and upper bounds for the value of its timestamp, and* o *indicates if is is an indeterminate event. Let* $\mathcal{U}_E = (\mathcal{U}_I \times \mathcal{P}_{NE}(\mathcal{U}_A) \times \mathcal{U}_T \times \mathcal{U}_T \times \mathcal{U}_O)$ *be the set of all simple uncertain events. Over the uncertain event* e *we define the projection functions* $\pi_{t_{min}}(e) = t_{min}$ *and* $\pi_{t_{max}}(e) = t_{max}$.

Definition 8 (Simple uncertain traces and logs). $\sigma \subseteq \mathcal{U}_E$ *is a* simple uncertain trace *if for any* $(e_i, A, t_{min}, t_{max}, o) \in \sigma$, $t_{min} < t_{max}$ *and all the event identifiers are unique.* $\mathcal{T}_U$ *denotes the universe of simple uncertain traces.* $L \subseteq \mathcal{T}_U$ *is a* simple uncertain log *if all the event identifiers in the log are unique.*

A necessary step to allow for analysis of simple uncertain traces is to obtain their *behavior graph*. A behavior graph is a directed acyclic graph that synthesizes the information regarding the uncertainty on timestamps contained in the trace.

Definition 9 (Behavior Graph). *Let* $\sigma \in \mathcal{T}_U$ *be a simple uncertain trace. A behavior graph* $\beta\colon \mathcal{T}_U \to \mathcal{U}_G$ *is the transitive reduction of a directed graph* $\rho(G)$, *where* $G = (V, E) \in \mathcal{U}_G$ *is defined as:*

- $V = \{e \in \sigma\}$
- $E = \{(v, w) \mid v, w \in V \wedge \pi_{t_{max}}(v) < \pi_{t_{min}}(w)\}$

The semantics of a behavior graph can effectively convey time and order information regarding the time relationship of the events in the corresponding uncertain trace in a compact manner. For a behavior graph $\beta(\sigma) = (V, E)$ and two events $e_1 \in \sigma$, $e_2 \in \sigma$, $(e_1, e_2) \in E$ if and only if e_1 is immediately followed by e_2 for some possible values of the timestamps for the events in the trace. A consequence is that if some events in the graph are pairwise unreachable, they might have happened in any order.

Definition 9 is clear and meaningful from a theoretical standpoint. It accurately describes a behavior graph and the semantics of its components. While useful to understand the purpose of behavior graphs, building them from process traces following this definition – that is, employing the transitive reduction – is slow and inefficient. This hinders the analysis of larger logs. It is possible, however, to obtain behavior graphs from traces in a quicker way.

4   Efficient Construction of Behavior Graphs

The procedure to efficiently build a behavior graph from an uncertain trace is described in Algorithm 1. For ease of notation, the algorithm textually indicates some conditions on the timestamp of an event. The keyword `continue` brings the execution flow to the next iteration of the loop in line 16, while the keyword **break** stops the execution of the inner loop and brings the execution flow on line 30. A certain event e is associated with one specific timestamp which we refer to as *certain timestamp*. Furthermore, an uncertain event e is associated with a time interval which is determined by two values: minimum and maximum timestamp of that event. An event e has a certain timestamp if and only if $\pi_{t_{\min}}(e) = \pi_{t_{\max}}(e)$. A timestamp t is the minimum timestamp of the event e if and only if $t = \pi_{t_{\min}}(e) \neq \pi_{t_{\max}}(e)$. A timestamp t is the maximum timestamp of the event e if and only if $t = \pi_{t_{\max}}(e) \neq \pi_{t_{\min}}(e)$.

We will consider here the application of Algorithm 1 on a running example, the trace shown in Table 2. Notice that none of the events in the running example display uncertainty on activity labels or are indeterminate: this is due to the fact that the topology of a behavior graph only depends on the (uncertain) timestamps of events.

The concept behind the algorithm is to inspect the time relationship between uncertain events in a more specific way, instead of adding many edges to the graph and then deleting them via transitive reduction. This is achieved by searching the possible successors of each event in a sorted list of timestamps. We then scan the list of timestamps with two nested loops, and we use the inner loop to search for successors of the event selected by the outer loop. It is important to notice that, since the semantics of the behavior graph state that events with overlapping intervals as timestamps should not be connected by a path, we draw outbound edges from an uncertain event only when, scanning the list, we encounter the timestamp at which the event has certainly occurred. This is the reason why outbound edges are not drawn from minimum timestamps (line 14) and inbound edges are not drawn into maximum timestamps (lines 24–28).

Algorithm 1: Efficient construction of the behavior graph

    Input   : The uncertain trace σ.
    Output : The behavior graph $\beta(\sigma) = (V, E)$.

1 $V \leftarrow \{e \in \sigma\}$;   // Set of vertices of the behavior graph
2 $E \leftarrow \{\}$;   // Set of edges of the behavior graph
3 $\mathcal{L} \leftarrow \langle\rangle$;   // List of timestamps and events
4 **for** $e \in \sigma$ **do**
5 | **if** e *has a certain timestamp* **then**
6 | |__ $\mathcal{L} \leftarrow \mathcal{L} \oplus (\pi_{t_{\min}}(e), e)$
7 | **else**
8 | | $\mathcal{L} \leftarrow \mathcal{L} \oplus (\pi_{t_{\min}}(e), e)$
9 | |__ $\mathcal{L} \leftarrow \mathcal{L} \oplus (\pi_{t_{\max}}(e), e)$
10 sort the elements $(t, e) \in \mathcal{L}$ based on the timestamps
11 $i \leftarrow 1$
12 **while** $i < |\mathcal{L}| - 1$ **do**
13 | $(t, e) \leftarrow \mathcal{L}[i]$
14 | **if** e *has a certain timestamp or* t *is the maximum timestamp of* e **then**
15 | | $j \leftarrow i + 1$
16 | | **while** $j < |\mathcal{L}|$ **do**
17 | | | $(t', e') \leftarrow \mathcal{L}[j]$
18 | | | **if** t' *is the minimum timestamp of* e' **then**
19 | | | | $E \leftarrow E \cup \{(e, e')\}$
20 | | | |__ **continue**
21 | | | **if** e' *has a certain timestamp* **then**
22 | | | | $E \leftarrow E \cup \{(e, e')\}$
23 | | | |__ **break**
24 | | | **if** t' *is the maximum timestamp of* e' **then**
25 | | | | **if** $(e, e') \notin E$ **then**
26 | | | | |__ **continue**
27 | | | | **else**
28 | | | | |__ **break**
29 | | |__ $j \leftarrow j + 1$
30 |__ $i \leftarrow i + 1$
31 **return** (V, E)

Table 2. Running example for the construction of the behavior graph.

Case ID	Event ID	Activity	Timestamp	Event type
872	e_1	a	05-12-2011	!
872	e_2	b	07-12-2011	!
872	e_3	c	[06-12-2011, 10-12-2011]	!
872	e_4	d	[08-12-2011, 11-12-2011]	!
872	e_5	e	09-12-2011	!
872	e_6	f	[12-12-2011, 13-12-2011]	!

If, while searching for successors of the event e, we encounter the minimum timestamp of the event e', we connect them, since their timestamps do not overlap. The search for successors needs to continue, since it is possible that other events occurred before the maximum timestamp of e' (lines 18–20). This happens for the events e_1 and e_3 in Table 2. As shown in Fig. 3, e_3 can indeed follow e_1, but the undiscovered event e_2 is another possible successor for e_1.

If we encounter a certain event e', we connect e with e' and we stop the search. A certain event e' will in fact preclude an edge from e to any event occurring after e' (lines 21–23). The trace in Table 2 shows this situation for events e_1 and e_2: once connected, nothing that occurs after the timestamp of e_2 can be a successor of e_1.

If we encounter the maximum timestamp of the event e' (line 24), there are two distinct situations to consider. Case 1: e was not already connected to e'. Then, either e is certain and occurred within the timestamp interval of e', or both timestamps of e and e' are uncertain and overlap with each other. In both situations, e should not be connected to e' and the search should continue (lines 25–26). Events e_3 and e_4 are an example: when the maximum timestamp of e_4 is encountered during the search for the successor of e_3, the two are not connected, so the search for a viable successor of e_3 continues. Case 2: e and e' are already connected. This means that we had already encountered the minimum timestamp of e' during the search for the successors of e. Since the whole time interval associated with the timestamp of e' is detected after the occurrence of e, there are no further events to consider as successors of e and the search stops (lines 27–28). In the running example, this happens between e_5 and e_6: when searching for the successors of e_5, we first connect it with e_6 when we encounter its minimum timestamp; we then encounter its maximum timestamp, so no other successive event can be a successor for e_5.

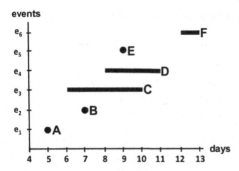

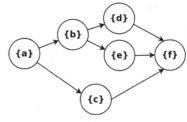

Fig. 3. A diagram visualizing the time perspective of the events in Table 2.

Fig. 4. The behavior graph of the trace in Table 2.

5 Asymptotic Complexity

Definition 9 provides a baseline method for the construction of the behavior graph consists of two main parts: the creation of the initial graph and its transitive reduction. Let us consider an uncertain trace σ of length $n - |\sigma|$ (with n events). Both the initial graph $G = (V, E)$ and the behavior graph $\beta(\sigma)$ have thus $|V| = n$ vertices. The initial graph is created by checking the time relationship between every pair of events; this is equivalent of checking if an edge exists between each pair of vertices of G, which is done in $\mathcal{O}(n^2)$ time.

The transitive reduction can be attained through many methods. Aho et al. [3] show a method to perform transitive reduction in $\mathcal{O}(n^3)$ time, better suited for dense graphs, and prove that the transitive reduction has the same complexity as the matrix multiplication. The Strassen algorithm [14] can multiply matrices in $\mathcal{O}(n^{2.807355})$ time. Subsequent improvements have followed suit: the asymptotically fastest algorithm has been described by Le Gall [10]. However, these improved algorithms are rarely used in practice, because of the large constant factors in their computing time hidden by the asymptotic notation, as well as very large memory requirements. The Strassen algorithm is useful in practice only for large matrices [7], and the Coppersmith-Winograd algorithm and successive improvements require an input so large to be efficient that they are effectively classified as galactic algorithms [9].

In light of these considerations, for the vast majority of event logs the best way to implement the construction of the behavior graph through transitive reduction runs in $\mathcal{O}(n^2) + \mathcal{O}(n^3) = \mathcal{O}(n^3)$ time in the worst-case scenario.

It is straightforward to find the upper bound for complexity of Algorithm 1. Lines 1–3 and line 11 run in $\mathcal{O}(1)$ time. The worst case scenario is when all events in a trace are uncertain. In that case, lines 4–5 build a list of length $2n$ with a single pass through the events in the trace, and thus run in $\mathcal{O}(n)$. Line 10 sorts the list, running in $\mathcal{O}(2n \log(2n)) = \mathcal{O}(n \log n)$. Lines 11–30 consist of two nested loops over the list, resulting in a $\mathcal{O}((2n)^2) = \mathcal{O}(n^2)$. The total running time for the novel method is then $\mathcal{O}(1) + \mathcal{O}(n) + \mathcal{O}(n \log n) + \mathcal{O}(n^2) = \mathcal{O}(n^2)$ time in the worst-case scenario.

6   Experiments

Both the baseline algorithm [12] and the novel algorithm for the construction of the behavior graph are implemented in Python, in the context of the PROVED project within the PM4Py framework. The experiments are designed to investigate the difference in performances between the two algorithms, and specifically how this difference scales with the increase of the size of the event log, as well as the number of events in the log that have uncertain timestamps.

For each series of experiments, we generate a synthetic event log with n many traces of length l (indicating the number of events in the trace). Uncertainty on timestamps is added to the events in the log. A percentage p of the events in the event log will have an uncertain timestamp, causing it to overlap with adjacent events. Finally, behavior graphs are obtained from all the traces in the event log with either algorithm, while the execution time is measured. All results are shown as an average of 10 runs of the corresponding experiment.

In the first experiment, we analyze the effect of the trace length on the overall time required for behavior graph construction. To this end, we generate logs with $n = 1000$ traces of increasing lengths, and added uncertain timestamps to events with $p = 0.4$. The results, presented in Fig. 5a, match our expectations: the computing time of the naïve algorithm scales much worse than the time of our novel algorithm, due to its cubic asymptotic time complexity. This confirms the findings of the asymptotic time complexity analysis discussed in Sect. 5. We can observe order-of-magnitude speedup. At length $l = 500$, the novel algorithm runs in 0.16% of the time needed by the naïve algorithm.

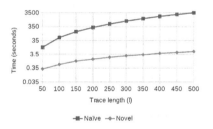
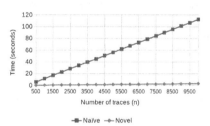

(a) Time in seconds for the creation of the behavior graphs for synthetic logs with $n = 1000$ traces and $p = 0.4$ of uncertain events, with increasing trace length.

(b) Time in seconds for the creation of the behavior graphs for synthetic logs with traces of length $l = 10$ events and $p = 0.4$ of uncertain events, with increasing number of traces.

Fig. 5. Results of the first and second experiments. The diagrams show the improvement in speed attained by our novel algorithm.

The second experiment verifies how the speed of the two algorithms scales with the log dimension in number of traces. We create logs with a trace length of $l = 50$, and a fixed uncertainty percentage of $p = 0.4$. The number of traces scales from $n = 500$ to $n = 10000$. As presented in Fig. 5b, our proposed algorithm outperforms the naïve algorithm, showing a relatively smooth behavior exposing a much smaller slope. As expected, the elapsed time to create behavior graphs scales linearly with the number of traces in the event log for both algorithms.

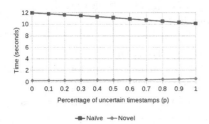

Event Log	p	Time (naïve)	Time (novel)
	0	1.17	0.15
Help Desk	0.4	1.11	0.17
	0.8	1.06	0.20
	0	31.69	4.09
Road Traffic	0.4	30.73	5.05
	0.8	29.45	5.79
	0	58.25	1.50
BPIC 2012	0.4	55.22	2.37
	0.8	51.79	3.33

(a) Time in seconds for the creation of the behavior graphs for synthetic logs with $n = 1000$ traces of length $l = 10$ events, with increasing percentages of timestamp uncertainty.

(b) Execution times in seconds for real-life event logs with increasing percentages of timestamp uncertainty.

Fig. 6. Effects of different percentages of uncertain timestamps in a trace on the execution time for both algorithms.

Finally, the third experiment inspects the difference in execution time for the two algorithms as a function of the percentage of uncertain events in the event log. Keeping the values $n = 1000$ and $l = 50$ constant, we scaled up the percentage p of events with an uncertain timestamp and measured computing time. As presented in Fig. 6a, the time required for behavior graph construction remains almost constant for our proposed algorithm, while it is decreasing for the naïve algorithm. This behavior is expected, and is justified by the fact that a worst-case scenario for the naïve algorithm is a trace that has no uncertainty on the timestamp: in that case, the behavior graph is simply a chain of nodes, thus the transitive reduction needs to remove a high number of edges from the graph. Notice, however, that for all possible values of p the novel algorithm runs is faster than the naïve algorithm: with $p = 0$, the new algorithm takes 1.91% of the time needed by the baseline, while for $p = 1$ this figure grows to 5.41%.

We also compared the elapsed time for behavior graphs construction on real-life event log, where we simulated uncertainty in progressively increasing percentage of events as described for the experiments above. We analyzed three event logs: an event log related to the help desk process of an Italian software company, a log related to the management of road traffic fines in an Italian municipality, and a log from the BPI Challenge 2012 related to a loan application process. The results, shown in Fig. 6b, closely adhere to the findings of the experiments on synthetic uncertain event data.

In summary, the results of the experiments illustrate how the novel algorithm hereby presented outperforms the previous algorithm for constructing the behavior graph on all the parameters in which the problem can scale in dimensions.

The third experiment shows that, like the baseline algorithm, our novel method being is essentially impervious to the percentage of events with uncertain timestamps in a trace. While for every combination of parameters we benchmarked the novel algorithm runs in a fraction of time required by the baseline method, the experiments also empirically confirm the improvements in asymptotic time complexity shown through theoretical complexity analysis.

7   Conclusions

The construction of the behavior graph – a fundamental structure for the analysis of uncertain data in process mining – plays a key role as processing step for both process discovery and conformance checking of traces that contain events with timestamp uncertainty, the most critical type of uncertain behavior. In this paper we improve the performance of uncertainty analysis by proposing a novel algorithm that allows for the construction of behavior graphs in quadratic time in the length of the trace. We argued for the correctness of this novel algorithm, the analysis of its asymptotic time complexity, and implemented performance tests for this algorithm. These show the speed improvement in real-world scenarios.

Further research is needed to inspect the capabilities of the novel algorithm. Future work includes extending the asymptotic time complexity analysis presented in this paper with lower bound and average case scenario analysis. Furthermore, behavior graphs are memory-expensive; we plan to address this through a multiset of graphs representation for event logs.

References

1. Van der Aalst, W.M.P.: Process Mining: Data Science in Action. Springer, Cham (2016)
2. Aho, A., et al.: Compilers: Principles, Techniques and Tools (2007)
3. Aho, A.V., Garey, M.R., Ullman, J.D.: The transitive reduction of a directed graph. SIAM J. Comput. $1(2)$, 131–137 (1972)
4. Al-Mutawa, H.A., Dietrich, J., Marsland, S., McCartin, C.: On the shape of circular dependencies in Java programs. In: 2014 23rd Australian Software Engineering Conference, pp. 48–57. IEEE (2014)
5. Bayes, T.: LII. An essay towards solving a problem in the doctrine of chances. By the late Rev. Mr. Bayes, FRS communicated by Mr. Price, in a letter to John Canton, AMFR S. Philos. Trans. R. Soc. Lond. (53), 370–418 (1763)
6. Berti, A., van Zelst, S.J., van der Aalst, W.M.P.: Process Mining for Python (PM4Py): bridging the gap between process- and data science. In: ICPM Demo Track (CEUR 2374), pp. 13–16 (2019)
7. D'Alberto, P., Nicolau, A.: Using recursion to boost ATLAS's performance. In: Labarta, J., Joe, K., Sato, T. (eds.) ALPS/ISHPC 2005-2006. LNCS, vol. 4759, pp. 142–151. Springer, Heidelberg (2008). https://doi.org/10.1007/978-3-540-77704-5_12
8. Dutta, S.: An event based fuzzy temporal logic. In: 1988 Proceedings of the Eighteenth International Symposium on Multiple-Valued Logic, pp. 64–71. IEEE (1988)

9. Le Gall, F.: Faster algorithms for rectangular matrix multiplication. In: 2012 IEEE 53rd Annual Symposium on Foundations of Computer Science, pp. 514–523. IEEE (2012)

10. Le Gall, F.: Powers of tensors and fast matrix multiplication. In: Proceedings of the 39th International Symposium on Symbolic and Algebraic Computation, pp. 296–303. ACM (2014)

11. Mariappan, M., Vora, K.: GraphBolt: dependency-driven synchronous processing of streaming graphs. In: Proceedings of the Fourteenth EuroSys Conference 2019, p. 25. ACM (2019)

12. Pegoraro, M., van der Aalst, W.M.P.: Mining uncertain event data in process mining. In: 2019 International Conference on Process Mining (ICPM), pp. 89–96. IEEE (2019)

13. Pegoraro, M., Uysal, M.S., van der Aalst, W.M.P.: Discovering process models from uncertain event data. In: Di Francescomarino, C., Dijkman, R., Zdun, U. (eds.) BPM 2019. LNBIP, vol. 362, pp. 238–249. Springer, Cham (2019). https:// doi.org/10.1007/978-3-030-37453-2_20

14. Strassen, V.: Gaussian elimination is not optimal. Numer. Math. **13**(4), 354–356 (1969)

Artificial Intelligence

Real-Time Detection of Unusual Customer Behavior in Retail Using LSTM Autoencoders

Oliver Nalbach[✉], Sebastian Bauer, Nanna Dahlem, and Dirk Werth

AWS-Institut für digitale Produkte und Prozesse gGmbH,
66123 Saarbrücken, Germany
{oliver.nalbach,sebastian.bauer,nanna.dahlem,dirk.werth}@aws-institut.de

Abstract. Personal customer care is one of the advantages of physical retail over its online competition, but cost pressure forces retailers to deploy staff as efficiently as possible resulting in a trend of staff reduction. For staff and managers it becomes harder to keep track of what is happening in a store. Situations that would benefit from intervention like cases of aimless customers, lost children or shoplifting go unnoticed. To this end, real-time tracking systems can provide managers with live data on the current in-store situation, but analysis methods are necessary to actually interpret these data. In particular, anomaly detection can highlight unusual situations that require a closer look. Unfortunately, existing algorithms are not well-suited for a retail scenario as they were designed for different use cases or are slow to compute. To resolve this, we investigate the use of long short-term memory autoencoders, which have recently shown to be successful in related scenarios, for real-time detection of unusual customer behavior. As we demonstrate, autoencoders reconcile the precision of reliable methods that have poor performance with a speed suitable for practical use.

Keywords: Anomaly detection · Retail · Unsupervised learning · Autoencoder · Long short-term memory

1 Introduction

Personal customer care is one of the remaining advantages of physical retail over e-commerce. Yet, increasing cost pressure forces retailers to deploy staff as efficiently as possible. As it becomes more difficult for managers and employees to keep track of individual situations in a store, important events may go unnoticed [3]. Examples include cases of aimless or clueless customers but also lost children or criminal behavior. This problem could be addressed by automated detection of unusual behavior: Customers unable to find a particular product will deviate from their typical route through the store to search for it. Shoplifters, too, differ from regular shoppers and spend more time in certain areas or near the checkout than is common [15]. These and similar situations lead to movement anomalies that retailers would benefit from to detect in real-time: Aimless customers

© Springer Nature Switzerland AG 2020
W. Abramowicz and G. Klein (Eds.): BIS 2020, LNBIP 389, pp. 91–102, 2020.
https://doi.org/10.1007/978-3-030-53337-3_7

could be identified and approached by a salesman. Suspicious behavior could be uncovered and observed to prevent theft. To achieve this, in the near future, visual tracking systems [6] will provide the required input in the form of live trajectories of customers moving in a store. However, the problem of real-time anomaly detection in this setting has not been covered in the literature yet and most existing approaches for related problems have shortcomings that prevent a simple transfer (Sect. 2). To this end, we propose the use of long short-term memory recurrent autoencoders (LSTM autoencoders) [10], a type of artificial neural network (ANNs). LSTM autoencoders can learn compact, lower dimensional encodings of a dataset and, as a by-product, can serve to identify outlier data points. By quantitative and qualitative analyses, we demonstrate that the method reconciles detection quality on-par with state-of-the-art methods, which are too slow for our use case, with a sufficiently high detection speed.

After reviewing related previous work (Sect. 2) to select a reference method, the specific problem setting and the algorithms are discussed in (Sect. 3). The latter are compared in a subsequent evaluation on real-world data from a large retail store (Sect. 4). We conclude the paper and discuss future work in Sect. 5.

2   Previous Work

Real-time anomaly detection in movement data of retail customers has not been addressed in the literature yet but is an important avenue of research to understand customer experience [13]. There is however related previous work: First, domain-specific literature on the analysis of retail customers in a broader sense. Second, literature on anomaly detection in movement patterns but for different application scenarios. Third, previous work on LSTM networks or autoencoders for sequential anomaly detection.

2.1   Analysis of Customer Behavior

Technologies like RFID or Bluetooth tags [17] have made it possible to acquire and analyze large data sets on customer behavior in offline stores. In their seminal paper, Larson et al. [11] cluster movement paths (traces) obtained by RFID tags on shopping carts. To deal with traces of varying length, each is re-sampled to a fixed number of 100 samples. The k-medoids algorithm is used to identify typical customer types. The authors also evaluate simple descriptive statistics for clustering, concluding that this approach is not sufficient due to its coarseness. While, in principle, clustering can be used for anomaly detection by checking samples against the identified clusters, the method of Larson et al. cannot be transferred to our use case: Real-time traces of customers in a store are by definition incomplete and deviate from clusters formed on completed traces and the representation by a fixed number of samples becomes spatially imprecise for longer traces. Anomalies characterized by a local deviation of behavior become undetectable. Yan and Zeng [22] analyze data acquired in the same way using a different representation: Traces are mapped to a pre-defined graph capturing the store layout and described as varying-length sequences of graph nodes. This makes them amenable to sequence clustering using a version of the longest common sub-sequence method [20]. In a similar way, Yada [21] analyzes sequences of visited shop sections to understand the behavior of customers purchasing particularly many items. Representing traces by symbolic sequences as done in both papers results in a loss of temporal information which is however important to identify abnormal behavior characterized by extended amounts of time spent in a particular area, such as in cases of shoplifting or searching customers. Another problem again is the spatial discretization which has the disadvantages discussed above. In their study, Sorensen et al. [18] identify behavioral patterns in a pool of data from 40 stores. The data is analyzed using statistics over manually defined features such as trip length, basket size or visited portion of the store. As concluded by Larson et al. [11], this representation is unsuitable for clustering and hence unpromising for outlier detection, too. To summarize, previous approaches resort to explicit temporal or spatial discretizations resulting in the loss of information that is relevant for anomaly detection; none of the methods is thus easily transferable to our scenario.

2.2 Anomaly Detection in Movement Data

If not for retail, anomaly detection has been applied to general surveillance settings. Owens and Hunter [16] detect unusual trajectories of objects in video streams. Object movement is characterized by a fixed-length vector containing the position of the object and moving averages of position, velocity and acceleration computed over a fixed number of samples. A self-organising map is used for the actual anomaly detection. Apart from being very coarse, the method, by design, only incorporates knowledge from a limited time window. Anomalies characterized by long-term behavior cannot be detected. Dogra et al. [1] target the same scenario differently: The tracking environment, i.e., the scene seen by a camera, is represented by a Region Association Graph in which nodes correspond to semantically homogeneous blocks and edges to region transitions. A labeled dataset is created by manually defining two kinds of anomalies and used to train a support vector machine. Besides the shortcomings of the explicit spatial discretization, this supervised approach assumes that all types of anomalies are known and can be defined in advance which is a strong assumption for complex scenarios. A promising method for online anomaly detection is the one by Laxhammar and Falkman [12], later extended by Guo and Bardera [5]. It circumvents many of the problems discussed above for other methods: It supports incomplete traces as they occur in a real-time setting, it does not require an explicit discretization in space or time as it works directly on trace data and it is unsupervised. Due to these useful properties, we choose it as the reference method for our evaluation in Sect. 4. We briefly discuss the algorithm in Sect. 3.2.

2.3 LSTM and Autoencoders for Sequential Anomaly Detection

The success of Deep Learning has also brought related concepts back to attention. One are recurrent neural networks (RNNs) which, unlike plain ANNs, can be applied to sequential data. An important extension of RNNs is long short-term memory [8], a structural modification to improve learning on long sequences. Another technique are autoencoders [7], a type of ANN that can be trained to learn compact encodings of a dataset. Intuitively, the autoencoder network is forced to compress input data to a fixed-length vector and then reconstruct the original input again. Autoencoders are suitable for anomaly detection: When a data sample is an anomaly, i.e., not well-represented in the original dataset, it has a high reconstruction error and can be identified that way. LSTM RNNs and autoencoders can be combined for sequential anomaly detection as demonstrated by Malhotra et al. [14] for outlier detection on multi-sensor datasets. Only little previous work exists on LSTM autoencoders for motion trajectories. Fernando et al. [2] apply a variant of the idea to trajectories defined with respect to the view of a single surveillance camera. Gatt et al. [4] detect abnormal behavior on a more detailed scale, also taking into account pose information. Both papers deal with scenarios different from ours as they address short-term behavior over a couple of seconds while we aim to detect anomalies in long-term behavior over the course of a shopping session which can take minutes to hours.

3   Real-Time Anomaly Detection for Retail

We will now describe our problem statement formally before outlining the two methods compared in this paper, the method of Laxhammar and Falkman [12] and our LSTM autoencoder-based approach.

3.1   Problem Statement

Our input is a set of traces $\mathcal{T} = \{\mathbf{t}_i \mid 0 \leq i < n_{\mathcal{T}}\}$ describing $n_{\mathcal{T}}$ completed shopping sessions. A trace captures the movements of a person inside the store over time, from entering until leaving again. Formally, trace $\mathbf{t}$ is a sequence $(\mathbf{t}^k)_{k=0}^{n_\mathbf{t}-1}$ with $\mathbf{t}^k = (t^k, x^k, y^k)$ where $t^i < t^j$ for $i < j$. Above, t^k is the time since entering the market and $(x^k, y^k) \in \mathbb{R}^2$ is the position at time t^k, defined with respect to the supermarket floor. Given $\mathcal{T}$, our aim is to decide for a new trace $\mathbf{t}'$ whether it constitutes an anomaly with respect to the traces in $\mathcal{T}$ or not. The retail setting implies three additional requirements. First, detection has to handle incomplete traces: Unusual behavior has to be detected as soon as it emerges so a manager can react in a timely manner. For the same reason, the detection has to offer real-time performance. A large store may accommodate hundreds of people at once. Only if all traces can be checked at a high rate the method is useful. Finally, no manual definition of "abnormal" behavior should be required. Even if we aim to detect specific anomalies of interest to shop managers, e.g., clueless customers or shoplifters, defining them in a way that is robust to variances e.g., in the store layout, is beyond question.

3.2   Method of Laxhammar and Falkman

We choose the method of Laxhammar and Falkman [12] as our reference as it meets all of the requirements stated above, is particularly easy to deploy due to

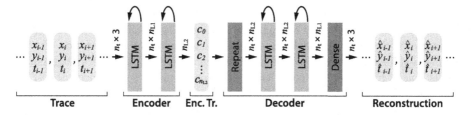

Fig. 1. The architecture of our LSTM autoencoder network. The horizontal arrows are labeled with the shape of the input/output of each step.

its limited parameter set, and has been designed for online surveillance scenarios similar to ours. The method consists of two steps: First, a distance $d_H(\mathbf{t}_i, \mathbf{t}_j)$ between traces $\mathbf{t}_i$, $\mathbf{t}_j$ is defined. Second, a non-conformity measure $\alpha(\mathbf{t})$ for a trace $\mathbf{t}$ is derived as the sum of the distances of $\mathbf{t}$ to its k nearest neighbors. Formally, $\alpha(\mathbf{t}) := \sum_{\mathbf{t}_i \in \mathrm{kNN}_\mathcal{T}(\mathbf{t})} d_H(\mathbf{t}, \mathbf{t}_i)$ where $\mathrm{kNN}_\mathcal{T}(\mathbf{t})$ yields the k nearest neighbors of $\mathbf{t}$ in $\mathcal{T}$ with respect to d_H. A trace $\mathbf{t}$ is considered an anomaly if its non-conformity measure exceeds a defined percentile of the non-conformity measures of the traces in $\mathcal{T}$. The distance proposed by Laxhammar and Falkman is the directed Hausdorff distance $d_H(\mathbf{t}_i, \mathbf{t}_j) = \max_{k=0}^{n_{\mathbf{t}_i}-1} \min_{l=0}^{n_{\mathbf{t}_j}-1} d(\mathbf{t}_i^k, \mathbf{t}_j^l)$, where d is the Euclidean distance. In particular, d_H behaves reasonably when comparing an incomplete real-time trace $\mathbf{t}'$ to a completed trace from $\mathcal{T}$: As long as the known part $\mathbf{t}'$ is similar to some sub-sequences of the traces in $\mathcal{T}$, its non-conformity will be low regardless of drastic differences in sequence length. For our evaluation, we use a windowed version of d_H proposed by Guo and Bardera [5] which provides results comparable to the original Hausdorff distance but has a lower computational complexity that is just linear, not quadratic, in the trace length. This is especially beneficial when dealing with long traces as in our scenario. We refer to the original paper [5] for the definition of d_H.

3.3 LSTM Autoencoder-based Detection

The structure of our LSTM autoencoder, a symmetrical neural network composed of encoder and decoder parts, is shown in Fig. 1 and was adapted from previous work [2,14]. The input (Fig. 1, left) is a trace $\mathbf{t}$ of length $n_\mathbf{t}$. The first LSTM layer computes a set of n_{L1} features for each sequence element, the second reduces this intermediate output to a vector of length n_{L2}—the encoded trace (Fig. 1, middle). In practice, we choose $n_{L1} = 200$ and $n_{L2} = 100$ (cf. Sect. 4). The encoded trace is replicated $n_\mathbf{t}$ times to form a sequence (Fig. 1, "Repeat"), followed by two LSTM layers outputting n_{L2} and n_{L1} features per sequence element, respectively. A final fully-connected layer (Fig. 1, "Dense") is applied to each element to reduce the n_{L1}-element feature vector to the original input dimensionality, yielding the reconstruction $\hat{\mathbf{t}}$. All layers use a tangens hyperbolicus activation function. The autoencoder is trained using the sum of elementwise absolute deviations as loss function, i.e., $L(\mathbf{t}, \hat{\mathbf{t}}) = \sum_{i=0}^{n_\mathbf{t}-1} \|\mathbf{t}^i - \hat{\mathbf{t}}^i\|_1$. For training, we augment $\mathcal{T}$, which only consists of complete traces, by randomly sampled shortened versions of them. For each $\mathbf{t} \in \mathcal{T}$, we add $n_{\text{sub}} = 10$ "sub-traces", i.e., prefixes of $\mathbf{t}$. Specifically, we apply stratified sampling over the duration of $\mathbf{t}$ to achieve a sufficient coverage of different sequence lengths. This data augmentation is crucial for the autoencoder to be able to reconstruct sub-traces, too. The final component of the method is the anomaly criterion: To determine whether trace $\mathbf{t}'$ is an anomaly, it is processed by the trained autoencoder to obtain the reconstruction loss $l(\mathbf{t}') = L(\mathbf{t}', \hat{\mathbf{t}}')$. Let $r(l(\mathbf{t}'), \mathcal{T})$ be the rank that l within the reconstruction losses computed over all traces in $\mathcal{T}$. We define the final anomaly score of $\mathbf{t}'$ as $\alpha(\mathbf{t}) = r(l(\mathbf{t}'), \mathcal{T})/n_{\mathcal{T}}$. If $\alpha(\mathbf{t}) > \epsilon$, $\mathbf{t}$ is considered an anomaly.

4 Evaluation

We perform two evaluations. First, a quantitative analysis on a synthetic, labeled dataset, comparing our approach to Laxhammar-Falkman [12] with respect to its detection and computational performance. To build the labeled dataset, we simulate both regular traces and different types of anomalies (Sect. 4.1). Second, a qualitative analysis on an unlabeled dataset recorded in a large German retail store assures that the method's classification aligns with human intuition on what is abnormal behavior. We implemented both methods in Python. The method of Laxhammar and Falkman was implemented according to the original papers [5,12]. The LSTM autoencoder was realized using Keras [9] and trained using the Adam optimizer. To speed up training, we parallelized the computation of the distance matrix for Laxhammar-Falkman and made use of the CPU-based parallelization in Keras to ensure a fair comparison. All timings given in Table 1 have been obtained using 8 parallel threads on an Intel Xeon Gold 5122 CPU.

4.1 Datasets

The dataset we base our evaluation on consists of traces from around $100,000$ individual shopping sessions recorded in a large German supermarket and has been kindly provided by the German Research Center for Artificial Intelligence (DFKI). The dataset was obtained using Bluetooth Low Energy beacons [17] mounted on shopping carts used in the store. It has to be noted that the use of shopping cart tracking data differs from our intended scenario in which people, not shopping carts, are tracked. In particular, customers may move independently from their cart from time to time. To compensate for this to some extent, we filter traces in which independent movement is predominant (cf. Sect. 4.1).

Pre-processing of the Unlabeled Dataset. We filter the original dataset for multiple reasons: To remove outliers originating from known limitations of the tracking technology such as wrong localisation we filter traces with gaps larger than 8 m between consecutive samples. To eliminate traces in which customers moved mostly independently, we filter those for which no movement occurred for more than 10 min. Finally, we remove traces shorter than 3 min and longer

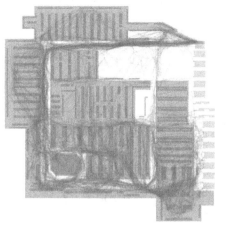 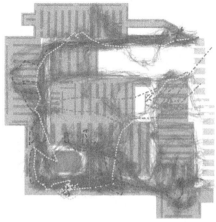

Real dataset (unlabeled)   **Synthetic dataset (labeled)**

Fig. 2. Samples from the two datasets we use. The left part shows traces from the unlabeled, real-world dataset. The right half depicts synthetic traces generated according to Sect. 4.1, three of which are anomalies (dashed lines).

than 180 min which are implausible and usually result from carts being used around the store for purposes other than shopping. In the end, we re-sample the remaining traces to a fixed sampling rate of 0.03 Hz.

Sampling of the Labeled Dataset. For the quantitative evaluation we generate a labeled data set of normal and anomalous traces. Normal traces (negatives) correspond to regular customers while anomalous ones (positives) represent behavior that should be detected. Specifically, we simulate two kinds of anomalies: cases of aimless customers searching for a particular product and shoplifting attempts. To model regular customers we sample multiple clusters of traces. All traces of a single cluster are derived from one particular random "seed trace" from the unlabeled dataset by randomly displacing the positions of the original trace and introducing additional, intermediate nodes according to a method by Technitis et al. [19]. The timestamps of new nodes are interpolated. In the end, each trace is re-sampled to 0.03 Hz again to ensure consistency. Figure 2 (right, solid lines) shows sample regular traces. The two kinds of anomalies are modeled as follows. Traces of searching customers are, too, based on seed traces from the regular dataset and constructed by inserting a new sub-sequence representing the searching process. The insertion position is chosen according to a Gaussian distribution centered around 30% of the progress of the original trace. The sub-sequence itself is modeled by a Brownian bridge. Our model for shoplifting traces is inspired by the descriptions found in [15]. A set of shelves is chosen and visited by the shoplifter on the shortest route possible, the walking speed is chosen randomly according to the distribution of the original dataset. The actual shoplifting event is inserted at around 50% of the progress and again

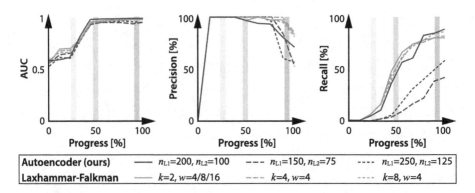

Fig. 3. AUC (left), precision (middle) and recall (right), computed plotted over relative progress for different variants of both methods.

modeled by a Brownian bridge. Another case of abnormal behavior is introduced towards the end of the trace by simulating an extended amount of time spent near the checkout area in which the thief watches for the right moment to leave. To generate additional variance, the same methods as for sampling trace clusters of regular shoppers, i.e., node displacement and new intermediate nodes, are applied. Figure 2 (right, dashed lines) depicts three exemplary anomalous traces.

4.2 Quantitative Evaluation

We first sample a labeled dataset according to Sect. 4.1 consisting of 6000 traces for training both methods and 1000 different ones for testing, each based on 10% seed traces. Assuming an average of about 1000 customers per day this corresponds to the data of a six-day week for training the method and an evaluation on a single day's traces. The prevalence of anomalies in the dataset was chosen to be 5% with 90% of searching customers and 10% shoplifting cases. Accordingly, for both methods we chose a detection threshold of $\epsilon = 0.95$ reflecting the prior assumption that 5% of all traces are anomalies. For both methods, we compare five metrics: the area under the receiver operating characteristic curve (AUC), precision and recall to assess the detection performance, as well as the time required for training and the average time to rate one trace to judge practical feasibility. From left to right, Fig. 3 depicts the change of AUC, precision and recall over the relative progress of the traces in the test set for different parameter settings of both methods. The plots were generated by trimming down traces to different relative lengths in steps of 10% and rating the resulting prefix sequences. In a practical application, it is desirable that detection of abnormal behavior happens with as little delay as possible. To judge this, the typical times of occurrence of anomalies in the synthetic dataset (Sect. 4.1) are highlighted using vertical bars in Fig. 3.

Comparing different parameter settings for Laxhammar-Falkman, we observe only minor effects when increasing the number of nearest neighbors k (Sect. 3.2). Since a smaller k means better computational performance (Table 1), we opt for $k = 2$. The effect of a larger window size w [5] is even so negligible that the resulting lines are indistinguishable. We thus choose $w = 4$ which again yields the best performance. For our LSTM autoencoder, we evaluated different combinations of layer sizes n_{L1}, n_{L2}, the three best of which are shown in Fig. 3. As all variants are fast to evaluate, we pick $n_{L1} = 200$ and $n_{L2} = 100$ which yields the most reliable detection.

Focusing on the best variants, i.e., $k = 2$, $w = 4$ for Laxhammar-Falkman and $n_{L1} = 200$, $n_{L2} = 100$ for the autoencoder, we notice a similar level and progression of their AUC (Fig. 3, left). As expected, no anomalies are detected before unusual behavior becomes apparent. However, with the onset of the first abnormal behavior in the anomalous traces (leftmost vertical bars) after around 30% progress, AUC increases rapidly. At about 50% progress (middle vertical bars), abnormal behavior has become apparent for all anomalies and AUC stabilizes at around 0.98 for both methods. Both methods show high precision which degrades towards the end of the traces. This can be attributed to both algorithms computing global anomaly criteria which degrade as the regular portion of the traces becomes larger which is after around 50% of progress for most traces. For example, traces of searching customers are correctly identified as anomalies while the search is ongoing but may be rated as regular once the trace continues regularly. The drop in precision is larger for the autoencoder. Recall measures how many anomalies can actually be detected and is even more important than precision: For a store manager, it is preferable to take a closer look at a few traces more than necessary than to miss any abnormal situation. Both methods show similar recall curves with the autoencoder having an edge over Laxhammar-Falkman. Overall, both methods achieve comparable results with individual trade-offs regarding the achieved precision vs. recall.

Table 1. Comparison of time required to train the method on our training set and average time required to rate a single trace afterwards.

	Laxhammar-Falkman					Autoencoder		
	$k = 2$ $w = 4$	$k = 4$ $w = 4$	$k = 8$ $w = 4$	$k = 2$ $w = 8$	$k = 2$ $w = 16$	$n_{L1} = 150$ $n_{L2} = 75$	$n_{L1} = 200$ $n_{L2} = 100$	$n_{L1} = 250$ $n_{L2} = 125$
Training	30 min	30 min	29 min	45 min	79 min	498 min	439 min	311 min
Testing	2182 ms	2242 ms	2202 ms	3646 ms	6626 ms	23 ms	24 ms	25 ms

The two methods differ significantly regarding their computational performance (Table 1). In practice, training time is of minor importance as long as it is possible to re-train or finetune on a daily basis. This is the case for all of the variants. Evaluation time is more critical. Here, Laxhammar-Falkman is slower by a factor of 100. Assuming about 100 customers in a market at a given time, this means the reference method can rate the traces of all customers only once every 200 s which means a significant delay that precludes timely reaction to anomalies. The autoencoder however can rate 100 traces at a rate of about 0.5 Hz which is sufficient for practical use. These numbers can be explained by the theoretical complexity of both methods: Laxhammar-Falkman is based on a

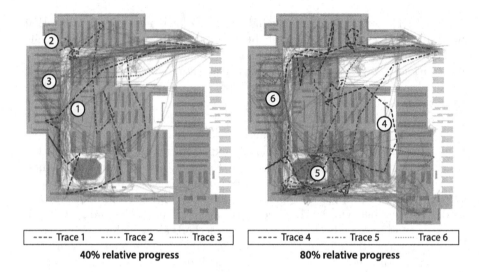

Fig. 4. Three sample traces from our real-world dataset rated as anomalies by our method (dashed lines) for two different progress levels.

k-nearest neighbor search requiring the pairwise distances d_{H} between all traces in $\mathcal{T}$. This results in a quadratic complexity of the training process in $n_{\mathcal{T}}$ and a complexity that is linear both in $n_{\mathcal{T}}$ and the respective trace length for rating a trace afterwards. As d_{H} is, in general, asymmetric, common optimization techniques for kNN that require the distance function to be a metric are not applicable. In contrast, the training time of the autoencoder depends on multiple factors and is governed by how fast the training loss stabilizes for a specific training set. In general, no theoretical complexity can be specified, but in practice the training time turns out to be in the order of a few hours for our example. The autoencoder's evaluation time is linear in the length of the trace and, different from Laxhammar-Falkman, independent of the size of the training set. This results in a large performance advantage over Laxhammar-Falkman.

4.3 Qualitative Evaluation

To show that our method actually yields plausible results on real-world data that align with human judgement, we examine the output of the best-performing autoencoder ($n_{\mathrm{L1}} = 200$, $n_{\mathrm{L2}} = 100$) trained on 6000 randomly chosen traces from the unlabeled dataset (Sect. 4.1) on a set of 100 different random traces. Figure 4 shows three sample anomalies identified by the autoencoder for both of two relative progress levels (40%, 80%). For reference, all test traces rated as regular are shown semi-transparently in the background. All other abnormal traces have been removed from the plots for improved visibility. We first examine the result for 40% progress: Trace 1, rated abnormal, differs from most traces by a diagonal walk across the market (label 1) and the fact that the customer comes close to the checkout area (bottom right corner) after a very short time already which could be due to time pressure experienced by the customer. Anomaly 2 is similarly unusual: While most customers follow a counter-clockwise route, the customer immediately visits the center of the store. Later on, a lot of time is spent in a small area (2) while the person is likely searching for a particular product. In the case of trace 3, the person first follows the common shopping route but then makes a 180 degree turn (3) and walks back right to the entrance area. The results for 80% of the relative progress are equally interpretable. Again there behaviors such as criss-crossing of the store (4), circling around a particular area for an extended time (5) and seemingly unplanned returns to a particular department of the market initially disregarded (6).

5   Conclusion and Future Work

We have proposed LSTM autoencoders for the detection of unusual customer behavior in retail. As has been demonstrated by quantitative and qualitative analysis, this novel approach is on-par with existing methods in terms of quality while being orders of magnitude faster, which makes it—unlike its competitors— viable in real world settings. Still, future research can be identified: We have only considered simple encoder-decoder structures consisting of two layers each. More complex, deeper architectures can be investigated to potentially improve detection in terms of precision and recall. Our anomaly criterion, too, is currently relatively simple. Instead of averaging the reconstruction loss over the full trace, a more complex criterion could give a higher weight to anomalous sub-sequences while disregarding regular parts of a trace, which should result in a higher precision. Finally, we have not taken into account any of the available semantic information, e.g., on the store layout, or other prior knowledge about factors influencing behavior. For example, different times of day, or days of the week, can be associated with different dominant customer groups. These factors are currently implicitly encoded in the dataset but could be made explicit.

Acknowledgments. This work is based on VICAR, a project partly funded by the German ministry of education and research (BMBF), reference number 01IS17085C. The authors are responsible for the publication's content.

References

1. Chebiyyam, M., Reddy, R.D., Dogra, D.P., Bhaskar, H., Mihaylova, L.: Motion anomaly detection and trajectory analysis in visual surveillance. Multimedia Tools Appl. **77**(13), 16223–16248 (2017). https://doi.org/10.1007/s11042-017-5196-6
2. Fernando, T., Denman, S., Sridharan, S., Fookes, C.: Soft+ hardwired attention: an LSTM framework for human trajectory prediction and abnormal event detection. Neural Netw. **108**, 466–478 (2018)
3. Forbes: Too few retail workers on the floor, too few retail sales and profits on p&l statement (2017). https://www.forbes.com/sites/pamdanziger/2017/12/16/too-few-retail-workers-on-the-floor-too-few-retail-sales-and-profits-on-pl-statement/

4. Gatt, T., Seychell, D., Dingli, A.: Detecting human abnormal behaviour through a video generated model. In: 2019 11th International Symposium on Image and Signal Processing and Analysis (ISPA), pp. 264–270. IEEE (2019)
5. Guo, Y., Bardera, A.: SHNN-CAD+: an improvement on shnn-cad for adaptive online trajectory anomaly detection. Sensors **19**(1), 84 (2019)
6. Hernandez, D.A.M., Nalbach, O., Werth, D.: How computer vision provides physical retail with a better view on customers. In: 2019 IEEE 21st Conference on Business Informatics (CBI), vol. 1, pp. 462–471. IEEE (2019)
7. Hinton, G.E., Zemel, R.S.: Autoencoders, minimum description length and Helmholtz free energy. In: Advances in Neural Information Processing Systems, pp. 3–10 (1994)
8. Hochreiter, S., Schmidhuber, J.: Long short-term memory. Neural Comput. **9**(8), 1735–1780 (1997)
9. Keras: The python deep learning library. https://keras.io
10. Kieu, T., Yang, B., Guo, C., Jensen, C.S.: Outlier detection for time series with recurrent autoencoder ensembles. In: 28th International Joint Conference on Artificial Intelligence (2019)
11. Larson, J., Bradlow, E., Fader, P.: An exploratory look at supermarket shopping paths. Int. J. Res. Market. **22**(4), 395–414 (2005)
12. Laxhammar, R., Falkman, G.: Online learning and sequential anomaly detection in trajectories. IEEE Trans. Pattern Anal. Mach. Intel. **36**(6), 1158–1173 (2013)
13. Lemon, K.N., Verhoef, P.C.: Understanding customer experience throughout the customer journey. J. Market. **80**(6), 69–96 (2016)
14. Malhotra, P., Ramakrishnan, A., Anand, G., Vig, L., Agarwal, P., Shroff, G.: LSTM-based encoder-decoder for multi-sensor anomaly detection (2016). arXiv preprint arXiv:1607.00148
15. NSW Justice: Shoplifting: Signs and prevention. http://www.crimeprevention.nsw.gov.au/Documents/RetailSecurityResource/04_Shoplifting-signs_and_prevention.pdf
16. Owens, J., Hunter, A.: Application of the self-organising map to trajectory classification. In: Proceedings of 3rd IEE International Workshop on Visual Surveillance, pp. 77–83 (2000)
17. Quuppa: Quuppa intelligent locating system. https://quuppa.com
18. Sorensen, H., Bogomolova, S., Anderson, K., Trinh, G., Sharp, A., Kennedy, R., Page, B.: Fundamental patterns of in-store shopper behavior. J. Retail. Consum. Serv. **37**, 182–194 (2017)
19. Technitis, G., Othman, W., Safi, K., Weibel, R.: From a to b, randomly: a point-to-point random trajectory generator for animal movement. Int. J. Geog. Inf. Sci. **29**(6), 912–934 (2015)
20. Vlachos, M., Kollios, G., Gunopulos, D.: Discovering similar multidimensional trajectories. In: Proceedings 18th International Conference on Data Engineering, pp. 673–684. IEEE (2002)
21. Yada, K.: String analysis technique for shopping path in a supermarket. J. Intel. Inf. Syst. **36**(3), 385–402 (2011)
22. Yan, P., Zeng, D.D.: Clustering customer shopping trips with network structure. In: ICIS 2008 Proceedings - 29th International Conference on Information Systems, p. 28 (2008)

A Novel Multi-agent-based Chatbot Approach to Orchestrate Conversational Assistants

Jan Felix Zolitschka[(⊠)] [iD]

University Ulm, Helmholtzstraße 22, 89081 Ulm, Germany
jan.zolitschka@uni-ulm.de

Abstract. Nowadays, chatbots have become more and more prominent in various domains. Nevertheless, designing a versatile chatbot, giving reasonable answers, is a challenging task. Thereby, the major drawback of most chatbots is their limited scope. Multi-agent-based systems offer approaches to solve problems in a cooperative manner following the "divide and conquer" paradigm. Consequently, it seems promising to design a multi-agent-based chatbot approach scaling beyond the scope of a single application context. To address this research gap, we propose a novel approach orchestrating well-established conversational assistants. We demonstrate and evaluate our approach using six chatbots, providing higher quality than competing artifacts.

Keywords: Conversational agent · Chatbot · Multi-agent-based system · Orchestration · Collaboration · Mediation · Divide and conquer

1 Introduction

In recent years, conversational agents, also called chatbots, are becoming an increasingly pervasive means of conceptualizing a diverse range of applications in various domains [1–4]. Furthermore, it is forecast that due to the usage of chatbots the annual cost savings in organizations will grow from $48.3 million in 2018 up to $11.5 billion until 2023 [5]. In the area of conversational artificial intelligence, many studies already tap the potential of chatbots and provide well-known approaches for chatbots interacting with humans as well as answering questions regarding open-domain (non-task-oriented) and closed-domain (task-oriented) topics [1, 2, 6–8]. Following a survey from Ramesh et al. [8], chatbot approaches have evolved from simple pattern matching (e.g. ALICE [9]) to modern complex knowledge- and retrieval-based approaches (e.g. MILABOT [10], ALQUIST [11], EVORUS [12] or ALANA [13]) with the aim of giving conversations more human-like shape in order to pass the Turing test. Nevertheless, designing a versatile chatbot, giving reasonable answers to a variety of possible requests, is a challenging task. In particular, the task of simultaneously being robust regarding various domains and answering domain-specific questions is extremely demanding [6]. As a consequence, the major drawback of most chatbots is their limited scope [14]. In order to increase the capabilities of a single chatbot, most approaches in literature manually build and add skills to existing chatbots [15, 16]. Although this works reasonably well, it leads to a

© Springer Nature Switzerland AG 2020
W. Abramowicz and G. Klein (Eds.): BIS 2020, LNBIP 389, pp. 103–117, 2020.
https://doi.org/10.1007/978-3-030-53337-3_8

high effort in generating and coordinating various and complementary skills [14]. Other approaches based on current state-of-the-art knowledge and retrieval models (e.g. deep neuronal networks [10, 11]), require massive datasets, skilled human resources as well as a huge amount of time and hardware to be trained, enhanced and optimized [10]. Besides, as chatbots have become more prominent, the size and complexity of chatbot systems is increasing, which, as for any software system, cannot increase indefinitely [17]. The area of multi-agent-based systems offers a wide range of approaches to solve problems in a cooperative manner following the "divide and conquer" paradigm [18, 19]. In other domains, multi-agent-based approaches are commonly used to model complex and emergent phenomena inspired by human behavior as, for instance, expert collaborations in organizations interacting with the aim of solving a customer request. Consequently, as existing chatbots already provide sound results regarding the scope of their application, it seems promising to design a multi-agent-based chatbot approach, inspired by expert collaboration in practice, scaling beyond the scope of a single chatbot. However, there is still a lack of chatbot approaches giving reasonable answers to a variety of possible requests. To address this research gap, we orchestrate multiple chatbots and propose a multi-agent-based chatbot approach, which is able to learn chatbots' capabilities and identify relevant chatbots capable of giving answers.

Guided by the Design Science Research (DSR) process due to Peffers et al. [20], the remainder of this paper is structured as follows: In the next section, we provide an overview of the related work and identify the research gap. In Sect. 3, we propose a multi-agent-based chatbot approach relying on capability-based middle-agents to orchestrate chatbots in a single conversational agent. In Sect. 4, we demonstrate and evaluate our approach based on six different chatbot datasets on which the approach could be successfully applied. Finally, we conclude with a brief summary, limitations and an outlook on future research.

2   Related Work and Research Gap

In the area of conversational artificial intelligence, many studies already tap the potential of conversational agents and provide well-known chatbots approaches. Nevertheless, there is still a lack of approaches solving user requests in a cooperative manner by reusing multiple well-established chatbots. In the following, informed by the related literature regarding interactions between humans and multiple conversational agents as well as agent collaboration in multi-agent-based systems, we identify the research gap.

2.1   Interactions Between Humans and Multiple Conversational Agents

There has been recent work on analyzing different kinds of interactions between humans and multiple chatbots. First of all, the communication between a user and multiple chatbots can be conducted by multiple *single-bot chats* (a user interacting with a single chatbot in a single chat, e.g. [9–11]), *multi-bot chats* (a user interacting with multiple chatbots in a shared chat [3, 14, 21]) or a *single bot chat orchestrating hidden chatbots* (a user interacting with a single chatbot in a single chat, which in turn interacts with multiple hidden chatbots [12, 13]). At a first glance, interacting with multiple chatbots in multiple single-bot chats or a shared chat seems to increase complexity. Thus, researchers have investigated the user experience of single- versus multi-bot chats [3, 14, 21]. Chaves and Gerosa [3] conducted a Wizard-of-Oz study, where subjects are deluded into thinking that they are interacting with chatbots. As a result, the participants reported more confusion using a multi-bot chat in comparison to single-bot chats. Besides, they identified no significant difference between conversations in single- and multi-bot chats. This is in line with Pinhanez et al. [14], who point out that there is no increase in collaboration and coordination costs while interacting in a multi-bot chat. In contrast, Maglio et al. [21] investigated interactions in an office setting and found out that participants needed less effort to control multiple hidden chatbots within a single chat compared to conversing with chatbots in an individual or shared chat, respectively. Hence, regarding different kinds of interactions, literature reveals promising potential in approaches orchestrating multiple hidden chatbots [3, 14, 21].

In this context, only a few researchers focus on human collaboration with multiple chatbots in terms of a single chatbot orchestrating hidden chatbots [12, 13, 22]. To do so, Papaioannou et al. [13], one of the top competitors of Amazon's Alexa prize, developed a chatbot named ALANA. Their approach is based on a contextual bot priority list and a ranking function trained on user feedback to choose a response from one out of seven (2017) or rather nine (2018) chatbots. Cui et al. [22] solely base their approach on a static priority list in order to choose responses out of four chatbots without any consideration of the customer's intent. In contrast, a chatbot called EVORUS [12] used not only six different chatbots but also crowd-sourced human workers when their approach was unable to answer. To do so, their approach collects feedback from the crowd and learns to select chatbots, which are most likely to generate high-quality responses depending on the given context [12]. Nevertheless, all mentioned approaches orchestrating hidden chatbots solely rely on static priority lists or require a huge amount of user-provided ratings and feedback, which is often not available, time-consuming or expensive to collect. To the best of our knowledge, besides these few examples, further literature does not focus on orchestrating the capabilities of multiple already existing chatbots. In particular, the collaboration between open- and closed-domain chatbots without the use of feedback is not yet addressed by existing literature.

2.2 Agent Collaboration in Multi-agent-based Systems

In particular, as the size and complexity of most systems cannot increase indefinitely, research in the area of multi-agent-based systems offers a wide range of approaches to address the challenge of jointly acting agents [17–19]. In general, multi-agent-based approaches are used to design complex and emergent phenomena inspired by human behavior by using a collection of autonomous and distributed entities, called agents, with individual decision-making. Each agent is designed as an individual software agent with well-defined and limited scope, which perceives its environment (e.g. prospective user messages) and determines its actions accordingly (e.g. reasonable answers) [18]. Several previous studies have proposed agent theories and architectures to provide multi-agent-based systems with a strong formal basis. Hence, different types of agent architectures can be applied depending on the complexity of the agents' deliberation process. The most widely applied types are reactive agents, which are determined by static action rules [23,

24], in contrast to deliberative agents, which use symbolic reasoning for planning their actions [25]. Furthermore, the predominantly used agent architecture is derived from the physical and internet economy [26] and subdivides agents into requester-agents demanding a service (e.g. a visitor requests an apple on a physical market), provider-agents supplying a service (e.g. an orchardist supplies apples to a physical market) and middle-agents as intermediaries (e.g. a salesman mediates fruits on a physical market) [19, 27]. Obviously, regarding the context of our study, particularly designing the user as requester-agent, existing chatbots as reactive provider-agents and middle-agents as intermediaries between users and chatbots seems promising to cope with the task of orchestrating multiple conversational agents.

2.3 Research Gap

Despite emerging scientific work in the field of chatbots [3, 10, 12–14, 28], we still observe a lack of research on how multiple well-established chatbots could be reused in a jointly coordinated manner to scale beyond the scope of a single chatbot application. Regarding related research in distributed artificial intelligence, the area of multi-agent-based systems already offers a wide range of approaches to solve problems in a cooperative manner [18, 19, 23–25]. However, orchestrating hidden well-established chatbots by means of multi-agent-based technology is a novel approach for the area of conversational agents and not yet investigated by previous literature so far. Merely, Hettige and Karunananda [28] take a first step by modeling the components (e.g. graphical interface, natural language processing or data access) of a chatbot as a multi-agent-based system called OCTOPUS, but do not integrate or rather orchestrate single hidden chatbots. Thus, we assume that investigating chatbots based on multi-agent-based architecture without relying on a huge amount of feedback, harbors enormous potential for research and copes with the current challenges in the context of collaborative chatbots. Indeed, to the best of our knowledge, so far none of the recent studies in conversation agents has considered orchestrating hidden chatbots in a single conversational agent while at the same time taking an integrated perspective by not only ranking chatbot answers based on priority lists or human feedback but rather combining research streams by embedding chatbots as provider-agents into a multi-agent-based architecture. Thus, we aim at designing a novel multi-agent-based chatbot approach combining conversational agents and multi-agent-based methods in a well-founded way which improves the versatility of chatbots giving reasonable answers.

3 Novel Multi-agent-based Chatbot Approach

Having stated the solution's objectives, following the DSR process by Peffers et al. [20], we set out to develop an approach to answer human questions with a single chatbot regarding open-domain and closed-domain topics by orchestrating and coordinating hidden chatbots (cf. Fig. 1). Since our research is concerned with the development of a novel approach, it constitutes a contribution of nascent design theory [29] and represents an example of work in interior mode [30]. In our research, we mainly employ a deductive, iterative knowledge creation strategy and develop our approach based on a series of hypotheses that we test and validate. In its entirety, the design and search process described in the following forms a single iteration of the DSR process [20, 31].

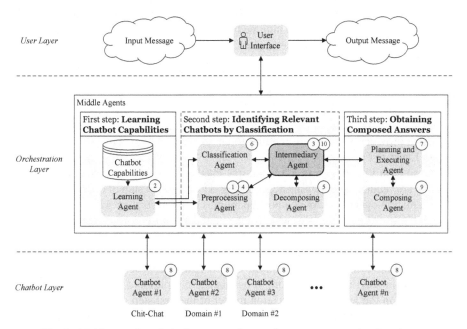

Fig. 1. Multi-agent-based chatbot approach to orchestrate conversational assistants

3.1 Basic Idea and Overview

In literature, multi-agent-based approaches are used to design complex and emergent phenomena inspired by human behavior, such as expert collaboration in organizations interacting with each other to solve a customer request [17–19]. Consequently, we base the architecture of our novel approach on the "divide and conquer" paradigm of multi agent based systems containing requester-, provider- and middle-agents as intermediaries [19].

Following the basic process of collaboration through mediation between requester-, provider- and middle-agents in multi-agent-based systems by Klusch and Sycara [19], the response of a user request can be identified by first, advertising the capabilities of all provider-agents to middle-agents; second, requesting responses from middle-agents; third, mediating requests against the knowledge on capabilities of registered provider-agents; fourth gathering responses from provider-agents; and fifth *composing* and *returning* the results to the requesting users. As outlined in Sect. 2.3, the main challenge of creating a versatile chatbot is the reusage of multiple well-designed chatbots in a jointly coordinated manner to scale beyond the scope of a single chatbot application. Therefore, in this paper, we focus on the integration of multiple chatbots in a multi-agent-based architecture by designing reasonable capability-based middle-agents as intermediaries orchestrating hidden chatbots in a single conversational agent (cf. Orchestration Layer in Fig. 1). Against this backdrop, we take an integrated perspective by not only ranking chatbot answers based on priority lists or human feedback [12, 13] but rather combining research streams by embedding chatbots as provider-agents into a multi-agent-based architecture with middle-agents as intermediaries (cf. Fig. 1).

Focusing on the design of reasonable capability-based middle-agents, we take a standardized agent communication language (e.g. KQML or FIPA ACL [32]) for granted to automatically process requests by enabling communication among different agents. Furthermore, we state a requester-agent with a user interface as given, which receives textual requests from and provides responses to a real user (cf. User Layer in Fig. 1). Beyond that, we also state multiple chatbots as given, whereby each chatbot is designed as reactive provider-agent by responding to an answer, when receiving a request (cf. Chatbot Layer in Fig. 1).

Focusing on the orchestration of multiple chatbots, our approach comprises three steps that are sequenced in ten substeps (cf. Fig. 1). In the first step of our approach, we design a middle-agent to learn chatbot capabilities or requests they are able to answer, respectively. In order to determine which of the chatbots is most appropriate for a prospective request, different supervised machine-learning models are trained (learning agent). In the second step of our approach, an intermediary agent identifies all chatbots, which have the potential to answer a user request. To do so, this agent coordinates other middle-agents to preprocess requests, decomposes them into topics and thus, classifies these topics based on the learned classification model of the first step. The third step of our approach builds upon the classified topics of the second step and determines an execution plan, which in turn is used to request appropriate chatbot(s) and gather responses. Subsequently, all gathered responses are composed by the composing agent and published by the intermediary agent as an answer to the user (cf. Fig. 1). In the following subsections, we present our three-step-approach to orchestrate hidden chatbots in a single conversational agent in more detail.

3.2 First Step: Learning Chatbot Capabilities

The aim of the first step is to learn the capabilities of all available chatbots by using a matching mechanism that supports the intermediary middle-agent in identifying an appropriate chatbot regarding prospective user requests (cf. first step in Fig. 1). To do so, literature in multi-agent-based systems states that the choice of a suitable matching mechanism between request and agent capabilities certainly depends on the structure and semantics of the descriptions to be matched [19, 27]. In our context, solvable requests and all prospective similar requests of a chatbot can be treated as capabilities of a chatbot. Therefore, both, user requests and capabilities of the chatbots, in terms of requests they are able to answer, are represented as free-text (e.g. "Is there an airport in Nairobi?"). Thus, we base the identification of appropriate chatbots on text analysis consisting of the common substeps *data preprocessing*, *classification*, and *classifier evaluation*, which have been proven to deliver reliable results in identifying relevant and similar content in text-based requests [33, 34].

More precisely, the task of identifying the chatbot with the highest probability to answer a request is framed as a problem of supervised learning, where a machine-learning model is trained to generate the desired output from training data, which consists of pairwise training inputs (chatbot capabilities) and expected outputs (appropriate chatbot). Following the "divide and conquer" paradigm of multi-agent-based systems and in line with Hettige and Karunananda [28], we propose to split the text analysis into two separate reactive agents performing natural language *preprocessing* (cf. preprocessing agent in Fig. 1), *classifier learning* as well as *classifier evaluation* (cf. learning agent in Fig. 1). In particular, as the preprocessing has to be reused in the subsequent step to apply the classification model, it is reasonable to separate natural language preprocessing from classifier learning in different agents.

In order to learn a suitable model, the *learning agent* is triggered, whenever the available chatbots' capabilities change (e.g. due to adding, removing or changing a chatbot). Following this, the *preprocessing agent* is triggered by the learning agent and applies natural language preprocessing to clean textual data (e.g. eliminate irrelevant and redundant information) and to reduce the number of terms in order to obtain the minimum of relevant terms to improve speed as well as accuracy of classification algorithms [35]. Subsequently, the learning agent applies text classification algorithms (e.g. SVMs or artificial neural networks [36]) to the preprocessed chatbots' capabilities to train supervised machine-learning models. Based on common classifier evaluation measures (e.g. Precision, Recall, Accuracy and F1 score [37]), the learning agent aims at finding the most accurate classifier whenever the set of available classifier models changes. To sum up, the first step constitutes a necessary preparation for solving new requests triggered by changed chatbots capabilities. By doing so, the first step preprocesses chatbot capabilities from all available chatbots by means of the preprocessing agent (first substep). Following this, the learning agent trains a set of classifiers and chooses the model with the highest evaluation measure in order to determine which chatbot is most appropriate for a prospective request (second substep). Thus, a sound capability-based classification of requests in the following step is enabled.

3.3 Second Step: Identifying Relevant Chatbots by Classification

The aim of the second step is to identify all chatbots, which are necessary for solving a new request (cf. second step in Fig. 1). As outlined in Sect. 3.1, we base our architecture on the "divide and conquer" paradigm of multi-agent-based systems, where intermediaries receive requests and cope with the task of orchestrating multiple hidden conversational agents to solve the request. According to literature, capability-based (also known as skill-based or service-oriented) middle-agents can be categorized into matchmaker, mediators and broker [35] depending on the extent of providing mediation services. As matchmaker middle-agents just provide a ranked list of relevant provider-agents and give the choice of selecting a provider-agent to the user, they are not appropriate in our setting. In contrast, mediator and broker middle-agents actively answer the request by forwarding requests to the most relevant provider-agents. However, as mediators are based on a fixed number of agents and need a static pre-integrated global model of the application scope, they seem to be inappropriate as an intermediary in a multi-agent-based chatbot approach. Accordingly, we define the intermediary agent as a reactive broker agent interfacing between the user, middle-agents and chatbot agents (cf. intermediary agent in Fig. 1). More precisely, after receiving a new request, the intermediary agent first contacts the preprocessing agent to apply natural language preprocessing. As one request could refer to more than one topic and thus, could require responses from more than one chatbot, it seems reasonable to identify all contained topics prior to the capability-based classifications of the requests (e.g. "Which species of zebra is known as the common zebra (topic of chatbot one) and does an occupational disability insurance pay in case of an accident in Nairobi (topic of chatbot two)?"). Furthermore, it is necessary to maintain the structure and dependencies of the topics in order to enable the determination of an execution plan in the following step. This is in line with Skorochod'ko [38], who suggested to divide texts into subsentences regarding their semantic overlap and represented the resulting structure as a connected graph. Hence, we define the decomposing agent (cf. decomposing agent in Fig. 1) as a reactive agent, which automatically partitions preprocessed requests into coherent topics in a graph structure based on a well-established statistical text segmentation algorithm (e.g. based on lexical cohesion, decision trees or semantic networks) [38, 39]. Afterward, the reactive classification agent (cf. classification agent in Fig. 1) uses the trained classifier of the first step to identify the most relevant chatbot for each topic. To sum up, the intermediary broker-agent is triggered by a new request from the user (third substep) and coordinates the preprocessing agent (fourth substep), decomposing agent (fifth substep) and classification agent (sixth substep). As a result of the second step, the new request is decomposed into topics and a corresponding graph structure. Finally, the most relevant chatbot is identified for each topic.

3.4 Third Step: Obtaining Composed Answers

The aim of the third step is to plan and execute topics in order to get responses from all relevant chatbots as well as composing and publishing a final answer to the user (cf. third step in Fig. 1). Regarding multi-agent-based literature, deliberative agents are well-established to reason and plan actions based on a symbolic representation of a planning problem [25, 40]. Moreover, literature has also proven approaches for symbolic representation to be successful as a formal basis for deliberation, such as the well-known representation formalism named STRIPS (Stanford Research Institute Problem Solver) [19, 40, 41]. Therefore, we define the planning and executing agent (cf. planning and executing agent in Fig. 1) as a deliberative middle-agent based on STRIPS as planning formalism, which creates an execution plan based on the graph structure and the classified topics of the second step. Obviously, requests in conversational systems are characterized by short texts, which are usually comprised of only a few sub-sentences or topics, respectively. These topics could be independent and therefore parallelizable in an execution plan (cf. zebra and insurance example in Sect. 3.3) or sequentially dependent in their graph structure (e.g. "What is the currency of Nairobi (first topic) and what are $200 in that currency (second dependent topic)?"). However, in the case of sequential dependency, the dependent part of the second topic has to be replaced with the answer received by the execution of the first topic (e.g. answer of the first topic: "Kenyan Shilling"; reformulated second topic: "What are $200 in Kenyan Shilling"). According to literature, the task of reformulating requests by substituting or adding words or phrases to the original request is framed as a problem of query reformulation (e.g. substituting and/or adding words or phrases to the original query or topic, respectively) [42]. Furthermore, answers ought to be combined to create a final answer, which can be determined by the conjunction of all gathered answers. Consequently, we define the composing agent (cf. composing agent in Fig. 1) as a reactive agent to reformulate sequentially dependent topics based on a query reformulation approach as well as to combine answers based on a conjunction of all gathered answers. Summing up, the third step is triggered by the intermediary agent and builds upon the graph structure and the classified topics of the second step. To obtain a final answer, the planning and execution agent determines an execution plan (seventh substep), which in turn is used to request appropriate chatbot(s) (eighth substep). The composing agent supports this process by reformulating topics and answers (ninth substep), and finally, the intermediary agent publishes the composed answer to the user (tenth substep).

4   Demonstration and Evaluation of the Novel Approach

As an essential part of the DSR process [20, 29, 31], we demonstrate and evaluate the practical applicability of our approach. First, we describe the chatbot datasets considered. Then, we demonstrate how the three steps of our novel multi-agent-based chatbot approach could be applied. Finally, we present the results of our application and compare them to baselines and a competing artifact.

4.1   Chatbot Datasets

In order to demonstrate the practical applicability and evaluate the effectiveness of our approach, we use six different chatbot datasets comprised of 116.060 distinct questions and answers. The chatbots can roughly be divided into three main categories with two representatives in each category. First, we use datasets from two popular chitchat chatbots: the Loebner Prize winner MITSUKU [43] (26.78% of all distinct questions) and a successor of ALICE [9] called ROSIE [44] (15.29%). Second, we use chatbots based on question-answer datasets derived from the well-known knowledge bases Quora [45] (13.43%) and Wikipedia [46] (8.28%). These chatbots are capable of answering general knowledge questions beyond the capabilities of the two chitchat chatbots. Third, we use two domain-specific chatbots based on a dataset of algebraic word problems provided by Ling et al. [47] (34.39%) and based on an insurance-specific dataset provided by a major German insurer [48] (1.83%).

As these chatbots are specialized in their scope in terms of distinct capabilities (e.g. only insurance-specific questions or questions available on Quora) and could be mutually enriched by each other's contributions, they foster the need for a chatbot approach composed of jointly acting chatbots. Moreover, the requests of all six chatbots can be easily combined by the conjunction of two different topics. Thus, our demonstration dataset is comprised of 116.060 single requests and 6.73 billion combined requests each comprised of two independent topics. While focusing on the conjunction of independent topics, the considered chatbot datasets provide an appropriate setting to demonstrate and evaluate the applicability of our novel approach orchestrating hidden chatbots in a single conversational agent.

4.2 Demonstration

In order to demonstrate the applicability of our approach, we use Aimpulse Spectrum [49] as a runtime environment for large multi-agent systems providing a platform to design and execute reactive as well as deliberative agents based on FIPA ACL [32] as standardized agent communication language. In order to apply our approach, we first create six reactive chatbot provider-agents, which are able to receive requests and provide an answer based on the chatbots and their datasets described in the previous section. Moreover, we implement the user as reactive requester-agent, forwarding requests to the intermediary agent and receiving answers from the intermediary agent. Based on the chatbot agents and the requester-agent in the multi-agent environment, we apply our approach. Following the first step of our approach, we use the requests of six chatbots to train a set of classifiers and thus learn the capabilities of each chatbot. Therefore, we rely on the vector space representation and apply the preprocessing agent insofar as terms of the requests had been cleared from stopwords and punctuations, transformed to lower case, reduced to their word stems, and weighted based on the relevance of individual terms by applying the well-established frequency measure tf-idf [33, 34]. Following this, the learning agent uses the most common classification algorithms decision tree classification, support vector machine (SVM), k-nearest neighbor (KNN), naïve Bayes and artificial neural network (ANN) [36] to learn a set of classifier models. To do so, the learning agent applies 10-fold cross-validation [50] for each classifier by using 90% of the 116.060 solvable requests for training and the remaining requests for classifier evaluation.

In order to ensure a rigorous evaluation and to quantify the quality of the results, the agent chooses the best available model based on the performance measure F1 score as this measure is widely used to assess the results of text analysis and condenses as well as balances the information entailed in the measures recall and precision [37]. The results of our first step reveal that the SVM classifier model constitutes the best available model in our setting to learn chatbot capabilities as it shows the highest average values regarding all evaluation metrics of the 10-fold cross-validation (cf. Table 1).

Table 1. Results of chatbot classification (maximum values are marked bold)

Approach	Recall (%)	Precision (%)	Accuracy (%)	F1 score (%)
KNN	80.15	83.83	81.33	79.76
Naïve Bayes	88.78	78.67	85.69	82.14
Decision Tree	88.72	90.14	92.04	88.64
ANN	94.53	95.06	96.38	94.78
SVM	**94.99**	**95.07**	**96.42**	**94.97**

For the second step, we base the text segmentation of the decomposing agent on the occurrence of question marks as the majority of topics contain a question mark at the end of their textual description. Even though using a question mark can be considered as a simple parametrization of our decomposing agent, in our setting, it seems to be a suitable indicator identifying the boundaries of topics. Consequently, the preprocessing prior to the decomposing agent must not perform a removal of punctuations, which contrasts with the preprocessing of the classification agent, which need to be identical to the preprocessing described in the first step to apply the classification model. Concerning the results of the text segmentation, we confirm that question marks are a suitable indicator in our context to identify the boundaries of topics. The majority of the 6.73 billion combined requests are well segmented into topic by the decomposing agent based on the occurrence of question marks (recall 100%, precision 98.33%, accuracy 98.33%, F1 score 99.16%). We found that only a few topics mislead our text segmentation approach by containing more than one question marks or a question mark followed by textual descriptions. In order to perform the third step and thus to determine a request-specific plan, an automatic deliberation is required. As the requests in our dataset are based on the conjunction of independent topics (cf. Sect. 4.1), all topics can be executed in parallel without any dependencies. Furthermore, and to create a final answer, the composing agent combines all independent answers based on a conjunction of all gathered answers as described in Sect. 3.4. Summing up, to create a composed answer regarding a new user request, the intermediary agent orchestrates other middle- and chatbot agents. Hence, multiple chatbots are reused in a jointly coordinated manner to scale beyond the scope of a single chatbot application.

4.3 Evaluation

To evaluate our approach as demanded by the DSR process [20], we follow the Framework for Evaluation in Design Science (FEDS) put forward by Venable, Pries-Heje, and Baskerville [51]. To this end, our evaluation strategy consists of comparing the results obtained by our novel multi-agent-based chatbot approach, a competing artifact and three baselines to the expected output. More precisely, we simulated a real user by randomly demanding requests out of the 6.73 billion combined requests as described in Sect. 4.1. While doing so, we applied 10-fold cross-validation [50] and skipped requests comprised of topics used while training the classifier in the first step. On this basis, we inspected the results of our approach relying on the average values of the well-established evaluation measures precision, recall, accuracy and F1 score [37]. To do so, we expected the conjunction of answers contained in the chatbot datasets (cf. Sect. 4.1) as correct output regarding a combined request. Moreover, we compared the results of our approach to the only competing artifact from chatbot literature (Huang et al. [12]) which can be applied in a multi-agent-based chatbot context without the presence of priority lists or a huge amount of user-provided ratings and feedback. Huang et al. [12] provide an appropriate approach, which is able to choose chatbots over time by ranking them based on the similarity of the new request and each chatbots' history (capabilities) as well as a specific prior probability of each chatbot. Although the prior probability is based on user feedback, Huang et al. state that the prior probability of chatbots without any feedback can be initially assigned to a constant value and updated with the presence of feedback [12]. Hence, we used the approach of Huang et al. [12] without any feedback as first competing artifact. Furthermore, we also compared our results to three multi-agent-based chatbot approaches as baselines, each selecting the answering chatbots based on a specific probability distribution. By doing so, the first sampling approach used an equalized distribution for choosing the answering chatbot (probability of 16.67% for choosing each chatbot); the second applied a distribution depending on each chatbots' dataset size (e.g. probability of 26.78% for choosing MITSUKU); and the third, skewed, baseline preferred the chatbot with the highest amount of requests (preferring the Chatbot dealing with algebraic word problems with a probability of 99.99%). Note that baselines and competing artifact are not able to decompose or compose requests and thus, our decomposing and composing agents are used to decompose and compose combined topics in our comparison. Regarding the comparison with baselines and competing artifact, shown in Table 2, our approach answered 94.78% of the user requests successfully. In contrast, the approach based on Huang et al. [12] reached a value of 73.78% successfully answered requests. With 33.82%, the skewed sampling approach reached the third highest value for successful answers, but the lowest F1 score of all approaches. However, the skewed sampling approach mainly just identified the correct answers regarding the scope of one chatbot, leading to high recall (98.33%) and precision (34.01%) for this chatbot but a critically low precision (0.11%) and recall (0.001%) regarding the scope of the other five chatbots. Finally, the equalized and size-dependent sampling performed worst by only obtaining 16.39% and 23.46% of the successful answers. Summing up, the results of our comparisons reveal that our approach provides higher quality than the competing artifacts and thus, indicate the suitability of our novel chatbot approach to scale beyond the scope of a single chatbot.

Table 2. Evaluation measures of our approach in comparison with baselines and competing artifacts (maximum values are marked bold)

Approach	Recall (%)	Precision (%)	Accuracy (%)	F1 score (%)
Equalized sampling	16.39	16.41	16.39	14.45
Size-dependent sampling	16.39	16.39	23.46	16.39
Skewed sampling	16.39	5.76	33.82	8.43
Huang et al. [12]	76.97	77.33	73.78	72.69
Novel approach	**91.11**	**93.11**	**94.78**	**92.12**

5   Conclusion, Limitations and Future Research

Despite emerging scientific work in the field of conversational agents, we still observe a lack of research on how multiple well-designed chatbots could be reused in a jointly coordinated manner to scale beyond the scope of a single chatbot application. Thus, we, proposed an approach consisting of three steps allowing us to decompose a user request into topics, classifying the answering chatbot for each topic and finally obtaining a composed answer. The evaluation of our approach using six chatbots provides promising results and illustrates the ability of our approach to scale beyond the scope of a single chatbot by orchestrating chatbots with different capabilities.

Thus, our work contributes to literature by taking an integrated perspective by not only ranking chatbot answers based on priority lists or human feedback but rather combining research streams by embedding chatbots as provider-agents into a multi-agent-based architecture. Furthermore, our novel approach paves the way for immensely versatile chatbot applications insofar as our chatbot approach is capable of scaling the scope of answerable questions and topics depending on the amount of integrated chatbots. In practice, for example, using our approach a company can combine freely available chitchat chatbots or certain specific chatbots with their domain-specific chatbot in order to enhance response options and capabilities. In addition, our approach might be used to address constraints such as "information overload" and "redundant information" mentioned in Stoeckli et al. [52]. Against this background, our novel multi-agent-based chatbot approach constitutes a promising first step in order to overcome current challenges in giving reasonable answers to a variety of possible requests.

Even though our research provides a sound integration of chatbots into a multi-agent-based architecture, there are some limitations, which can serve as starting points for future research. First, we only considered the decomposing, planning and composing of single-turn requests and independent topics in our demonstration and evaluation. While in a first step, it seemed appropriate to take such a perspective, future studies can enhance the agents of the second and third step in order to demonstrate and evaluate our novel approach in terms of sequential dependent topics. Moreover, in our demonstration and evaluation, we consider only a simple parametrization and compare our approach with competing artifacts applicable in a multi-agent-based chatbot context. Thus, we encourage further research to evaluate our approach using more complex parametrizations and competing artifacts in non-multi-agent-based contexts. Finally, it would be of interest to analyze the feedback of real chatbot users in order to evaluate the proposed approach in a real-world application as well as integrate dynamic learning strategies that consider context information and personal preferences of users.

Summing up, we believe that our study is a first, but an indispensable step towards multi-agent-based chatbots. We hope our work will stimulate further research in this exciting area and encourage scientists as well as practitioners to reuse existing chatbots by applying the "divide and conquer" paradigm of multi-agent-based systems.

References

1. Ahmad, N.A., Che, M.H., Zainal, A., et al.: Review of chatbots design techniques. IJACSA **181**(8), 7–10 (2018)
2. Klopfenstein, L.C., Delpriori, S., Malatini, S., et al.: The rise of bots: a survey of conversational interfaces, patterns, and paradigms. In: Proceedings of the 12th Conference on Designing Interactive Systems, pp. 555–565 (2017)
3. Chaves, A.P., Gerosa, M.A.: Single or multiple conversational agents? An interactional coherence comparison. In: Proceedings of the 36th CHI (2018)
4. Masche, J., Le, N.-T.: A review of technologies for conversational systems. In: Proceedings of the 5th ICCSAMA, pp. 212–225 (2017)
5. Dhanda, S.: How chatbots will transform the retail industry. Juniper Research (2018)
6. Abdul-Kader, S.A., Woods, J.C.: Survey on chatbot design techniques in speech conversation systems. IJACSA **6**(7), 72–80 (2015)

7. Chen, H., Liu, X., Yin, D., et al.: A survey on dialogue systems: recent advances and new frontiers. ACM SIGKDD Explor. Newslett. **19**(2), 25–35 (2017)
8. Ramesh, K., Ravishankaran, S., Joshi, A., Chandrasekaran, K.: A survey of design techniques for conversational agents. In: Kaushik, S., Gupta, D., Kharb, L., Chahal, D. (eds.) ICICCT 2017. CCIS, vol. 750, pp. 336–350. Springer, Singapore (2017). https://doi.org/10.1007/978-981-10-6544-6_31
9. Wallace, R.S.: The anatomy of ALICE. In: Epstein, R., Roberts, G., Beber, G. (eds.) Parsing the Turing Test, pp. 181–210. Springer, Dordrecht (2009). https://doi.org/10.1007/978-1-4020-6710-5_13
10. Serban, I.V., Sankar, C., Germain, M., et al.: A deep reinforcement learning chatbot (2017)
11. Pichl, J., Marek, P., Konrád, J., et al.: Alquist: the Alexa prize socialbot. In: Proceedings of the 1st Alexa Prize (2017)
12. Huang, T.-H.K., Chang, J.C., Bigham, J.P.: Evorus: a crowd-powered conversational assistant built to automate itself over time. In: Proceedings of the 36th CHI (2018)
13. Papaioannou, I., Curry, A.C., Part, J.L., et al.: Alana: social dialogue using an ensemble model and a ranker trained on user feedback. In: Proceedings of the 1st Alexa Prize (2017)
14. Pinhanez, C.S., Candello, H., Pichiliani, M.C., et al.: Different but equal: comparing user collaboration with digital personal assistants vs. teams of expert agents (2018)
15. Janarthanam, S.: Hands-On Chatbots and Conversational UI Development. Packt Publishing, Birmingham (2017)
16. Chandar, P., et al.: Leveraging conversational systems to assists new hires during onboarding. In: Bernhaupt, R., Dalvi, G., Joshi, A., Balkrishan, D., O'Neill, J., Winckler, M. (eds.) INTERACT 2017. LNCS, vol. 10514, pp. 381–391. Springer, Cham (2017). https://doi.org/10.1007/978-3-319-67684-5_23
17. Jennings, N.R.: Commitments and conventions: the foundation of coordination in multi-agent systems. Knowl. Eng. Rev. **8**(3), 223–250 (1993)
18. Jennings, N.R.: An agent-based approach for building complex software systems. Commun. ACM **44**(4), 35–41 (2001)
19. Klusch, M., Sycara, K.: Brokering and matchmaking for coordination of agent societies. a survey. In: Omicini, A., Zambonelli, F., Klusch, M. (eds.) Coordination of Internet Agents, pp. 197–224. Springer, Heidelberg (2001). https://doi.org/10.1007/978-3-662-04401-8_8
20. Peffers, K., Tuunanen, T., Rothenberger, M.A., et al.: A design science research methodology for information systems research. JMIS **24**(3), 45–77 (2007)
21. Maglio, P.P., Matlock, T., Campbell, C.S., Zhai, S., Smith, B.A.: Gaze and speech in attentive user interfaces. In: Tan, T., Shi, Y., Gao, W. (eds.) ICMI 2000. LNCS, vol. 1948, pp. 1–7. Springer, Heidelberg (2000). https://doi.org/10.1007/3-540-40063-X_1
22. Cui, L., Huang, S., Wei, F., et al.: Superagent. A customer service chatbot for e-commerce websites. In: Proceedings of the 55th Annual Meeting of the ACL, pp. 97–102 (2017)
23. Arentze, T., Timmermans, H.: Modeling the formation of activity agendas using reactive agents. Environ. Plan. B **29**(5), 719–728 (2002)
24. Ehlert, P., Rothkrantz, L.J.M.: Microscopic traffic simulation with reactive driving agents. In: 4th Proceedings of IEEE Intelligent Transportation Systems, pp. 861–866 (2001)
25. Rao, A.S., Georgeff, M.P.: BDI agents. In: 1st ICMAS, pp. 312–319 (1995)
26. Barua, A., Whinston, A.B., Yin, F.: Value and productivity in the internet economy. Computer **33**(5), 102–105 (2000)
27. Decker, K., Sycara, K., Williamson, M.: Middle-agents for the internet. In: Proceedings of the 15th IJCAI, pp. 578–583 (1997)
28. Hettige, B., Karunananda, A.S.: Octopus: a multi agent chatbot. In: Proceedings of the 8th International Research Conference, pp. 41–47 (2015)
29. Gregor, S., Hevner, A.R.: Positioning and presenting design science research for maximum impact. MIS Q. **37**, 337–355 (2013)

30. Baskerville, R., Baiyere, A., Gregor, S., et al.: Design science research contributions: finding a balance between artifact and theory. JAIS **19**, 358–376 (2018)
31. Hevner, A.R., March, S.T., Park, J., et al.: Design science in information systems research. MIS Q. **28**, 75–105 (2004)
32. Labrou, Y., Finin, T., Peng, Y.: Agent communication languages: the current landscape. Intell. Syst. Appl. **14**(2), 45–52 (1999)
33. Park, S., An, D.U.: Automatic e-mail classification using dynamic category hierarchy and semantic features. IETE Tech. Rev. **27**(6), 478–492 (2010)
34. Li, N., Wu, D.D.: Using text mining and sentiment analysis for online forums hotspot detection and forecast. DSS **48**(2), 354–368 (2010)
35. Storn, R., Price, K.: Differential evolution – a simple and efficient heuristic for global optimization over continuous spaces. J. Global Optim. **11**(4), 341–359 (1997). https://doi.org/10.1023/A:1008202821328
36. Russell, S.J., Norvig, P.: AI. A Modern Approach. Pearson Education, London (2010)
37. Sokolova, M., Japkowicz, N., Szpakowicz, S.: Beyond accuracy, F-score and ROC: a family of discriminant measures for performance evaluation. In: Sattar, A., Kang, B. (eds.) AI 2006. LNCS (LNAI), vol. 4304, pp. 1015–1021. Springer, Heidelberg (2006). https://doi.org/10.1007/11941439_114
38. Skorochod'ko, E.F.: Adaptive method of automatic abstracting and indexing. In: Proceedings of the 5th Information Processing Congress, pp. 1179–1182 (1972)
39. Beeferman, D., Berger, A., Lafferty, J.: Statistical models for text segmentation. Mach. Learn. **34**(1–3), 177–210 (1999). https://doi.org/10.1023/A:1007506220214
40. Wooldridge, M., Jennings, N.R.: Intelligent agents: theory and practice. Knowl. Eng. Rev. **10**(2), 115–152 (1995)
41. Fikes, R.E., Nilsson, N.J.: STRIPS: a new approach to the application of theorem proving to problem solving. Artif. Intell. **2**(3–4), 189–208 (1971)
42. Dang, V., Croft, B.W.: Query reformulation using anchor text. In: Proceedings of the 3rd WSDM, pp. 41 50 (2010)
43. Mitsuku Dataset. https://github.com/pandorabots/Free-AIML. Accessed 06 Dec 2019
44. Rosie Dataset. https://github.com/pandorabots/rosie. Accessed 06 Dec 2019
45. Quora Dataset. https://www.kaggle.com/c/quora-question-pairs. Accessed 06 Dec 2019
46. Wikipedia Dataset. https://www.kaggle.com/rtatman/questionanswer-dataset. Accessed 06 Dec 2019
47. Ling, W., Yogatama, D., Dyer, C., et al.: Program induction by rationale generation: learning to solve and explain algebraic word problems. In: Proceedings of the 55th Annual Meeting of the ACL, pp. 158–167 (2017)
48. Bedué, P., Graef, R., Klier, M., et al.: A novel hybrid knowledge retrieval approach for online customer service platforms. In: Proceedings of the 26th ECIS (2018)
49. Aimpulse Spectrum. https://developer.aimpulse.com. Accessed 23 Aug 2019
50. Kohavi, R.: A study of cross-validation and bootstrap for accuracy estimation and model selection. In: Proceedings of the 14th IJCAI, vol. 14, no. 2, pp. 1137–1145 (1995)
51. Venable, J., Pries-Heje, J., Baskerville, R.: FEDS: a framework for evaluation in design science research. Eur. J. Inf. Syst. **25**, 77–89 (2016)
52. Stoeckli, E., Uebernickel, F., Brenner, W.: Exploring affordances of slack integrations and their actualization within enterprises-towards an understanding of how chatbots create value. In: Proceedings of the 51st HICSS (2018)

Computer Vision for the Ballet Industry: A Comparative Study of Methods for Pose Recognition

Margaux Fourie[✉] and Dustin van der Haar[✉]

University of Johannesburg, Kingsway Avenue and University Rds,
Auckland Park, Johannesburg, South Africa
{margauxf,dvanderhaar}@uj.ac.za

Abstract. The presence of computer vision technology is continually expanding into multiple application domains. An industry and an art form that is particularly attractive for the application of computer vision algorithms is ballet. Due to the well-codified poses, along with the challenges that exist within the ballet domain, automation for the ballet environment is a relevant research problem. The paper proposes a model called BaReCo, which allows for ballet poses to be recognised using computer vision methods. The model contains multiple computer vision pipelines which allows for the comparison of approaches that have not been widely explored in the ballet domain. The results have shown that the top-performing pipelines achieved an accuracy rate of 99.375% and an Equal Error Rate (EER) of 0.119% respectively. The study additionally produced a ballet pose dataset, which serves as a contribution to the ballet and computer vision community. By combining suitable computer vision methods, the study demonstrates that successful recognition of ballet poses can be accomplished.

Keywords: Computer vision · Pose recognition · Dance · Ballet industry · Automation

1 Introduction

Movements of the human body can be captured using various types of cameras and sensors. As a result of the increased availability of new capture technologies as well as Graphics Processing Units (GPUs), computer vision technology is increasingly used in order to perform body movement and pose recognition [1,2]. Those involved in human motion and pose-based fields such as medical, sports and performing arts can benefit from additional technological assistance, which contributes to the improvement in field-related tasks. Therefore, new ways of applying computer vision to these fields, in order to enhance and contribute to the lives of those involved, need to be explored [3].

Ballet is a particularly noteworthy form of dance as it has a history reaching back to the 16th century and is, therefore, a well respected, established art

© Springer Nature Switzerland AG 2020
W. Abramowicz and G. Klein (Eds.): BIS 2020, LNBIP 389, pp. 118–129, 2020.
https://doi.org/10.1007/978-3-030-53337-3_9

form [4]. It has become foundational for a large majority of other dance forms [5]. The precise structure of ballet is what makes it an attractive art form to investigate and further explore in terms of its relevance to technological fields.

Initially, the scientific domain of technology may seem to be completely separate and unrelated to the artistic domain of ballet. However, this study explored the intricacies of both ballet and technology to reveal how applicable and suitable it may be to bring the two fields together. Artists and scientists have similar aims in their work as both intend to produce work that is novel and original. As a result, the collaboration between an artistic field such as ballet and a scientific field such as computer technology may be valuable to address challenges in both fields [6]. Specific challenges in the ballet environment include individual training correction as well as the accurate documentation of choreographic works.

Ballet pose recognition and correction is an activity executed by ballet dancers and teachers frequently during training sessions. Pose recognition is also relevant to ballet choreography where sequences of postures are involved in the construction of dances. Research has been done on how technology, such as computer vision can be applied to the ballet environment. However, it is still a growing application field with room for the exploration of a range of technological automation approaches. The research problem of the study involves the need for a standardised approach that can recognise multiple ballet poses automatically. This study is, therefore, an initial step towards further research to aid in the tasks of ballet training and choreography.

The paper aims to address the problem with the use of a captured dataset and proposing a model that combines multiple traditional and novel computer vision methods to achieve ballet pose recognition. Background on the problem is provided first, along with current work conducted in related fields. The experiment setup is described next, followed by a discussion on the model. The results are then presented, and the paper ends with a conclusion which highlights key findings and future work.

2   Problem Background

Ballet is a dance form where the dancer's movements are composed out of predefined poses [7]. The clarity with which a dancer performs these poses needs to be enforced to maintain the aesthetics required by the ballet art form. The application of computer vision to the domain of ballet has become a relevant area of research due to growth in the computer vision field over recent years.

Ballet technique is the driving force behind the definite structure of the dance form and is concerned with the creation of strong body lines, poses and fluid movements. Foundational concepts of ballet technique include turnout (outward rotation of the legs), alignment (vertical and horizontal lines of the shoulders and hips) as well as stretched legs and feet [7]. To achieve these classical ideals, it is essential that ballet teachers accurately convey the fundamental principals of technique to students during training, specifically as they relate to different poses.

The environment in which ballet training typically occurs is a studio class-room where a ballet teacher instructs and corrects about 8–10 students [8]. Expertise and knowledge in ballet are traditionally passed on verbally to future dancers during these coaching sessions [9]. However, a challenge that arises in this environment is that the teacher cannot keep his/her eyes on every student simultaneously. It is therefore often essential for the serious ballet student to receive one-on-one coaching [10]. Furthermore, without proper training and a grasp of technique, ballet dancers risk developing bad habits or injuries [11].

The creative process of constructing dance sequences is known as choreography, and in ballet, a series of the codified movements are combined to produce the resulting visual expression [12]. The documentation of ballet choreography is important for the preservation and protection of created dances as well as enabling future dancers to study the created works [13]. There is a need for finding new methods to address the challenges in the choreographic domain and this is where the use of technology becomes relevant.

A few studies in current research that address problems in the ballet environment through the use of technology include wearable technology systems, choreography systems, as well as pose recognition systems. These related works show that pose recognition is an essential first step towards technology-based training and documentation of ballet poses. Ballet-specific research and work in related domains are presented in the subsections that follow.

2.1   Wearable Technology and Other Related Domains

One category of capturing sensors that can be leveraged to collect data on dancers is known as wearable technology. Research by Gupta et al. involves the instruction of beginner adult ballet students through a wearable full-body garment [14]. The garment would allow for the clarification of core movements demonstrated by a teacher wearing the garment which lights up the essential limbs being used. The establishment of the basics and only later moving into the core details of motions is a well-known and effective method to coach ballet dancers regardless of their experience level. Despite the advantages, wearable technologies in ballet have potential restrictions as these garments are typically costly and often require special technical construction [14]. A more practical approach to assist with training would be to make use of vision-based pose data.

A recent relevant vision-based system in the domain of sports was aimed at the creation of a posture analysis tool for basketball free-throw shooting. This basketball free-throw study indicates that it is possible to accurately predict whether or not a basketball throw will be successful based on OpenPose skeleton key-point data as well as correct beginner players [15]. The study demonstrates that it is feasible to make use of the OpenPose [16] library in settings that deal with different body postures.

2.2   Choreography Systems

Dancs et al. investigated the choreographic side of ballet by studying the concept of having a technological tool to help choreographers. The proposed system had the aim to recognise and record a choreographer's movements automatically [17].

Computer vision algorithms proved to be useful for such a system where the consecutive movements of a dancer were to be recorded. The study had the goal to enhance the area of ballet using a Microsoft Kinect to find the joints of the body. The classification algorithms used by the study included, amongst others, the Nearest Neighbor (NN) as well as the Support Vector Machine (SVM) classification algorithms [17]. The approach by Dancs et al. produced promising results with the main algorithms generating accuracy results over 90%.

The limitations of the work proposed by Dancs et al. include the processing speed, which may be affected by any small changes in the system. The authors mentioned that, despite the classification model's recognition being fast, any additions to the model would have to consider how the processing speed would be affected [17]. Another disadvantage of this system is linked with the hardware limitations of the Kinect Sensor. According to a study by Hong et al., the Kinect sensor is unable to detect turnout and crossed feet positions properly [18]. Since turnout and crossed feet occur quite frequently in ballet poses, this challenge is relevant to other related research in the ballet pose recognition domain. Therefore, it may be worthwhile for researchers to explore alternative ways to extract skeleton features from Kinect-captured images without relying on the built-in Kinect skeleton tracking abilities.

2.3   Pose Recognition Systems for Ballet

A known concern in many sport or artistic disciplines that require any level of specialised skill and physical training is that it comes at the cost of being educated in the particular discipline. Saha et al. address this concern with their concept of fuzzy image matching for ballet poses by making the idea of e-learning in ballet easier [19]. The stages of the project involved skin colour segmentation and straight-line approximation in order to reduce the initial image of a dancer to a skeleton and eventually a stick figure representation.

The advantages of pose recognition implementations for ballet are multiple. These types of pose recognition systems lower the level at which ballet students need to rely on the physical presence of a teacher to train. E-learning of ballet through automated pose recognition therefore enables students to practise at any time in any suitable space [20].

One issue pointed out by Banerjee et al. in progressions of the above research was that the proposed algorithm relied on having the dancer wearing only particular colours in order for the skin colour segmentation pre-processing step to be effective [21]. A disadvantage of these systems, therefore, include that they rely heavily on the environmental constraints to be in place for pre-processing to take place effectively.

The research conducted by Saha et al. progressed over the years as different approaches were tried and tested by the researchers for ballet pose recognition. The results of three current related pose recognition studies, including those by Saha et al. can be seen in Table 1.

Table 1. Similar system results of studies conducted in 2014 and 2015 by Kyan et al. and Saha et al. [20, 22, 23]

No.	Similar system	Algorithm description	Accuracy
1	Posture Recognition in Ballet [20]	Fuzzy discrete membership function for matching	85%
2	Topomorphological Approach to Ballet Posture Recognition [23]	Matching of Radon transforms	91.35%
3	CAVE VR with MS Kinect [22]	L2 norm for similarity matching	92%

The results presented in Table 1 highlight achievements for current pose recognition systems, which indicate there is an opportunity to expand and improve on previous work. Along with the limited availability of ballet-datasets, this opportunity warrants the creation of a sufficiently sized ballet pose dataset as well as the implementation of multiple computer vision methods. These aspects of dataset creation as well as method implementations are addressed by this study and described in the following section.

3   Experiment Setup

To obtain consistent and scientific results, this study required certain constraints to be in place during the data capturing phase. The first constraint is that capturing should take place within a ballet studio with the appropriate floor surface for the safe execution of ballet poses. In the case of this study, the floor surface consists of ballet dance mats. Another environmental constraint includes that no mirrors or clutter should be present in the background of the capture space. Furthermore, the lighting conditions for capturing should be at a suitable level, not being too bright or too dim. In terms of role constraints, the study requires participants to be advanced level dancers that could execute the determined poses clearly. Another constraint involves the clothing that is necessary for a capture session. Standard black ballet attire has to be worn to maintain consistency and provide the assurance that there are no unnecessary variations or noise in the captured data.

A total of eight ballet poses are selected for the study, and a primary dataset was created with thirty real ballet dancers performing each respective pose in a studio. During data collection, the participants were instructed to perform the eight different ballet poses of varying difficulty. The poses include Demi-Plié, Second Position, Tendu, Sussous, Retiré, Développé, Arabesque and Penché. In order to capture the data for this study, a Microsoft Kinect sensor and a GoPro camera were used. These sensors allow for the collection of image, depth and video data.

The Microsoft Kinect images of the poses are all captured at a resolution of 640 by 480 pixels to build the dataset for this study. Once all the pose data of each dancer is gathered, it is necessary to arrange the dataset in such a way that it would be easy to feed the data to the relevant computer vision methods. The data is split into a training and testing set with 80% of the data used for training, and 20% used for testing. The dataset contains 7198 images in total, with an even distribution of images amongst different classes for both the training and testing sets. The gathered depth data was not used in this study due to the initial focus being on achieving ballet pose recognition through various methods on normal image data. The depth data will, therefore, be utilised in future work. A link to sample images in the dataset is available at: http://bit.ly/2vMZ1gq.

Once the data has been captured, and the necessary constraints are in place for the study, the relevant computer vision methods for different pipeline phases can be considered. The captured dataset is, therefore, the starting point for each of the model's pipeline implementations which will be presented next.

4   Model

For ballet pose recognition to take place using computer vision methods, this paper proposes the BaReCo model. This model consists of three broad pipeline categories, namely traditional, OpenPose and Artificial Neural Network pipelines. The variations for each of the broader categories are presented in this section, ultimately introducing eight individual computer vision pipelines. The methods involved in these pipelines were chosen based on their presence in related research [16,17,23] as well as their general relevance to computer vision pose-based problems. Furthermore, this study contributes by applying computer vision methods to the ballet environment that have not yet been explored by related research. Images from the captured dataset were used as input for each of the pipelines.

4.1   Traditional Pipeline

The traditional pipeline implementation for this study involves five stages, namely capturing, pre-processing, localisation, feature extraction and classification as illustrated in Fig. 1. For the pre-processing stage of this pipeline grayscaling and histogram equalisation was chosen. The histogram of oriented gradients (HOG) method was used for localisation to isolate the dancer's body in each frame. This method was also used for the feature extraction phase to gather key features for classification. The classification stage of the pipeline introduces three options namely a Support Vector Machine (SVM), a Random Forest (RF) and a Gradient Boosted Tree (GBT).

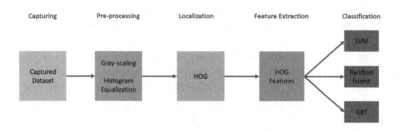

Fig. 1. Traditional pipeline architecture

4.2 OpenPose Pipeline

The OpenPose pipeline for this study consists of three main phases, including capturing, feature extraction and classification, which is illustrated in Fig. 2. OpenPose is a recent and useful approach which uses a multi-stage Convolutional Neural Network (CNN) for extracting human skeleton key-point data from an image [16]. The classification phase of this pipeline uses the same three classifier variations that the previously described traditional pipeline used.

4.3 VGG16

The third category of pipelines the BaReCo model made use of is a deep learning approach which is concerned with Deep Neural Networks. The VGG16 Convolutional Neural Network (CNN) is a deep learning algorithm utilised for the model which is illustrated in Fig. 3. Since accurate results have been achieved using CNNs for various computer vision problems [24], it is suitable to utilise the approach for the implementation of this study.

4.4 Faster Region-Based Convolutional Neural Network

A family of algorithms that are extremely effective for performing object detection and localisation tasks are known as Region-Based Convolutional Neural Networks (R-CNNs). One of these algorithms is known as the Faster R-CNN algorithm which is more efficient than its predecessors [26,27] because it makes use of Region Proposal Networks (RPNs) for determining regions of interest within an image. The Faster Region-Based CNN approach is considered to be an end-to-end deep learning object detection pipeline. Accurate object detection effectively assists in visual recognition tasks, which makes the Faster R-CNN an

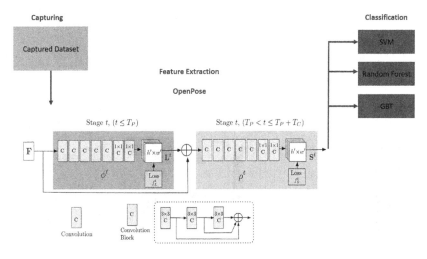

Fig. 2. OpenPose pipeline architecture with the multi-stage CNN architecture of Open-Pose [16]

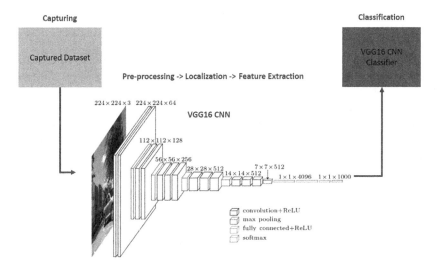

Fig. 3. VGG16 pipeline architecture [25]

appropriate method choice for the pose recognition problem of this study. A visual illustration of the Faster R-CNN pipeline for the study is presented in Fig. 4.

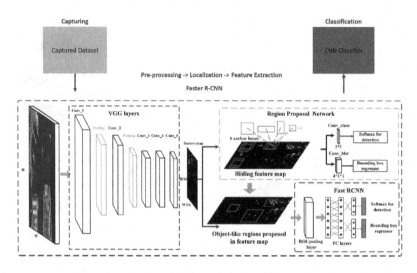

Fig. 4. Faster R-CNN pipeline architecture [27]

5   Results

Table 2 presents the results of this study and highlights the top-performing pipelines. It can be observed that the OpenPose and Random Forest Approach achieves the best results out of all the pipelines with the best scores for the considered metrics. This suggests that the OpenPose data had a positive effect on classification and that the Random Forest classifier is an appropriate model for pose recognition. Another favourable result is that of the VGG16 CNN pipeline, which has the lowest EER of 0.119% and a high accuracy score of 98.472%.

Table 2. Summary of the results obtained by this study as percentages.

Pipeline variation	Accuracy	EER	Recall	Precision	F1	Specificity
OpenPose+RF	**99.375**	0.268	**99.375**	**99.399**	**99.377**	**99.911**
VGG16 CNN	98.472	**0.119**	98.472	98.639	98.466	99.782
OpenPose+GBT	99.305	1.261	99.306	99.320	99.307	99.901
Traditional+HOG+SVM	99.199	0.139	99.178	99.016	99.096	99.887
OpenPose+SVM	99.097	0.308	99.097	99.125	99.098	99.871
Traditional+HOG+RF	98.958	0.486	99.048	98.135	98.523	99.860
Traditional+HOG+GBT	98.379	0.757	98.254	97.781	97.991	99.776
Faster R-CNN	96.049	0.237	95.381	96.577	95.561	99.433

The metrics gathered for the pipelines of this study were promising with accuracy scores above 95% for all the classifiers, which indicates that correct classifications were made for the majority of the involved poses. One interesting observation that can be seen in the confusion matrices in Fig. 5 is that a common pose misclassification that occurred was the incorrect prediction of a Tendu as a Développé as well as an Arabesque as a Penché. A reason for this may be due to the fact that these poses are closely related to similar body orientations and arm lines. The Receiver Operating Characteristic (ROC) curves of three implemented pipelines are presented in Fig. 6 which also indicate that recognition of the Développé and Tendu poses were in some cases not as accurate as other poses. However, the ROC curves generally show that the classifiers performed well in recognising the relevant ballet poses.

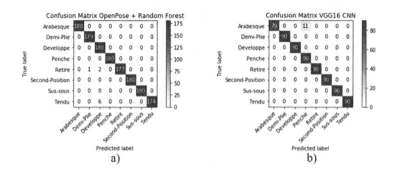

Fig. 5. Confusion matrices of the OpenPose + Random Forest and VGG16 pipelines

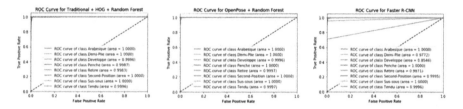

Fig. 6. The Receiver Operating Characteristic (ROC) curves of three implemented pipelines

6   Conclusion and Future Work

This study has validated the use of computer vision in the ballet domain to achieve pose recognition successfully. Some key findings of the study indicate that closely related poses are the potential cause for errors in recognition. The results also reveal that the use of novel deep learning techniques such as OpenPose and artificial neural networks, along with traditional classification approaches, yield promising results which could drive automation forward in the ballet industry.

The most accurate pipeline of this study made use of OpenPose key-point data. This study has, therefore, successfully utilised the OpenPose approach for the domain of ballet poses. The successful results of the OpenPose pipelines in this study indicate that the combination of deep learning for feature extraction with traditional classifiers is a feasible approach for ballet pose recognition. The results attained by this study is largely due to the quality of the created dataset, which contributes to research in the ballet pose recognition domain as no similar dataset is openly available.

Future work for the study includes making improvements to the current implementation, as well as the application of other computer vision approaches on image as well as video data. Future research may, therefore, expand from static pose recognition to movement-based recognition and tracking. For ballet training there is also value in building upon the work of this study by looking at deviations from the standard ballet poses in order to correct dancer poses. Further computer vision approaches that may be of interest in future work include different Convolutional Neural Network architectures, N-shot learning as well as Recurrent Neural Networks.

The developed solution ultimately has the potential to affect the realm of ballet training and choreography in a technologically focused world. In a broader sense, this study indicates that exciting explorations can be made in artistic industries as their potential is found within the field of computer vision.

Acknowledgements. This research benefitted, in part, from support from the Faculty of Science at the University of Johannesburg.

References

1. Nishani, E., Çiço, B.: Computer vision approaches based on deep learning and neural networks: deep neural networks for video analysis of human pose estimation. In: 2017 6th Mediterranean Conference on Embedded Computing (MECO), pp. 1–4. IEEE (2017)
2. Yao, B., Hagras, H., Alhaddad, M.J., Alghazzawi, D.: A fuzzy logic-based system for the automation of human behavior recognition using machine vision in intelligent environments. Soft Comput. **19**(2), 499–506 (2015). https://doi.org/10.1007/s00500-014-1270-4
3. Kale, G.V., Patil, V.H.: A study of vision based human motion recognition and analysis. Int. J. Ambient Comput. Intell. (IJACI) **7**(2), 75–92 (2016)
4. Di Orio, L.: Ballet: Method to Method. Dance Informa American Edition (2013)
5. New York Film Academy: Ballet and modern dance: using ballet as the basis for other dance techniques (2014)
6. Clay, A., Domenger, G., Conan, J., Domenger, A., Couture, N.: Integrating augmented reality to enhance expression, interaction & collaboration in live performances: a ballet dance case study. In: 2014 IEEE International Symposium on Mixed and Augmented Reality-Media, Art, Social Science, Humanities and Design, ISMAR-MASH'D, pp. 21–29. IEEE (2014)
7. Royal Academy of Dancing: The Foundations of Classical Ballet Technique. Royal Academy of Dancing (1997)
8. Kassing, G., Jay, D.M.: Beginning Ballet Technique. Human Kinetics (1998)
9. Trajkova, M., Cafaro, F.: E-ballet: designing for remote ballet learning. In: Proceedings of the 2016 ACM International Joint Conference on Pervasive and Ubiquitous Computing Adjunct - UbiComp 2016, pp. 213–216 (2016)

10. Dance Spirit: Working one-on-one: what to expect from private lessons (2014)
11. Grieg, V.: Inside Ballet Technique. Dance Books (1994)
12. Speck, S., Cisneros, E.: Ballet for Dummies. Wiley, Hoboken (2003)
13. Snyder, A.F.: Securing our dance heritage: issues in the documentation and preservation of danceatio (1999)
14. Gupta, M., Hallam, J., Keen, E., Lee, C., McKenna, A.: Ballet hero: building a garment for memetic embodiment in dance learning. In: Proceedings of the 2014 ACM International Symposium on Wearable Computers Adjunct Program - ISWC 2014 Adjunct, pp. 49–54 (2014)
15. Nakai, M., Tsunoda, Y., Hayashi, H., Murakoshi, H.: Prediction of basketball free throw shooting by OpenPose. In: Kojima, K., Sakamoto, M., Mineshima, K., Satoh, K. (eds.) JSAI-isAI 2018. LNCS (LNAI), vol. 11717, pp. 435–446. Springer, Cham (2019). https://doi.org/10.1007/978-3-030-31605-1_31
16. Cao, Z., Hidalgo, G., Simon, T., Wei, S.-E., Sheikh, Y.: OpenPose: real-time multi-person 2D pose estimation using part affinity fields. arXiv preprint arXiv:1812.08008 (2018)
17. Dancs, J., Sivalingam, R., Somasundaram, G., Morellas, V., Papanikolopoulos, N.: Recognition of ballet micro-movements for use in choreography. In: 2013 IEEE/RSJ International Conference on Intelligent Robots and Systems, pp. 1162–1167 (2013)
18. Hong, G.-S., Park, S.-W., Park, S.-H., Nasridinov, A., Park, Y.-H.: A ballet posture education using IT techniques: a comparative study. In: Proceedings of the Sixth International Conference on Emerging Databases: Technologies, Applications, and Theory, pp. 114–116. ACM (2016)
19. Saha, S., Banerjee, A., Basu, S., Konar, A., Nagar, A.K.: Fuzzy image matching for posture recognition in ballet dance. In: 2013 IEEE International Conference on Fuzzy Systems (FUZZ-IEEE) (2013)
20. Saha, S., Konar, A., Janarthanan, R.: Posture recognition in ballet dance a case study on fuzzy uniform discrete membership function. In: Proceedings of the 2014 International Conference on Control, Instrumentation, Energy and Communication (CIEC), pp. 708–711 (2014)
21. Banerjee, A., Saha, S., Basu, S., Konar, A., Janarthanan, R.: A novel approach to posture recognition of ballet dance. In: 2014 IEEE International Conference on Electronics, Computing and Communication Technologies (CONECCT) (2014)
22. Kyan, M., et al.: An approach to ballet dance training through ms kinect and visualization in a cave virtual reality environment. ACM Trans. Intell. Syst. Technol. (TIST) 6(2), 23 (2015)
23. Saha, S., Konar, A.: Topomorphological approach to automatic posture recognition in ballet dance. IET Image Process. 9(11), 1002–1011 (2015)
24. Géron, A.: Hands-On Machine Learning with Scikit-Learn and TensorFlow: Concepts, Tools, and Techniques to Build Intelligent Systems. O'Reilly Media Inc., Newton (2017)
25. Frossard, D.: VGG in TensorFlow (2016)
26. Ren, S., He, K., Girshick, R., Sun, J.: Faster R-CNN: towards real-time object detection with region proposal networks. In: Advances in Neural Information Processing Systems, pp. 91–99 (2015)
27. Deng, Z., Sun, H., Zhou, S., Zhao, J., Lei, L., Zou, H.: Multi-scale object detection in remote sensing imagery with convolutional neural networks. ISPRS J. Photogram. Remote Sens. 145, 3–22 (2018)

Specific Language Impairment Detection Through Voice Analysis

Kayleigh Joy Slogrove⑩ and Dustin van der Haar⁽⊠⁾ ⑩

University of Johannesburg, Johannesburg, South Africa
kay.slogrove@gmail.com, dvanderhaar@uj.ac.za

Abstract. Specific Language Impairment is a communication disorder regarding the mastery of language and conversation that impacts children. The system proposed aims to provide an alternative diagnosis method that does not rely on specific assessment tools. The system will accept a voice sample from the child and then detect indicators that differentiate individuals with specific language impairment from that voice sample. These indicators were based on the timbre and pitch characteristics of sound. Three different feature spaces are calculated, followed by derived features, with three different classifiers to determine the most accurate combination. The three feature spaces are Chroma, Mel-frequency cepstral coefficients (MFCC), and Tonnetz and the three classifiers are Support Vector Machines, Random Forest and a Recurrent Neural Network. MFCC, representing the timbre characteristic, was found to be the most accurate feature vector across all classifiers and Random Forest being the most accurate classifier across all feature spaces. The most accurate combination found was the MFCC feature vector with the Random Forest classifier with an accuracy level of 99%. The MFCC feature vector has the most features that are extracted giving the reason for the high accuracy. However, this accuracy decreases when the recorded word is three syllables or longer. The system proposed has proven to be a valid method that can detect SLI.

Keywords: Chroma · Machine learning · Mel frequency cepstral coefficient · Pitch · Random Forest · Recurrent Neural Network · Sound · Specific Language Impairment · Support Vector Machines · Timbre · Tonnetz · Voice

1 Introduction

The diagnosis of Specific Language Impairment (SLI) requires that the language ability of the child meets four criteria. The primary condition is that the language difficulties need to have been present for the child between the ages of two and twelve years old [1]. For SLI, there is no scale for diagnosis, unlike other neurological disorders such as dyslexia. The diagnosis is binary, either the child has SLI or they do not. Misdiagnosis of SLI occurs when psychologists, who typically diagnose SLI, do not include all assessment possibilities at the time of diagnosis. If the child's difficulties with the language are not covered in the assessment, they may be misdiagnosed as healthy [2].

© Springer Nature Switzerland AG 2020
W. Abramowicz and G. Klein (Eds.): BIS 2020, LNBIP 389, pp. 130–141, 2020.
https://doi.org/10.1007/978-3-030-53337-3_10

This paper presents an assessment tool for SLI diagnosis using voice analysis. This will eliminate the need for specific assessment tools per assessment, therefore lowering the number of misdiagnosis of SLI being mistaken for another neurological condition as well as false positive/negative diagnoses. The proposed assessment tool will detect the presence of SLI within the spoken voice by extracting features that represent the fundamental characteristics of sound: timbre and pitch. Timbre is the harmonic tone of the voice that allows the voice to be distinct to each person and uniquely identified by others. Timbre identifies the quality of a particular sound [3]. Pitch is the characteristic that allows a person to judge whether the particular tone is higher or lower than another [4].

The remainder of this paper is laid out as follows: Sect. 2.1 discusses Specific Language Impairment and its diagnosis, Sect. 2.2 discusses the current tests and systems are used for the diagnosis and detection of SLI, Sect. 2.3 identifies systems that use the methods and algorithms that the proposed system will implement. Section 3 discusses the research methodology of the proposed system, explaining the methodology, data samples and population. Section 4 explains the model and the specific algorithms that will be used within each phase of the model. Section 5 analyses the results produced by the system.

2 Related Work

2.1 Specific Language Impairment

Specific Language Impairment is classified as a communication disorder and, more specifically, as a language disorder [1]. This disorder is diagnosed when a child struggles with all forms of language, spoken, written, sign language etc. Children diagnosed with SLI are diagnosed very early, between the ages of two and four years old as symptoms present themselves clearly during that timeframe. These symptoms are demonstrated by the child's inability to talk at the appropriate age. The children affected by SLI have no difficulties with comprehension of words spoken to them, however, they struggle with responding with the right vocabulary and tone. There are four criteria that need to be met when diagnosing a child with SLI. The first is that the child has difficulties with the use of language, meaning that the child has a limited vocabulary, has difficulty constructing sentences and is unable to apply their known vocabulary to explain a series of events within a conversation. The second criteria is that all their abilities relating to the language is below the level that should be maintained for that age. Thirdly, these difficulties must be present within the early developmental period, between two and twelve years old. The last condition states that these difficulties cannot be the result of or attributed to another disability, neurological condition or disorder [1].

2.2 Current Systems that Detect SLI

The current method used to diagnose SLI starts with a screening process. This process begins by gathering information from people who are in direct and constant communication with the child, these people typically being parents and educators [5]. After the information is gathered, a hearing test is conducted to ensure that hearing difficulty are not the cause of the language problems. Formal/informal assessments are then conducted in order to recommend further diagnosis possibilities. If the screening process indicates a possible SLI diagnosis, the child will then be recommended to undergo a comprehensive assessment. These assessments are conducted by educational psychologists or pathologists. In the comprehensive assessment, an analysis of the child's history is performed which involves their medical history, family's history, the educators' and parents' concerns and the common language the child uses, dialect spoken at home. Furthermore, an oral test is conducted as well as a spoken language test which includes phonology, semantics, morphology, syntax and pragmatics [5].

To address the before-mentioned limitation of SLI diagnosis, Grill and Tuckova proposed a system to determine the applicability of using voice analysis to diagnose SLI with artificial intelligence classification [6]. They recorded children with diagnosed SLI as well as children that were not diagnosed with SLI. They used MFCCs as a feature extraction algorithm and then calculated derived features from the MFCCs. The derived features included mean, standard deviation, skewness, kurtosis etc. These features were passed to an artificial neural network for classification and reached an accuracy of 85.14%. Although there was no live use of their system, they did prove that it is possible to detect SLI from voice analysis [6].

Georgopoulos, Malandraki and Stylious proposed to diagnose SLI through the use of Fuzzy Cognitive Maps [7]. Fuzzy Cognitive Maps are the combined methodologies of Fuzzy Logic and Neural Networks. They proposed the use of this model, in addition to diagnosis, to demonstrate the connection between SLI, dyslexia and autism. By demonstrating this relationship, they aimed to show the high misclassifications of SLI as either dyslexia or autism. The model used was able to make the distinction between the three disorders, fulfilling their secondary aim. The network created by the Fuzzy Cognitive Map was used for classification of each disorder, however, Georgopoulos et al. were unable to classify disorders based on the map and made that a point for further research [7].

2.3 Similar Systems for Voice Analysis

The system developed by Yeo, Al-Haddad and Ng is used to correctly identify individual animals when they are 'speaking' [8]. The features that were extracted were pitch, intensity and repetition rate. Mel-frequency cepstral coefficients (MFCC) were used to create a feature vector that was passed through to classification which was based on Dynamic Time Warping. The process created by Yeo et al. proved a more accurate method for animal identification while improving on the samples extracted making them more compact and less redundant. A limitation that was noted, was that several different species of animals have similar methods of communication with regards to the noises made. When two different species made the same noise, the accuracy level of the system decreased as the system classified both as one species [8].

Kumar and Muthukumaraswamy used MFCC as a feature extraction algorithm for voice recognition systems. Before the MFCC feature vector is extracted, the captured samples undergo both framing and windowing with noise elimination as the final step within the preprocessing stage [9]. Vector Quantization is the classification algorithm used within this system with a probability classification result returned. With the additional algorithms of the preprocessing phase, the false acceptance rate is significantly lower than other voice recognition security systems. However, to achieve the highest accuracy, the user would have to double the amount of input voice samples. In addition to this limitation, it was noted that the accuracy level for males was significantly higher than it was for females proven with an accuracy of 81.25% for males and 67.5% for females. [9].

A system designed by Liu, Yin, Jiang, Kan, Zhang, Chen, Zhu and Wang identifies bowel sounds for bowel disease diagnosis [10]. After the MFCC feature vector has been extracted, it is passed to an LSTM neural network. This system was aimed at eliminating the shortcomings of similar systems and by accomplishing this, Liu et al. were able to diagnose the various bowel diseases with a greater level of accuracy. Although steps were taken to eliminate as much noise as possible from the samples captured, this limitation skewed the classification outcome and lowered the accuracy level [10].

A system designed by Korba, Bourouba and Rafik has the aim of improving current noise elimination algorithms [11]. Korba et al. proposed to implement Moving-Average Filtering as their noise elimination algorithm followed by applying a Gaussian Mixture Model to reduce the speaker's variability. It was found that, although the proposed algorithm improved accuracy, the order of the features extracted heavily impacted the accuracy of the overall system [11].

3   Experiment Setup

3.1   Research Methodology

The research design is quantitative with a positivist philosophy where a model is created based on the hypothesis that it is able to detect SLI based on speech. The detection of specific learning impairment within a voice sample will be determined according to the previously stated symptoms.

The research methods that have been used are the literature review, the model and then a developed prototype. The literature review is an analysis of all the systems that have been designed that are similar to the proposed system. The results of the literature review are used to develop the model for the proposed system. The model outlines the requirements for the proposed system and identifies the pipelines and the phases within each pipeline. Within each phase, individual tasks and algorithms are selected to be conducted. A prototype is then implemented according to the phases determined by the model.

3.2   Data Sampling

The data set chosen is a secondary data set developed by Jana Tučková and Vladimír Komárek [6]. The data set is called the Lanna Speech Database for Children with SLI. The data is split into two categories, Healthy (H) and Patients (P). There were forty-four children used for healthy samples and fifty-four for patients. For each child, there are seven subsections of voice samples. The first and second subsections contains voice recordings of the child saying the vowels and consonants, respectively, in individual wav files. The consonants chosen were 'D', 'G', 'H', 'K', 'L', 'M', 'R', 'T' and 'X' as they are consonants the children with SLI struggle with pronouncing. The third subsection contains voice recordings of one-syllable words with two-four letters. Examples include 'Be', 'Pro', 'Prst' and 'Vla'. The next subsection is voice recordings of three-four letter, two-syllable words such as 'Kolo', 'Pivo' and 'Pap'. The fifth subsection contains voice recordings of children saying words with three syllables. The sixth subsection contains voice recordings of words with four syllables such as 'Motovella'. The last subsection contains recordings of words with five syllables. All recordings are stored in the wav format. The duration of the voice recordings ranges from less than one second to one second. The quality of each sample is 705 kbps. In total there are 3853 voice recordings that will be used to train the classifiers proposed in Sect. 4. 70% of data set was used for training the model and the remainder was split with the ratio 2:1 between validation and testing samples respectively.

3.3 Population

The direct stakeholders for this system would be the educational psychologists and the children that are being tested. The educational psychologists would be the ones administering the SLI tests on the children and would be the primary users as they would control the system. The children would be providing the input into the system as a voice recording to be tested.

4 Model

This section outlines the various phases of the proposed system. Figure 1 depicts the core pipeline with alternative algorithms for each phase.

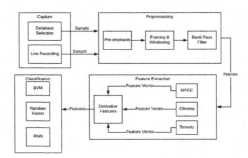

Fig. 1. Figure showing the pipeline of the system with the alternative algorithms in each phase.

4.1 Capture

There are two methods of capturing the voice sample. The user can select a pre-recorded file from the test section of the database, or the user can choose to perform a live recording. The child will be required to speak into the microphone while saying the chosen letter/word. The recording is then saved as a wav file and passed to the next phase of the pipeline.

4.2 Preprocessing

Pre-emphasis. The captured sample is first passed through a pre-emphasis filter. Pre-emphasis filters amplify the higher frequencies within the sample without impacting the lower frequencies. Pre-emphasis is performed with the filter coefficient being set to 0.97.

Framing. The sample is then cut into 25 ms frames with a 10 ms step between frames. This is performed for easier feature extraction and, therefore, easier classification.

Windowing. As framing has the potential to cut the sine wave of the sample into sections that do not end at 0, a windowing function is then performed on the frames. The windowing algorithm selected is the Hamming Window.

Band Pass Filter. Once the sample has been windowed, a bandpass filter is then applied to the sample. The bandpass filter identifies the frequencies that fall within the specified range and removes the frequencies that fall outside this range. This filter is a combination of the low pass filter and the high pass filter. The range for a human voice is 300 Hz–3.1 kHz. The filter is applied according to equations below, where f_h is the upper frequency, f_l is the lower frequency range and n is a sequence of evenly spaced numbers between 0 and 4/0.08. The signal passed to the filter is represented by x and the returned filtered sample is represented by w.

$$z = \sin\left(\frac{\pi\left(2f_h\left(\frac{n-\frac{4}{0.08}-1}{2}\right)\right)}{\pi\left(2f_h\left(\frac{n-\frac{4}{0.08}-1}{2}\right)\right)}\right) \tag{1}$$

$$lp = \frac{z}{sum(z)} \tag{2}$$

$$y = \sin\left(\frac{\pi\left(2f_l\left(\frac{n-\frac{4}{0.08}-1}{2}\right)\right)}{\pi\left(2f_l\left(\frac{n-\frac{4}{0.08}-1}{2}\right)\right)}\right) \tag{3}$$

$$hp = \frac{y}{sum(y)} \tag{4}$$

$$w(t) = x(t) \cdot (lp \cdot hp) \tag{5}$$

4.3 Feature Extraction

Mel Frequency Cepstral Coefficients. Once the sample has been processed, it is passed to feature extraction. MFCC is used to represent the timbre characteristic of voice. To calculate the MFCC, first Discrete Fourier Transform is performed on each frame according to Eq. 8, where the signal is represented by x(n) and S is the returned signal.

$$S(k) = \sum_{n}^{N} x(n)e^{-j2\pi kn/N} \tag{6}$$

After the Discrete Fourier Transform is performed on each frame, a periodogram-based power spectral estimate is calculated for each frame according to the formula below.

$$P(k) = \frac{1}{N}|S(k)|^2 \tag{7}$$

The Mel Filterbanks are then calculated according to the previously defined range. These filterbanks are then multiplied with the power spectrums. These steps are shown in Formulas 8 and 9 respectively, with H(k) representing the filterbanks calculated and m being the current filterbank being calculated.

$$H_m(k) = \begin{cases} 0 & k < f(m-1) \\ \frac{k-f(m-1)}{f(m)-f(m-1)} f(m-1) \le k \le f(m) \\ \frac{f(m+1)-k}{f(m+1)-f(m)} f(m) \le k \le f(m+1) \\ 0 & k > f(m+1) \end{cases} \tag{8}$$

$$E(k) = H(k) \cdot P(k) \tag{9}$$

The log of each energy filterbank is then calculated following by taking the Discrete Cosine Transform for each filterbank. The result of the Discrete Cosine Analysis is stored as the MFCC.

Chroma. An alternative algorithm for feature extraction is calculating the chroma features of the sample. Chroma features depict the usage of each pitch class for the sample. To calculate the chroma features, the sample is first passed through Short-Time Discrete Transformation. After each frame has been passed through the transformation, the log is taken for each frame giving the chroma features according to the following equation, where P is the sequence of pitches in numeric form (p being an element of P where $p \in [0:127]$) and m representing the current frame.

$$C(m, p) = \sum_{k \in P(p)} |S(m, k)|^2 \tag{10}$$

Tonnetz. A second alternative feature extraction algorithm is calculating the tonal centroid features. These features describe the pitch space between the network of relationships between each of the musical pitches. To calculate the tonal centroid, first, the chroma features are calculated according to the calculations above. The tonal centroid for the chroma vector is then calculated according to the equations below, where C_f is the chroma vector, b is the integer pitch class where $b \in [0, \beta - 1]$ and β is the bins per octave.

$$TC(d) = \frac{1}{\sum_b |C_{f(b)}|} \sum_{b=0}^{\beta-1} \Phi(d, b) C_f(b) \tag{11}$$

$$\Phi = [\phi_0, \phi_1 \ldots \phi_{\beta-1}] \tag{12}$$

$$\phi_b = \begin{bmatrix} r_1 \sin(b\frac{7\pi}{6}) \\ r_1 \cos(b\frac{7\pi}{6}) \\ r_2 \sin(b\frac{3\pi}{2}) \\ r_2 \cos(b\frac{3\pi}{2}) \\ r_3 \sin(b\frac{2\pi}{3}) \\ r_3 \cos(b\frac{2\pi}{3}) \end{bmatrix} \tag{13}$$

Derivative Features. For each feature space calculated, for each filter bank/pitch class/relationship, nineteen derivate features are determined from the feature space. These features are then passed through for classification training and predictions. The first two derivative features calculated are the positions of the minimum and maximum values. The mean, standard deviation, skewness and kurtosis are then calculated. Following this, the slope and offset of the linear approximation of the contour are determined. The three interquartile values are calculated as well as the ranges between these values. The 1% and 99% percentile are calculated, followed by the value of the difference between them. Lastly, two percentages are calculated according to Formula 15 with x representing 75% and 90% respectively

$$percents = all\ values > (x * (99\%\ percentile - 1\%\ percentile) + 1\%\ percentile \tag{15}$$

4.4 Classification

Support Vector Machine. The first classifier used is an SVM classifier. SVMs convert a nonlinear function to linear to create the biggest distance possible between the classes using Euclidean distance [12]. A Radial Basis Function kernel was chosen for the proposed model with the influence of each node set to 0.0001. These parameters were selected from the result of a parameter study conducted through a grid search where different kernels and gamma values were tested.

Random Forest. One alternative classification algorithm is the Random Forest classifier that develops multiple decision trees at the time of training [13]. The number of trees chosen for the proposed model was 200. The parameters that were chosen were based on the output of a parameter study where the number of trees, the maximum depth of each tree, criterion and maximum features of each tree were considered.

Recurrent Neural Network. The second alternative classification algorithm is an RNN with one LSTM layer and three time-distributed hidden layers [14]. The input layer contained equal number of nodes to the parameters passed in while the output layer contained two nodes. This architecture was selected as the result of an architecture study.

5   Results

The metrics used to evaluate each feature vector and classifier were precision, recall, F1 score, accuracy, the area under the curve (AUC), sensitivity and specificity. These metrics were calculated and displayed in Table 1. The confusion matrix is also displayed with the format shown in Formula 16. The values were calculated by the researcher.

$$
\begin{array}{cc} & H \; P \\ H & \\ P & \end{array} \tag{16}
$$

From Table 1, it is clear to see that the MFCC feature vector has the highest scores across all classifiers. MFCC represent the harmonic tone of a person's voice, the harmonic tone determines the quality of sound from one person to another. This harmonic tone is a data rich feature vector as can be seen by the filter banks calculated. MFCC has the most features calculated with thirteen filter banks and nineteen features calculated per filter bank. The chroma feature space has twelve pitch classes while tonnetz only has six relationships to calculate derived features from. Furthermore, Random Forest is the most accurate classifier across all feature spaces. The potential reasoning for this is due to the linearity of the derived features. This allows for easy identification for split points for the Random Forest classifier. Recurrent Neural Networks have the highest accuracy when working with sequential/non-linear data which was tested with an alternative pipeline where the extracted features were not derived before being passed to the classification algorithm. This pipeline is depicted in Fig. 2. However, when the features were derived to linear features, the RNN performed worse than the Random Forest classifier. The ROC curves for the three MFCC classifiers are shown in Fig. 3.

The ROC curves show the True Positive Rate versus the False positive rate and depicts how easily the classifier can distinguish between the two classes, children with SLI and children without SLI. From Figure two, it can be seen that Random Forest can distinguish the difference between the two classes the most accurately. This confirms that Random Forest was the most accurate classifier found for the MFCC feature space and that the MFCC feature space was the most accurate feature space. When studying the confusion matrices, it was determined that two thirds of the misclassified samples were made up of words between three and seven syllables. The remaining third is made of samples where either one letter is spoken or a word is lower than three syllables. Therefore, it can be said that letters and shorter words are more accurate. This can be further researched to determine the possibility of classifying on sentences compared to single words.

Table 1. Table depicting the metrics for each feature vector and classifier.

Features	Metric	SVM	Random Forest	RNN
MFCC	Precision	0.94	**0.99**	0.97
	Recall	0.93	**0.99**	0.94
	F1 Score	0.93	**0.99**	0.96
	Accuracy	0.94	**0.98**	0.96
	AUC	0.98	**1.00**	0.99
	Sensitivity	0.96	**0.98**	0.98
	Specificity	0.898	**0.99**	0.94
	Confusion matrix	423 16 34 298	**431 8 4 328**	435 9 19 328
Tonnetz	Precision	0.70	**0.71**	0.56
	Recall	0.68	**0.69**	0.64
	F1 Score	0.68	**0.69**	0.60
	Accuracy	0.70	**0.70**	0.63
	AUC	0.73	**0.78**	0.69
	Sensitivity	0.83	**0.84**	0.63
	Specificity	0.52	**0.53**	0.64
	Confusion matrix	353 86 133 199	**370 69 127 205**	281 163 119 208
Chroma	Precision	0,73	**0.78**	0.65
	Recall	0.73	**0.78**	0.69
	F1 Score	0.73	**0.78**	0.67
	Accuracy	0.73	**0.78**	0.71
	AUC	0.79	**0.86**	0.78
	Sensitivity	0.76	**0.83**	0.72
	Specificity	0.69	**0.72**	0.69
	Confusion matrix	329 110 100 232	**344 95 105 227**	320 124 101 226

The highest accuracy of the proposed model was for the MFCC feature vector with the Random Forest classifier at 99%. From Table 2, the most accurate similar system reached an accuracy of 94.60% with an RNN. From this, it is clear to see that the proposed model is more accurate than other MFCC-based systems. Further research on this topic could involve employing a spectrogram to replace the extracted features.

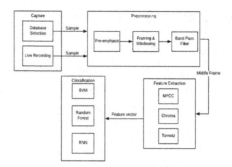

Fig. 2. Figure detailing the phases of the alternative pipeline

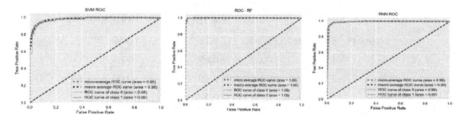

Fig. 3. The ROC curves for the MFCC feature space across all classifiers. From left to right, SVM classifier, Random Forest classifier and RNN classifier.

Table 2. Table depicting the accuracy of similar systems with their respective classifiers

Classifier	Accuracy	Sensitivity	Specificity
Vector Quantization [9]	73.3%	Not given	Not given
Recurrent Neural Network [10]	94.60%	62.10%	96.2%
Gaussian Mixture Model [11]	74.31%	Not given	Not given
Artificial Neural Network [6]	85.14%	Not given	Not given

6 Conclusion

A method of detecting SLI with a voice sample was proposed in this paper. Three feature extraction algorithms were compared along with three classification algorithms. The three feature extraction algorithms aimed at representing the characteristics of sound, timbre and pitch. It was determined that MFCC, representing the timbre characteristic, was the most accurate feature vector with Random Forest being the most accurate classifier. As there were derived features from the MFCC calculations, the features used for classifications were linear features. The linearity of the features is a possible reason for the higher accuracy of the Random Forest classifier over the Recurrent Neural network which is known for high accuracy of speech data. The limitations of the system include a loss of accuracy when a word longer than three syllables is inputted as a sample. Future research can include the third dimension of sound, intensity, as a feature to be extracted. From the results calculated, the proposed method for diagnosing Specific Language Impairment will work as expected to correctly diagnose children with Specific Language Impairment.

References

1. American Psychiatric Association: Diagnostic and Statistical Manual of Mental Disorders, 5th edn. American Psychiatric Publishing (2013)
2. Grimm, A., Schulz, P.: Specific language impairment and early second language acquisition: the risk of over-and underdiagnosis. Child Ind. Res. **7**(4), 821–841 (2014). https://doi.org/10.1007/s12187-013-9230-6
3. Miriam Webster: Timbre. https://www.merriam-webster.com/dictionary/timbre. Accessed 25 Nov 2019
4. Klapuri, A.: Signal Processing Methods for Music Transcription. Springer, Boston (2006). https://doi.org/10.1007/0-387-32845-9
5. American Speech-Language-Hearing Association: Spoken Language Disorders. https://www.asha.org/PRPSpecificTopic.aspx?folderid=8589935327&section=Assessment. Accessed 10 Sept 2019
6. Grill, P., Tučková, J.: Speech databases of typical children and children with SLI. PLoS One **11**(3), 1–21 (2016)
7. Georgopoulos, V.C., Malandraki, G.A., Stylios, C.D.: Development of intelligent method for differential diagnosis of specific language impairment. In: Proceedings of the 23rd Annual EMBS International Conference. IEEE, Istanbul (2001)
8. Yeo, C.Y., Al-Haddad, S.A.R, Ng, C.K.: Animal voice recognition for identification (id) detection system. In: 2011 IEEE 7th International Colloquium on Signal Processing and Its Applications, pp. 198–201. IEEE (2011)
9. Kumar, A.N.A., Muthukumaraswamy, S.A.: Text dependent voice recognition system using MFCC and VQ for security applications. In: International Conference on Electronics, Communication and Aerospace Technology, pp. 130–136. IEEE, Coimbatore (2017)
10. Liu, J., et al.: Bowel sound detection based on MFCC feature and LSTM neural network. In: 2018 IEEE Biomedical Circuits and Systems Conference (BIOCAS). IEEE, Cleveland (2018)
11. Korba, M.C.A., Bourouba, H., Rafik, D.: Text-independent speaker identification by combining MFCC and MVA features. In: 2018 International Conference on Signal, Image, Vision and Their Applications (SIVA). IEEE, Guelma (2018)
12. Statistica Help: Support Vector Machines Introductory Overview. https://documentation.statsoft.com/STATISTICAHelp.aspx?path=MachineLearning/MachineLearning/Overviews/SupportVectorMachinesIntroductoryOverview. Accessed 05 Oct 2019
13. Random Forest Classifier. https://www.globalsoftwaresupport.com/random-forest-classifier/. Accessed 05 Oct 2019
14. Recurrent Neural Networks. https://leonardoaraujosantos.gitbooks.io/artificial-inteligence/content/recurrent_neural_networks.html. Accessed 05 Oct 2019

ICT Project Management

The Role of Openness and Extension Modularization in Value Capture for Platform-Based Digital Transformation

David Soto Setzke[(✉)], Markus Böhm, and Helmut Krcmar

Technical University of Munich, Boltzmannstraße 3, 85748 Garching, Germany
{david.soto.setzke,markus.boehm,helmut.krcmar}@tum.de

Abstract. Digital transformation is radically changing the way companies conduct business and compete in established markets. In particular, a growing number of companies are switching from predominantly product-focused to platform-based business models. However, it remains unclear how these platforms should be designed to enable platform owners to maximize value capture. In this study, we investigated the interactions between platform openness and extension modularization and their influence on value capture in the context of digital transformation. To do so, we combined a case survey strategy with a configurational approach using fuzzy-set Qualitative Comparative Analysis. We found that there is no single condition necessary to achieve a high degree of value capture. Furthermore, our results show the importance of closedness and tight coupling of platforms and their applications. Finally, we confirmed the importance of interface conformance to high value capture. In addition, our results contribute to both theory and practice and provide implications for future research into the role of digital platforms in digital transformation.

Keywords: Digital transformation · Digital platforms · Configuration theory

1 Introduction

Digital technologies are radically changing today's business environments and ecosystems. The widespread availability of these technologies has given birth to a multitude of start-ups, entering markets that were traditionally dominated by established companies. Recently, these companies have begun digital transformation initiatives to implement new business models, increase internal efficiency, and enhance customer experience [1]. In particular, more and more companies have begun switching from predominantly product-focused to service-based models centered on digital platforms [2]. Thus, the source of value creation is shifting from traditional dyadic, one-on-one relationships with partners to more complex, interconnected ecosystems with digital platforms at their centers [3]. Many established companies cannot draw on prior experience regarding the design and maintenance of digital platforms and thus, many platform-based endeavors have failed [4]. One particular challenge for platform owners lies in capturing part of the value that is created on a platform and its ecosystem by complementors [5].

W. Abramowicz and G. Klein (Eds.): BIS 2020, LNBIP 389, pp. 145–156, 2020.
https://doi.org/10.1007/978-3-030-53337-3_11

Extant literature suggests that design choices such as the openness of the platform and the modularization of its extensions or applications influence the degree of value that can be captured [5]. For example, an open platform might increase the number of available applications and the potential value that can be captured [6]. On the other hand, a rather closed platform may be more easily governed by its owners. This may be realized through the enforcement of stricter regulations controlling exactly how value may be captured by platform owners [7]. However, extant literature has only theoretically hypothesized the nature of these relationships. To the authors' best knowledge, no empirical study has yet investigated the specific interaction between platform design and value capture. In order to close this research gap, our guiding research question is as follows: *In the context of digital transformation, how do design choices, such as platform openness and extension modularization, influence the degree of value capture to the platform?*

To answer this question, we adopted a configurational viewpoint. We conducted a case survey of digital platform case studies using scholarly and practice-oriented sources and subsequently analyzed the results with fuzzy-set Qualitative Comparative Analysis (fsQCA) [8, 9]. We identified both single necessary conditions and sufficient configurations of conditions for achieving high degrees of value capture.

2 Platform-Based Digital Transformation

2.1 Value Capture and Platform Design

The widespread availability of digital technology has had a profound impact on companies and their environments. New technologies, such as the blockchain or the Internet of Things, have radically changed the processes of established companies and their business models as they try to defend their shares of traditional markets that are currently being conquered by start-ups [1, 10]. In particular, many companies are adopting digital platform-based business models and service offerings [3]. Accordingly, some scholars have observed a transformational shift from product-centric to service-centric offerings that are based on digital platforms. Since most established companies are novices regarding the design and maintenance of digital platforms, they face a multitude of challenges [2]. One particular challenge lies in capturing the monetary value that is co-created on these platforms [5]. Extant research suggests that the potential value capture is profoundly influenced by digital platform design choices [5, 7]. Therefore, we focus on design choices made by the platform owner regarding relationships with third-party developers and applications, namely, platform openness and extension modularization. For the remainder of this paper, the terms "application" and "extension" are interchangeable.

Platform Openness. Openness is a crucial feature of any platform design. It can generally be defined according to the categories of accessibility and transparency [11]. *Accessibility* indicates a platform's degree of discrimination regarding different actors and their access to the ecosystem [11]. A platform owner can vary the degree of accessibility to control who is and who is not allowed and under what conditions they have access to the platform [12]. For example, a platform owner may choose to only provide boundary resources such as application programming interfaces (APIs) or software development kits (SDKs) to those third-party developers that pay a certain licensing fee or deny them access if their applications receive consistently bad ratings [13]. *Transparency* indicates the degree to which users understand "what is happening and why" [11]. In particular, this determines whether governance-related decisions are made transparently and comprehensively for all users. A platform owner may, for example, provide extensive documentation regarding their boundary resources or communicate the conditions that third-party apps need to fulfill to be listed in a platform application marketplace [14].

Platform openness may have both a positive and negative influence on value capture: a more open platform may attract a larger number of developers and fewer restrictions may simultaneously increase the number of platform applications along with potential value to be captured [6]. However, the platform owner may have difficulty enforcing the capture of such value in the absence of strict regulation [7]. A closed platform, on the other hand, due to stricter rules regarding the inclusion of third-party developers may decrease the number of applications, potentially resulting in less value to be captured. Still, stricter control may increase the quality of permitted apps, ultimately resulting in a larger user base and increased value [15].

Extension Modularization. *Extension modularization* describes the client-side design choices made by a digital platform. Following Tiwana [16], extension modularization can be assessed using two subcategories: loose coupling and interface conformance. Both can be applied either to each single extension or to the entire ecosystem [16]. *Loose coupling* refers to whether platforms and their applications have minimal dependence on each other and the necessity of these dependencies [17]. In particular, loose coupling implies that a change in the digital platform does not generate a ripple effect that requires all third-party applications to make accommodating changes and vice versa [16]. Thus, the system remains relatively stable. On tightly coupled platforms, on the other hand, applications heavily depend on the platform and are severely affected whenever changes occur [18]. Loosely coupled platforms may provide a higher degree of freedom to application developers, thus increasing the stability of applications and ensuring a constant degree of value capture. Still, tightly coupled platforms may also result in higher value capture since they give platform owners the possibility to exert stricter control on its relationship with third-party developers, similar to more closed platforms. Furthermore, tightly coupled platforms typically emphasize "an increased understanding of each other's needs, a close relationship, a low degree of information asymmetry, and the ability to tailor products or services to strategic needs" [18]. These factors may eventually reinforce and improve the relationship between platform owner and third-party developer, facilitating a higher degree of value capture.

Interface conformance, the second subcategory of extension modularization, relates to the degree to which extensions conform "to the interface specifications explicitly specified by the platform owner," such as APIs or proprietary protocols [16]. Therefore, it measures whether applications interact with the platform through clearly specified, stable, and well-documented interfaces [19]. Based on extant literature, we assume that interface conformance has a positive effect on value capture in general since it fosters high quality in its applications, resulting in higher user adoption or acceptance and, ultimately, a higher degree of value capture available to the platform owner [16].

2.2 Research Model

We investigate how different design choices influence the degree of value capture achieved by digital platforms in the context of digital transformation. Regarding value capture, we refer to monetary value and define value capture as "'the appropriation and retention […] of payments made by consumers in expectation of future value from consumption' that one member of a value system can claim for itself" [5]. We argue that the interactions between the predictor variables are of particular interest and have not been investigated yet. During our comprehensive review of the existing literature, we have found that these predictors may have both positive and negative effects on outcomes, depending on the specific context. Operating under the assumption that there are a number of different paths that may lead to success, we have adopted a configurational perspective. In our study, a platform owner achieving a high degree of **value capture** represents success. As predictor variables, we added two design choices characteristic of digital platforms to our research model: **platform openness** and extension modularization. We further divided extension modularization into two subcategories: the degree of **coupling** and **interface conformance**. These three explanatory factors along with our chosen outcome represent our research model. We summarize this in Fig. 1.

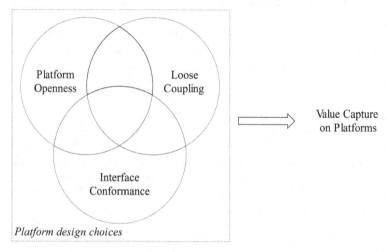

Fig. 1. Configurational research model

3 Methodology

We believe that an in-depth perspective based on case studies is an appropriate approach to explore the complex causal relationships in our research model. Since it is highly resource-intensive to conduct in-depth case studies, we made use of the large number of existing case studies already available in scientific databases, thus adopting a case survey strategy [8]. Furthermore, we adopted a configurational perspective since we assumed that there are various paths leading to success and complex interactions between the specific factors in our research model. Therefore, we created a coding scheme based on fuzzy values for conducting the case survey and analyzing the resulting data matrix with fsQCA [9]. Before we started collecting case studies, we defined appropriate inclusion criteria [8]. In our analysis, since we focused on digital platforms that were specifically introduced during an established company's digital transformation process, we defined inclusion criteria to ensure accordingly appropriate coverage. Thus, we included cases only if (1) the narrative sufficiently described the introduction of a digital platform into an existing company's business model, (2) the digital platform was introduced by an established company as part of a digital transformation strategy, and (3) the digital platform offered third-party developers the possibility to develop their own complementary applications using resources provided by the platform owner. Considering these criteria, we developed an appropriate search string and then applied it to searches of various databases. We selected the following databases for our search: EBSCOhost, Emerald Insights, Google Scholar, Web of Science, and Scopus. After an initial screening of the results, we applied our inclusion criteria. Any case study that did not fulfill these criteria was excluded from the set. Afterward, a set of 20 case studies remained, which formed part of our analysis.

After collecting our set of 20 cases, we began the coding process. We developed a coding scheme based on the research model introduced earlier, comprising the dimensions of platform openness, loose coupling, interface conformance, and value capture. We followed established guidelines for the use of qualitative data such as case studies or archival data in qualitative comparative analysis [20]. Thus, for each dimension, we defined a "theoretical ideal" that represented "the best imaginable case in the context of the study that is logically and socially possible" [21]. Furthermore, we used a fuzzy five value scheme to code each case against the theoretical ideal, as recommended by Iannacci and Cornford [22], when underlying data may be "too weak to support fine-grained distinctions." Two researchers independently coded all cases using the defined coding scheme. Subsequently, both researchers prepared a coding report that contained detailed information about the case, the chosen fuzzy values, as well as quotes from the text supporting the choice of said fuzzy values. The researchers then compared their reports to each other. Disagreements in coding lead to both researchers re-reading the cases and reconciling disagreements through a consensus approach as recommended by Larsson [8]. The resulting fuzzy values for each case and the theoretical ideals for each dimension are summarized in Table 1.

After coming to an agreement regarding the coding of all cases, the researchers proceeded to analyze the agreed-upon fuzzy values. For this step, we employed fsQCA, a configurational research method. While regression-based approaches usually postulate that a certain predictor variable is both necessary and sufficient for a given outcome, configuration theory posits that predictors may be either sufficient or necessary, both, or neither. To further reinforce this difference, configuration theory uses the term "condition" rather than "predictor". To evaluate the quality of solutions, fsQCA is based on two "parameters of fit": consistency and coverage. Consistency captures "how consistently empirically observed configurations are linked to the outcome and thus provide information regarding the model's validity" [23]. We also considered an alternative measure of

Table 1. Fuzzy membership values for openness and application modularization

Case/Dimension	Openness [11]	Extension modularization [16, 17]		Value capture [5, 26]
		Loose coupling [16, 17]	Interface conformance [16, 17]	
	(Ideal type: Platform resources are available and accessible for everyone without restrictions; documentation and governance rules are transparent)	(Ideal type: Platform and applications are loosely coupled, with a small number of necessary and no unnecessary interdependencies)	(Ideal type: Applications interact with the platform using interface standards and protocols that are clearly specified, standardized, and stable)	(Ideal type: Platform owner is able to capture most of the value that is co-created on the platform by third-party developers and the platform owner)
APIBank [27]	1	0.49	1	0.25
DBS [28]	0	1	1	1
Intuit [29]	0.75	0.49	0.75	1
CommCo [30]	0.49	0.75	0.25	0.25
Advantech [31]	1	1	1	1
Cisco [32]	0	0.25	1	1
Siemens [33]	1	0.75	1	1
ABB [34]	0.75	0.75	1	0.25
GE [35]	0.75	1	0.75	1
Telekom [36]	1	0.75	0.75	0.25
Telenor [37]	0.75	0.49	1	1
China Mobile [38]	0	0.49	1	0.75
Signify [39]	1	1	1	1
Royal Philips [40]	1	1	1	1
Aker BP [41]	0.75	1	1	1
Maersk [42]	1	1	1	1
Trafiklab [43]	0.75	1	0.75	0
Volvo [44]	0	0.25	0.25	1
Lego [45]	0.75	1	1	0
Aadhaar [46]	0.75	0.75	1	0.25

consistency: Proportional Reduction in Consistency (PRI), which eliminates the influence of cases that are subsets of both the outcome and its negation [24, 25]. Coverage represents the percentage of cases of a certain outcome that are explained by the chosen configurational model [9].

To conduct our analysis, we used an R library developed by Thiem and Dusa [47] that supported all steps of fsQCA. As recommended by Schneider and Wagemann [48], we first performed a necessity analysis to identify any single conditions that are necessary for achieving the outcome. We then employed the commonly accepted consistency threshold of 0.9. In the next step, we used the R library to construct a truth table of all potential combinations of conditions and outcomes. Each truth table row was assigned a consistency measure indicating how well the cases represented the given table row. Next, we applied thresholds for consistency (0.8) and frequency (1). Thresholds for (PRI) consistency reduced the truth table by eliminating all rows with consistency values that did not reach the given threshold. With 0.8 for raw consistency and 0.75 for PRI consistency, we exceeded the lower bound of 0.75 for both consistency measures [24, 48]. After further reducing and simplifying the truth table, three solution types remained: the complex, parsimonious, and intermediate. As recommended by Fiss [49], we focused on the intermediate and parsimonious solutions and used them to identify core and peripheral conditions. Core conditions have a strong causal relationship with the relevant outcome, while peripheral conditions surround core conditions and have a weaker causal relationship with the outcome [49].

4   Results

The results of our necessity analysis reveal interface conformance as a candidate condition for high value capture and loose coupling for low value capture (see Table 2). However, in both instances, relevance and coverage are rather low which puts their nature of necessity in doubt. Thus, we proceeded to analyze the results of sufficiency to further investigate the necessity of our candidate conditions and to identify sufficient conditions.

Table 2. Results of necessity analysis

	High value capture			Low value capture		
	Consistency	Relevance	Coverage	Consistency	Relevance	Coverage
Openness	0.66	0.61	0.69	0.83	0.43	0.37
Loose coupling	0.78	0.53	0.72	0.96	0.34	0.38
Interface conformance	0.91	0.35	0.73	0.88	0.17	0.3
~Openness	0.39	0.93	0.85	0.29	0.74	0.27
~Loose coupling	0.32	0.98	0.95	0.29	0.83	0.37
~Interface conformance	0.13	0.96	0.7	0.21	0.93	0.5

Commonly applied consistency threshold: 0.9 [25]

Our sufficiency analysis yielded three distinct configurations for achieving high degrees of value capture on digital platforms (see Table 3). For the negation of the outcome, i.e., achieving low degrees of value capture, we could not identify configurations that passed our thresholds of consistency. Regarding the positive outcome, we observed a robust consistency level of 0.87, well above the commonly accepted threshold of 0.75 [48]. The coverage level of 0.47 demonstrates that our solution explained almost half of the observed outcome. The first configuration represents closed platforms with a low degree of low coupling, which was also a core condition. Interface conformance is irrelevant for this configuration, which confirmed our earlier assumption that it was not a necessary condition. Volvo represents an example for this case [44]. The second configuration shows closed platforms with a high degree of interface conformance. Both conditions are core conditions while the degree of coupling does not play a major role in the achievement of the outcome. This configuration is represented by, for example, DBS [28]. The third and last configuration depicts platforms with a low degree of loose coupling and a high degree of interface conformance. In this configuration, platform openness is irrelevant. An example for this configuration can be found in the case of APIBank [27].

Table 3. Results of sufficiency analysis

Conditions	High value capture		
	1	2	3
Platform openness	⊗	⊗	
Loose coupling	⊗		⊗
Interface conformance		●	●
Raw consistency	1.00	0.86	0.94
PRI consistency	1.00	0.83	0.91
Raw coverage	0.23	0.34	0.29
Unique coverage	0.04	0.14	0.09
Solution raw consistency	0.87		
Solution PRI consistency	0.83		
Solution coverage	0.47		

Black circle: presence of a condition; crossed-out circle: absence of a condition; empty row: may be either present or absent; large circle: core condition; small circle: peripheral condition; raw consistency threshold: 0.8; PRI consistency threshold: 0.75; frequency threshold: 1.

5   Discussion

5.1   Cross-configurational Patterns

Our results show different interesting patterns across the identified configurations. First, it is interesting to note that we could not find any necessary conditions for high value capture. This implies that there is no single success condition for achieving high value capture. On the contrary, there are various configurations that all lead to the desired outcome. Furthermore, we also confirmed the positive effect of interface conformance, since the presence of this condition forms part of two sufficient configurations. Still, it is not a relevant factor in the first identified configuration. Interestingly, we could not ascertain high degrees of platform openness nor loose coupling as being sufficient elements. On the contrary, closed platforms and tight coupling appear to be parts of sufficient configurations, mostly even as core conditions with high empirical relevance. For example, a closed platform with tight coupling is a sufficient configuration for high value capture. Thus, we assume that platform owners can maximize value capture by keeping a platform closed and exercising strict control over what third-party applications may be developed on it. Furthermore, a tight coupling approach seems promising for maintaining closer relationships with developers and maximizing value capture. While closedness and tight coupling in combination are sufficient, either condition may also be combined with interface conformance to achieve high value capture.

5.2   Contributions and Limitations

This study contributes to existing research on digital platforms and digital transformation in several ways. First, we answered calls for research taking a configurational perspective instead of the traditional econometric and qualitative approaches [50] of previous studies. Furthermore, we responded to a call for research by de Reuver, Sørensen and Basole [51] that proposed further analyses of design choices for digital platforms. Our study is a first exploration of the interrelationship between value capture and design choice for platforms in the context of digital transformation. Thus, we showed several avenues for future research. Finally, practitioners may use our findings to optimize the design of their own platforms to maximize value capture.

Our research is not free from limitations that need to be addressed. One such limitation is the relatively low value of solution coverage, indicating a potentially low empirical relevance of the solution. However, as Schneider and Wagemann [25] note, the "empirical importance expressed by coverage is not the same as the theoretical or substantive relevance of a sufficient condition." Thus, even solutions with low coverage can still be interesting for the purpose of theory building. We argue that this specifically applies to our findings since this is the first study that uses configuration theory to explore the interaction between openness and application modularization and their effects on value capture. Future research could include further conditions that are theoretically relevant to identify additional configurations.

Further limitations result due to our use of a case database. First, the number of cases is relatively low (20). This restricts the number of conditions that could be included in our research model since a high number of conditions would result in a high number of logical remainders. Since theoretical knowledge regarding our chosen research model is rather low, it is difficult to eliminate these remainders. Second, the case studies were originally conducted with very different goals. The degree to which they provide information about the conditions of our study may vary. However, we tried to account for this fact by having two independent coders. In summary, we call for further research by conducting in-depth case studies and increasing the set of studies that may be investigated to analyze the phenomenon of value capture. This increased set could also be used to test the results that were derived in this paper.

6 Conclusion

This study aimed to identify successful platform configurations with respect to the design choices of openness and extension modularity, in the context of digital transformation. Therefore, we conducted a case survey combined with an fsQCA approach, using a set of scholarly and practice-oriented case studies. Our analysis revealed three distinct configurations of successful design choices. In particular, our results demonstrate that no single condition is necessary and that there are several paths to success. Our analysis also revealed three configurations of platforms with high degrees of value capture and showed the effectiveness of closed platform design using a tight coupling approach. Our study contributes to theory and practice by providing a first exploration into the influence of platform design on value capture for platforms in digital transformation.

References

1. Vial, G.: Understanding digital transformation: a review and a research agenda. J. Strateg. Inf. Syst. **28**, 118–144 (2019)
2. Parker, G.G., Van Alstyne, M.W., Choudary, S.P.: Platform Revolution: How Networked Markets are Transforming the Economy and How to Make Them Work for You. WW Norton & Company, New York (2016)
3. Hein, A., et al.: Digital platform ecosystems. Electron. Markets **30**, 87–98 (2020). https://doi.org/10.1007/s12525-019-00377-4
4. Cusumano, M.A., Yoffie, D.B., Gawer, A.: The Business of Platforms: Strategy in the Age of Digital Competition, Innovation, and Power. HarperCollins Publishers, Manhattan (2019)
5. Schreieck, M., Wiesche, M., Krcmar, H.: The platform owner's challenge to capture value – insights from a business-to-business IT platform. In: 38th International Conference on Information Systems, Seoul, South Korea (2017)
6. Ondrus, J., Gannamaneni, A., Lyytinen, K.: The impact of openness on the market potential of multi-sided platforms: a case study of mobile payment platforms. J. Inf. Technol. **30**, 260–275 (2015). https://doi.org/10.1057/jit.2015.7
7. Van Alstyne, M.W., Parker, G., Choudary, S.P.: Reasons platforms fail. Harvard Bus. Rev. **31**, 2–6 (2016)
8. Larsson, R.: Case survey methodology: quantitative analysis of patterns across case studies. Acad. Manag. J. **36**, 1515–1546 (1993)
9. Ragin, C.C.: Redesigning Social Inquiry: Fuzzy Sets and Beyond. University of Chicago Press, Chicago (2008)

10. Riasanow, T., Soto Setzke, D., Böhm, M., Krcmar, H.: Clarifying the notion of digital transformation: a transdisciplinary review of literature. J. Competences Strategy Manag. **10**, 5–36 (2019)
11. West, J., O'Mahony, S.: The role of participation architecture in growing sponsored open source communities. Ind. Innov. **15**, 145–168 (2009)
12. Soto Setzke, D., Böhm, M., Krcmar, H.: Platform openness: a systematic literature review and avenues for future research. In: 14th International Conference on Wirtschaftsinformatik, Siegen, Germany (2019)
13. Boudreau, K.: Open platform strategies and innovation: granting access vs. devolving control. Manag. Sci. **56**, 1849–1872 (2010)
14. Dal Bianco, V., Myllärniemi, V., Komssi, M., Raatikainen, M.: The role of platform boundary resources in software ecosystems: a case study. In: IEEE/IFIP Conference on Software Architecture, pp. 11–20. IEEE, Sydney (2014)
15. Müller, R.M., Kijl, B., Martens, J.K.J.: A comparison of inter-organizational business models of mobile app stores: there is more than open vs. closed. J. Theor. Appl. Electron. Commer. Res. **6**, 13–14 (2011)
16. Tiwana, A.: Evolutionary competition in platform ecosystems. Inf. Syst. Res. **26**, 266–281 (2015)
17. Orton, J.D., Weick, K.E.: Loosely coupled systems: a reconceptualization. Acad. Manag. Rev. **15**, 203–223 (1990)
18. Hein, A., Böhm, M., Krcmar, H.: Tight and loose coupling in evolving platform ecosystems: the cases of airbnb and uber. In: Abramowicz, W., Paschke, A. (eds.) BIS 2018. LNBIP, vol. 320, pp. 295–306. Springer, Cham (2018). https://doi.org/10.1007/978-3-319-93931-5_21
19. Sanchez, R., Mahoney, J.T.: Modularity, flexibility, and knowledge management in product and organization design. Strateg. Manag. J. **17**, 63–76 (1996)
20. de Block, D., Vis, B.: Addressing the challenges related to transforming qualitative into quantitative data in qualitative comparative analysis. J. Mixed Methods Res. **13**(4), 503–535 (2018)
21. Basurto, X., Speer, J.: Structuring the calibration of qualitative data as sets for qualitative comparative analysis (QCA). Field Methods **24**, 155–174 (2012)
22. Iannacci, F., Cornford, T.: Unravelling causal and temporal influences underpinning monitoring systems success: a typological approach. Inf. Syst. J. **28**, 384–407 (2018)
23. Greckhamer, T., Furnari, S., Fiss, P.C., Aguilera, R.V.: Studying configurations with qualitative comparative analysis: best practices in strategy and organization research. Strateg. Organ. **16**, 482–495 (2018)
24. Park, Y., El Sawy, O.A., Fiss, P.C.: The role of business intelligence and communication technologies in organizational agility: a configurational approach. J. Assoc. Inf. Syst. **18**, 648–686 (2017)
25. Schneider, C.Q., Wagemann, C.: Set-Theoretic Methods for the Social Sciences: A Guide to Qualitative Comparative Analysis. Cambridge University Press, Cambridge (2012)
26. Bouncken, R.B., Fredrich, V., Kraus, S.: Configurations of firm-level value capture in coopetition. Long Range Plann. (2019, in print)
27. Schreieck, M., Wiesche, M.: How established companies leverage IT platforms for value co-creation – insights from banking. In: 25th European Conference on Information Systems, Guimarães, Portugal (2017)
28. Sia, S.K., Soh, C., Weill, P.: How DBS bank pursued a digital business strategy. MIS Q. Exec. **15**, 105–121 (2016)
29. Hagiu, A., Altman, E.J.: Intuit QuickBooks: From Product to Platform. Harvard Business School Publishing (2014)
30. Saarikko, T., Jonsson, K., Burström, T.: Software platform establishment: effectuation and entrepreneurial awareness. Inf. Technol. People **32**, 579–602 (2019)

31. Bai, G., Zhao, L., Wang, Z.E.: Advantech: evolution of its IoT ecosystem strategy. Emerald Emerg. Markets Case Stud. **8**, 1–28 (2018)
32. Li, Y.: The technological roadmap of Cisco's business ecosystem. Technovation **29**, 379–386 (2009)
33. Collis, D.J., Junker, T.: Digitalization at Siemens. Harvard Business School (2017)
34. Sandberg, J., Holmström, J., Lyytinen, K.: Digital transformation of ABB through platforms: the emergence of hybrid architecture in process automation. In: Urbach, N., Röglinger, M. (eds.) Digitalization Cases. MP, pp. 273–291. Springer, Cham (2019). https://doi.org/10.1007/978-3-319-95273-4_14
35. Iansiti, M., Lakhani, K.R.: Digital ubiquity: how connections, sensors, and data are revolutionizing business. Harvard Bus. Rev. **92**, 1–11 (2014)
36. Kuebel, H., Limbach, F., Zarnekow, R.: Business models of developer platforms in the telecommunications industry - an explorative case study analysis. In: 47th Hawaii International Conference on System Sciences, Waikoloa, HI, USA (2014)
37. Dasi, A., Elter, F., Gooderham, P.N., Pedersen, T.: New business models in-the-making in extant MNCs: digital transformation in a telco. In: Pedersen, T., Devinney, T.M., Tihanyi, L., Camuffo, A. (eds.) Breaking up the Global Value Chain: Opportunities and Consequences, vol. 30, pp. 29–53. Emerald Publishing Limited, Bingley (2017)
38. Zhang, J., Liang, X.-J.: Business ecosystem strategies of mobile network operators in the 3G era: the case of China mobile. Telecommun. Policy **35**, 156–171 (2011)
39. Hilbolling, S.: Organizing ecosystems for digital innovation. Ph.D. Vrije Universiteit Amsterdam, Amsterdam, Netherlands (2018)
40. Mocker, M., Ross, J.: Digital transformation at Royal Philips. In: 39th International Conference on Information Systems, San Francisco, CA, USA (2018)
41. Laborie, F., Røed, O.C., Engdahl, G., Camp, A.: Extracting value from data using an industrial data platform to provide a foundational digital twin. In: Offshore Technology Conference (2019)
42. Lal, R., Johnson, S.: Maersk: Betting on Blockchain. Harvard Business School Publishing (2018)
43. Ofe, H.A., Sandberg, J.: Platform establishment: navigating competing concerns in emerging ecosystems. In: 52nd Hawaii International Conference on System Sciences, Maui, HI, USA (2019)
44. Svahn, F., Mathiassen, L., Lindgren, R.: Embracing digital innovation in incumbent firms: how volvo cars managed competing concerns. MIS Q. **41**, 239–253 (2017)
45. Törmer, R.L.: How the LEGO group is embarking on architectural path constitution to transform its information infrastructure into a digital platform. In: 20th International Conference on Enterprise Information Systems, vol. 2, pp. 649–656. SCITEPRESS, Funchal (2018)
46. Mukhopadhyay, S., Bouwman, H., Jaiswal, M.P.: An open platform centric approach for scalable government service delivery to the poor: the Aadhaar case. Gov. Inf. Q. **36**, 437–448 (2019)
47. Thiem, A., Dusa, A.: QCA: a package for qualitative comparative analysis. R J. **5**, 87–97 (2013)
48. Schneider, C.Q., Wagemann, C.: Standards of good practice in qualitative comparative analysis (QCA) and fuzzy-sets. Comp. Sociol. **9**, 397–418 (2010)
49. Fiss, P.C.: Building better causal theories: a fuzzy set approach to typologies in organization research. Acad. Manag. J. **54**, 393–420 (2011)
50. El Sawy, O.A., Malhotra, A., Park, Y., Pavlou, P.A.: Research commentary—seeking the configurations of digital ecodynamics: it takes three to tango. Inf. Syst. Res. **21**, 835–848 (2010)
51. de Reuver, M., Sørensen, C., Basole, R.C.: The digital platform: a research agenda. J. Inf. Technol. **33**, 124–135 (2018)

Influence of Relationship Capability on Project Performance of Post-merger IS Integration

Julie Yu-Chih Liu[(✉)], Hung-Chin Hsu, and Jun-Lin Lin

Department of Information Management, Yuan Ze University,
Chung-Li, 320, Taoyuan City, Taiwan
{imyuchih,jun}@saturn.yzu.edu.tw, s1056218@mail.yzu.edu.tw

Abstract. Mergers and acquisitions (M&A) highly rely on the information technology systems integration (ISI) of two companies to realize the operation and resource synergy. Given the vast cost of M&A and the high failure rate of the post-merger ISI, various strategies, enforcement mechanisms, and potential determinants for the ISI success have been proposed. The capability of IS managers is one of the recognized factors that are critical to ISI success. This study extends the theory of relationship quality to explain how the relationship ability of IS personnel influences the outcomes of the ISI in the M&A context. The empirical survey of IS personnel shows that relationship ability improves not only relationship quality but also teamwork quality of the ISI. The findings shed light on the different functions of relationship capability in the managerial design and imply the significance of relational building ability as an important skill of IS personnel for achieving post-merger ISI success.

Keywords: Mergers and acquisitions · IS integration · Teamwork quality · Relationship quality · Management performance

1 Introduction

Mergers and acquisitions (M&A) are regarded as important corporate strategies to increase the competitive advantages and market growth of organizations. M&A enables saving cost, synergy benefit, and business diversification [1] The potential generation of these values explains the phenomenon of an increasing trend of M&A. The global deal struck in M&A totaled $3.4 trillion in 2014, up about 30% from 2013 [2], and reached more than US $3.7 trillion in 2016 [3]. Top executives in the M&A corporations often expect the integration of the resources from both sides can promote economies of scale or create organizational value [4]. Abound evidence in the literature shows that the expected benefits from the merger are often not achieved [5]. More than 60% of all M&A in the private sectors failed to create financial value for its shareholders as anticipated [6].

According to McKinsey's industry report[1], the expected benefits of M&A are around 45–60%, directly dependent on the effectiveness of IS integration. Acquirers imperatively rely on the integration of information technology systems (IS) of two companies to

---

[1] Report from McKinsey, global management consulting firm.

© Springer Nature Switzerland AG 2020
W. Abramowicz and G. Klein (Eds.): BIS 2020, LNBIP 389, pp. 157–166, 2020.
https://doi.org/10.1007/978-3-030-53337-3_12

realize overall efficiency and expected benefits from M&A. Post-merger IS integration (ISI) support information sharing, acquire business operation synergies, facilitate regulatory compliance, and reduce the monitor costs [7]. Ineffective ISI would cause business instability, customer dissatisfaction, human costs, and personal loss, resulting in the expected benefits delayed or unrealized. For example, in 2012, the merge of United with Continental Airlines failed to effectively integrate two company's reservation system, resulting in website shutting down, airport kiosks banned, and flights delayed or canceled [8]. Many cases have shown that the inability to smoothly and effectively integrating IS will cause short- or long-term damage to business running and reputation. Thus, effective ISI is a crucial factor in achieving expected benefits from M&A implementation [9, 10].

However, the ISI has a higher failure rate in the M&A context. Up to 90% of ISI projects fail to deliver the outcomes that meet the expectation [11]. The post-merger ISI includes not only the integration of systems (functions, software packages, and hardware) but also the coordinated integration of the organization (operational procedures and information technology departments). Studies show that the ISI success depends on the organizational property, organizational consolidation management, and integration activities, information synergy, IS compatibility, and personnel changes [12]. The managers need to integrate two IS develop teams, deal with different organizational cultures, adjust business processes[13]. For many M&A cases, heterogeneous IT systems and distinct business processes complicate their ISI [7]. The intertwine of technical and social issues makes the ISI troublesome. Unforeseen challenges usually surface at the process of ISI, and the unsolved obstruction delays the anticipated outcome [13].

On the other hand, organizational identity, cultural conflict and turnover harm individual performance [5]. Employees need to face prompt uncertainty and have less attachment to the new organization compared with the original smaller subunits at the beginning[14]. Because of underestimating these issues, top executives' often have the unpractical expectation of post-merger ISI [1].

Given the significant benefits of resource consolidation and the high failure rate of post-merger ISI, various strategies and enforcement mechanisms for post-merger ISI have been proposed [12, 15]. The strategies, including absorption, coexistence, best of the breed, renewal [15], focus on how to consolidate two companies' ISs. The strategies do not always achieve the desired results, and which implies that social issues, such as enterprise heterogeneity, relationship, and interest conflicts, could be the real factors of raising challenges. Researchers also dedicated to exploring the potential variables that can explain the outcome of post-merger ISI. They investigated the variables from both organizational issues and IS ones [11, 15, 16]. Among the IS issues, IS capacities are regarded as an important factor in determining the success of the ISI. The ability of IS managers on the planning of ISI and the communication between the IS and user influence the ISI progress and user satisfaction [17].

However, prior research has not paid attention to the possible influence of the relationship capacity of IS personnel on the ISI performance. This omission is surprising as the IS relationship capability has been recognized as a significant determinant of team performance [18]. The relationship quality is more important in the M&A context because the IS teams for two corporations confront more complex relationship building when having dilemma roles between competition and cooperation. Therefore, this research aims to empirically investigate whether IS relationship capability influences the outcome of ISI in the M&A context. Specifically, two primary questions were examined: Whether IS team members' relationship capability influence the relationship quality of the post-merger ISI team? Whether IS team members' relationship capability influence the teamwork quality of the post-merger ISI team? (Fig. 1)

Fig. 1. Proposed research model

2 Hypotheses

IS personnel and users usually have different opinions or suggestions to the ISI approach or strategy adopted by the new company. IS personnel generally favor to preserve the existing systems and resistance to the new one with different IT technique or business process [19]. Their requirement or comment on the ISI strategy and the burden coming from the ISI influence their willingness to participate, support, and cooperate. Complicated this further is the uncertainty of the structure, position, and authority in IS. The uncertainty raise IS personnel's interest in conflict and distrust. Conflict is one of the significant risk factors of IS project delay, cost overrun or user dissatisfaction [20]. The relationship capability of IS personnel in communication capability, cultural compatibility, information sharing, flexibility, conflict management would potentially release the risk from team members' conflict.

In the context of M&A, the IS departments of both corporations often have a considerable change in terms of personnel, structure and tasks, such as personnel transfer, change of authority. The team members in M&A ISI will have a higher possibility for role change and uncertainty than those teams working for ISI in a single organization. An unfamiliar environment and potential change in position and responsibility cause high future uncertainty to employees. The IS personnel may encounter the social identity problem [5]. This potentially generates stress, anxiety and conflict [5]. Particularly, the IS supervisors of two parties are cooperators for IS development but also a competitor for authority and position in the corporation. The role disagreement will trap the IS personnel into contradictory phycology and ill interaction. The contradictory phycology reduces team members' commitment and dedication to work, increasing risk factors for project development and leading to an unexpected outcome [13].

Building a good relationship is a significant approach for individuals from different parties to create a common goal [21]. Good relationship among team members can provide task flexibility during project development and enhance the opportunities and quality of follow-up cooperation. Good relationships capability depend on multiple skills, including communication, cultural compatibility, information sharing, conflict management, and more [22]. These abilities may release the IS team members' stress, anxiety and conflict coming from social identity and contradictory phycology, and increase their trust, interaction, mutual understanding and cooperation, which are parts of major dimensions of team relationship quality.

Some evidence has shown a significant correlation between *relationship quality* and team performance in the outsourcing context [18]. The information sharing among team members facilitates developers effectively obtaining the knowledge about the systems to be integrated. Team members' trust in each other can reduce their conflict toward the system design or functions during the integration process. Mutual understanding can help them to target the expertise of team members coming from the merged sectors. The relationship quality leads to successful long-term relations that are operationally beneficial to overcome the difficulty of integrating the systems developed by and used for other sectors. Research indicates that the relationship between project managers and stakeholders expend an influence on the project processes and outcomes and ultimate project success [23]. Based on this viewpoint, to the ISI project team, the relationship quality may reduce the harm coming from their cultural conflict and organizational identity, as well as the uncertainty and less attachment to the new team. Therefore, we hypothesize that

H1: IS relationship capability is positively associated with the relationship quality of post-merger ISI team.
H2: IS relationship quality is positively associated with post-merger ISI project success.

Teamwork quality (TWQ) is an essential factor in achieving project success in IS development teams [24]. Researchers regard TWQ as the quality of interactions among group members for completing their team works [25]. TWQ is conceptualized a multidimensional construct of group behavior, consisting of coordination, mutual support, effort, the balance of contribution, and cohesion, initiated by Hackman [26] and derived from McGrath [27].

Research has found a significant relationship between IS management ability and TWQ [28]. Communication capability enables effective communication among team members. Effective communication is a channel of information sharing and exchange for specific ISI details. Information sharing and exchange facilitate team members to learn from each other and perform their tasks [29]. Positive communication and interaction increase the opportunity of the team member to find out the potential risk or problems at the early stage and provide effort and support to resolve the challenge of ISI.

As the post-merger ISI team coming from two corporations, the difference in organizational structure, culture and business and the conflict in individual benefit, position, and authorization raise the difficulty for managers to develop TWQ. The relationship-building ability of IS staffs includes communication capability, cultural compatibility, information sharing, flexibility, conflict management. The ability can obviously promote their coordination, mutual support, effort, and cohesion in the team. Communication ability is an intra-team and inter-team interpersonal skill that has been regarded as an important element of teamwork quality [30]. Abound evidence has shown a strong association between TWQ and team performance [28].

To integrate information systems successfully, team members must put effort and learn from other team members toward their expertise and knowledge about the original systems to together effectively and efficiently understand the design and details of the systems. On the other hand, putting effort and learning other team members is essential to understand the systems developed by these team members. Mutual support enables team members together to overcome the complication of system integration to achieve the project goal. Researchers have empirically demonstrated the relationship between the interaction quality and information system project team [31]. Several studies have also shown TWQ is critical to software project success [24]. Therefore, we hypothesize that

H3: *IS relationship capability is positively associated with the teamwork quality of post-merger ISI team.*
H4: *IS teamwork quality is positively associated with post-merger ISI project success.*

3 Methods

3.1 Sample

The target sample includes IS developers, users and project managers who are belong to the ISI teams in M&A corporations. A cross-sectional internet survey was conducted. A total of 44 valid responses were obtained (Table 1).

Table 1. Demographics (N = 44)

Variables	Categories	#	%
Gender	Male	32	73
	Female	12	27
Age	<30 years old	13	30
	31–40	11	25
	41–45	10	23
	>46	10	23
Years in IT/IS	<1 year	8	18
	1–3	9	20
	4–6	7	16
	7–9	2	5
	>10 years	18	41
Fist M&A	Yes	33	75
	No	11	25

3.2 Measures

All of the considered variables were measured using a five-point scale. The conceptual definition and operational definitions of relationship capability and relationship quality are based on Swar's definitions [18]. We adopt the construct of teamwork quality (TWQ) defined by Hoegl and et al. [24]. All the constructs in the proposed model are 2nd-order variables.

The dimensions of relationship capability considered in this study include communication, cultural compatibility, confidentiality flexibility, information sharing, conflict handling. Communication is measured by the degree of proactive sharing or exchanging meaningful and timely information between team members from two parties in formal or informal. Cultural compatibility is defined by the coexistence with beliefs on values of each other, behaviors, goals, and policies, which could be the perspective of importance, appropriateness, and righteousness. Confidentiality refers to the degree of avoiding provide data or information to be disclosed to unauthorized or unwanted people. Flexibility is measured by the extent to which team members willing to make adaptations as circumstances change for two parties. Information sharing is measured by the degree of the critical or proprietary task information and know-how that are transferred. Conflict handling refers to the degree of disagreement on planning, execution, activities, resources allocation, and goals between the team members from two partners.

Relationship quality considered in this study includes cooperation, trust and mutual understanding. Cooperation refers to spiritly working together by two parties on project activities to achieving mutual benefits. Trust is measured by the extent of which a party is expected to act predictably, fulfill its obligations, and honestly behave even if existing the possible opportunity to take advantage of the others. Mutual understanding is defined as the understanding of behaviors, goals, and policies between the team members from two parties. The ISI success is adapted from the work of Swar et al. [18] and measured by user satisfaction and the usefulness of the ISI outcome. The example indicators are users are satisfied with the outcome of the ISI and you feel the integrated IS useful.

Table 2 shows descriptive statistics for the 2nd-order constructs, where M3 and M4 respectively represent skewness and kurtosis. The measures have proper distribution for the performance measures since the means and medians are similar, standard deviation (StdDev) is small, skewness is less than two, and kurtosis less than five.

Table 2. Descriptive statistics

Variables	Mean	StdDev	M3	M4
Relationship capability	4.08	0.84	2.27	−1.17
Relationship quality	4.05	0.81	0.47	−0.7
Teamwork quality	4.16	0.81	1.89	−1.13
IS integration success	3.93	0.88	0.17	−0.56

4   Data Analysis

We adopted the partial least squares (PLS) analysis to test the measurement and structural models using the smart-PLS tool. PLS places minimal demands on sample size and distribution of residuals [32]. Based on a simple rule of sample size in PLS-SEM, the sample size should be at least ten times the maximum number of inner or outer model links pointing at any latent variable in the model. Thus, the sample size should be at least forty to prevent biased results. The standardized coefficients can illustrate the relative strength of the statistical relationships among the variables. Reliability and validity are demonstrated through measures of internal reliability, convergent, and discriminant validity. Convergent validity was assessed by examining the loading of each item on the corresponding factors. The internal reliability of the examined constructs is obtained by estimating composite reliability and Cronbach's Alpha. Composite reliability and Cronbach's Alpha of 0.7 or greater is acceptable. Results are shown in Table 3, demonstrating the criteria of the validity and reliability tests are met.

Table 3. Internal reliability

Variables	Cronbach's α	Com. Rel.	AVE
Communication capability	0.96	0.96	0.56
Conflict handling	0.73	0.84	0.64
Confidence maintaining	0.9	0.94	0.83
Coordination	0.93	0.96	0.88
Culture compatibility	0.69	0.83	0.62
Flexibility	0.91	0.94	0.84
Information sharing	0.9	0.93	0.76
Coordination	0.93	0.96	0.88
Trust	0.87	0.92	0.79
Mutual understanding	0.82	0.89	0.73
Mutual support	0.92	0.94	0.74
Effort	0.90	0.94	0.84
Cohesion	0.94	0.95	0.70
Learning	0.87	0.92	0.79
ISI success	0.81	0.89	0.73

Table 4 shows the path coefficients of the structure model. The links from relationship capability to relationship quality and teamwork quality are respectively 0.875 ($p < 0.001$) and 0.935 ($p < 0.001$). The links from relationship quality and teamwork quality to ISI success are respectively 0.296 ($p < 0.05$) and 0.598 ($p < 0.001$). Relationship quality and teamwork quality together explain 75% of the variance of ISI success. The analysis results show that the path coefficient between relationship capability and teamwork quality is even higher than that between relationship capability and relationship quality. Moreover, compared with relationship quality, teamwork quality has a better explanation in the variance of ISI success. Discriminant validity was tested by examining the correlation coefficients of each item within and among constructs.

Table 4. Path coefficient

Independent variables	Dependent variables		
	RQ	TWQ	ISI success
Relationship capability	0.875*** (H1)	0.935*** (H3)	
Relationship Quality (RQ)			0.296* (H2)
Teamwork Quality (TWQ)			0.598*** (H4)
R^2	0.774	0.879	0.750
ΔR^2	0.769	0.876	0.738

*$P < 0.05$ **$P < 0.01$ ***$P < 0.001$

5 Conclusion

The present survey found IS personnel's relationship capability not only significantly influence relationship quality but also teamwork quality of the ISI team. The effect of teamwork quality on the ISI success was higher than for that of relationship quality. The contributions of this study are threefold. First, the study confirms the theoretical basis of relationship quality theory in post-merger ISI. Despite the emphasis on relationship quality in the outsourcing IS project, in the M&A and the outsourcing surveys alike, the influences of relationship quality on ISI and outsourcing success were similar. Second, it extends the perspective on ISI success off system strategy angle, shedding light on the different functions of relationship capability in facilitating managerial design. Third, the analysis results echo the proclaim of other M&A researchers that IS capabilities play an important role in the post-merger ISI success. In addition to IS managers' process and planning abilities, this study adds relationship capability as one of the essential elements of IS personnel competence.

Some future works need to be done. The first is to increase the sample size for testing the proposed model. Although sample size 44 is larger than 10 times the number of links, it may produce the results with bias because of the 2nd-order constructs used in this research. This work will increase the sample size in the near future to reconfirm the findings of this work. The second is to conduct a mediation analysis for both mediators (i.e., relationship quality and teamwork quality) after obtaining a sufficient sample size.

Acknowledgments. This work was supported by the MOST (Grant No. 108-2410-H-155 -038).

References

1. Hedman, J., Sarker, S.: Information system integration in mergers and acquisitions: research ahead. Eur. J. Inf. Syst. **24**(2), 117–120 (2015)
2. Raice, C.: Broken deals rein in a strong M&A market. Wall Street J. (2015). https://www.wsj.com/articles/broken-deals-rein-in-a-strong-m-a-market-1420155330. Accessed 29 Nov 2019
3. Mergers & Acquisitions review- full year 2015. http://share.thomsonreuters.com/general/PR/MA_4Q_2016_E.pdf. Accessed 28 Nov 2019
4. Tanriverdi, H., Uysal, V.B.: When IT capabilities are not scale-free in merger and acquisition integrations: how do capital markets react to IT capability asymmetries between acquirer and target? Eur. J. Inf. Syst. **24**(2), 145–158 (2015)
5. Weber, Y.: The role of organizational identity in post-merger integration. In: Thrassou, A., Vrontis, D., Weber, Y., Shams, S.M.R., Tsoukatos, E. (eds.) The Synergy of Business Theory and Practice. PSCBRIAEAB, pp. 91–107. Springer, Cham (2019). https://doi.org/10.1007/978-3-030-17523-8_5
6. Haleblian, J., Devers, C.E., McNamara, G., Carpenter, M.A., Davison, R.B.: Taking stock of what we know about mergers and acquisitions: a review and research agenda. J. Manag. **35**(3), 469–502 (2009)
7. Tanriverdi, H., Uysal, V.B.: Cross-business information technology integration and acquirer value creation in corporate mergers and acquisitions. Inf. Syst. Res. **22**(4), 703–720 (2011)
8. Mouawad, J.: For united, big problems at biggest airline. New York Times (2012). http://www.nytimes.com/2012/11/29/business/united-is-struggling-two-years-after-its-merger-with-continental.html. Accessed 29 Nov 2019
9. Yetton, P., Henningsson, S., Bjorn-Andersen, N.: Ready to acquire': IT resources for a growth-by-acquisition strategy. MIS Q. Exec. **12**(1), 19–35 (2013)
10. Stylianou, A.C., Jeffries, C.J., Robbins, S.S.: Corporate mergers and the problems of IS integration. Inf. Manag. **31**(4), 203–213 (1996)
11. Chang, Y.B., Cho, W.: The risk implications of mergers and acquisitions with information technology firms. J. Manag. Inf. Syst. **34**(1), 232–267 (2017)
12. Henningsson, S., Kettinger, W.J.: Understanding information systems integration deficiencies in mergers and acquisitions: a configurational perspective. J. Manag. Inf. Syst. **33**(4), 942–977 (2016)
13. Lohrke, F.T., Frownfelter-Lohrke, C., Ketchen Jr., D.J.: The role of information technology systems in the performance of mergers and acquisitions. Bus. Horiz. **59**(1), 7–12 (2016)
14. Van Knippenberg, D., Van Knippenberg, B., Monden, L., de Lima, F.: Organizational identification after a merger: a social identity perspective. Br. J. Soc. Psychol. **41**(2), 233–252 (2002)
15. Wijnhoven, F., Spil, T., Stegwee, R., Fa, R.T.A.: Post-merger IT integration strategies: an IT alignment perspective. J. Strateg. Inf. Syst. **15**(1), 5–28 (2006)
16. Baker, E.W., Niederman, F.: Integrating the IS functions after mergers and acquisitions: analyzing business-IT alignment. J. Strateg. Inf. Syst. **23**(2), 112–127 (2014)
17. Robbins, S.S., Stylianou, A.C.: Post-merger systems integration: the impact on IS capabilities. Inf. Manag. **36**(4), 205–212 (1999)
18. Swar, B., Moon, J., Oh, J., Rhee, C.: Determinants of relationship quality for IS/IT outsourcing success in public sector. Inf. Syst. Front. **14**(2), 457–475 (2012). https://doi.org/10.1007/s10796-010-9292-7
19. Jiang, J.J., Klein, G., Chen, H.-G.: The effects of user partnering and user non-support on project performance. J. Assoc. Inf. Syst. **7**(1), 6 (2006)
20. Liu, J.Y.-C., Chiu, G.C.-T.: Influence of project partnering on stakeholder role ambiguity and project manager risk perception in information system projects. Proj. Manag. J. **47**(6), 94–110 (2016)

21. Lee, J.N., Kim, Y.G.: Effect of partnership quality on IS outsourcing success: conceptual framework and empirical validation. J. Manag. Inf. Syst. **15**(4), 29–61 (1998)
22. Alaranta, M., Henningsson, S.: An approach to analyzing and planning post-merger IS integration: insights from two field studies. Inf. Syst. Front. **10**(3), 307 (2008). https://doi.org/10.1007/s10796-008-9079-2
23. Sutterfield, J.S., Friday-Stroud, S.S., Shivers-Blackwell, S.L.: A case study of project and stakeholder management failures: lessons learned. Proj. Manag. J. **37**(5), 26–35 (2006)
24. Hoegl, M., Weinkauf, K., Gemuenden, H.G.: Interteam coordination, project commitment, and teamwork in multiteam R&D projects: a longitudinal study. Organ. Sci. **15**(1), 38–55 (2004)
25. Lindsjørn, Y., Sjøberg, D.I., Dingsøyr, T., Bergersen, G.R., Dybå, T.: Teamwork quality and project success in software development: a survey of agile development teams. J. Syst. Softw. **122**, 274–286 (2016)
26. Hackman, J.R.: The design of work teams. In: Lorsch, J.W. (Ed.) Handbook of Organizational Behaviour. The Design of Work Teams, pp. 315–342. Prentice-Hall, Englewood Cliffs (1987)
27. McGrath, J.E.: Social Psychology: A Brief Introduction. Holt, Rinehart and Winston, New York (1964)
28. Faraj, S., Sproull, L.: Coordinating expertise in software development teams. Manag. Sci. **46**(12), 1554–1568 (2000)
29. Hsu, J.S.-C., Shih, S.-P., Chiang, J.C., Liu, J.Y.-C.: The impact of transactive memory systems on IS development teams' coordination, communication, and performance. Int. J. Proj. Manag. **30**(3), 329–340 (2012)
30. Stevens, M.J., Campion, M.A.: Staffing work teams: development and validation of a selection test for teamwork settings. J. Manag. **25**(2), 207–228 (1999)
31. Wang, E.T., Chen, H.H., Jiang, J.J., Klein, G.: Interaction quality between IS professionals and users: impacting conflict and project performance. J. Inf. Sci. **31**(4), 273–282 (2005)
32. Kroonenberg, P.M.: Latent variable path modeling with partial least squares. J. Am. Stat. Assoc. **85**(411), 909–911 (1990)

Applications

Detecting Tax Evaders Using TrustRank and Spectral Clustering

Priya Mehta[1], Jithin Mathews[1], Dikshant Bisht[1], K. Suryamukhi[1], Sandeep Kumar[2], and Ch Sobhan Babu[1(✉)]

[1] Indian Institute of Technology Hyderabad, Sangareddy, India
{cs15resch11007,cs15resch11004,cs17mtech11027,cs17mtech01002,
sobhan}@iith.ac.in
[2] Plianto Technologies, Sangareddy, India
cs15mtech11017@iith.ac.in

Abstract. Indirect taxation is a significant source of livelihood for any nation. Tax evasion inhibits the economic growth of a nation. It creates a substantial loss of much needed public revenue. We design a method to single out taxpayers who evade indirect tax by dodging their tax returns. Towards this, we derive six correlation parameters (features), three ratio parameters from tax return statements submitted by taxpayers, and another parameter based on the business interactions among taxpayers using the TrustRank algorithm. Then we perform spectral clustering on taxpayers using these ten parameters (features). We identify taxpayers located at the boundary of each cluster by using kernel density estimation, which are further investigated to single out tax evaders. We applied our method on the iron and steel taxpayer's data set provided by the Commercial Taxes Department, Government of Telangana, India.

Keywords: Cluster analysis · TrustRank algorithm · Spectral clustering · Social network analysis · Tax evasion · Goods and Services Tax

1 Introduction

Indirect taxation is a significant source of livelihood for any country. An indirect tax (for example, GST and sales tax) is collected by an intermediary (such as retailer and manufacturer) from the consumer who bears the ultimate burden of the tax. The intermediary forwards the tax he/she collected to the Government by filing tax returns at regular intervals. Indirect taxation is governed by carefully designed rules and regulations to which an intermediary (taxpayer/dealer) is expected to adhere. At the outset, indirect tax payment may seem as a liability on the intermediary, but in reality, the intermediary acts as a conduit to the flow of tax from the consumer of the goods/services to the Government. These taxes provide much-needed revenue for the growth of the nation. In this article, we work towards handling tax evasion happening in *Goods and Services Tax (GST)*, which is an indirect taxation system followed in India from July 2017 [20].

© Springer Nature Switzerland AG 2020
W. Abramowicz and G. Klein (Eds.): BIS 2020, LNBIP 389, pp. 169–183, 2020.
https://doi.org/10.1007/978-3-030-53337-3_13

1.1   Goods and Services Tax

Goods and Services Tax (GST) is a comprehensive, multi-stage, destination-based tax that is levied on every value addition [20]. In GST, the tax is collected incrementally at each stage of the production of the goods based on the value addition happened in that stage. This tax is levied at each stage of the supply chain such a way that the tax paid on purchases (*Input tax* or *Input tax credit*) will be given as set-off for the tax levied on sales (*Output tax* or *Liability*).

Figure 1 shows how incremental tax is collected at each stage of the supply chain. In this example, the manufacturer purchases goods from the supplier for a value of $100 and pays $10 as tax at a 10% GST rate. The supplier then pays this tax to the Government. In the next stage, the retailer purchases processed goods from a manufacturer for a value of $120 by paying tax $12. The manufacturer pays ($12−$10 = $2) to the Government, the difference between the tax collected from the retailer and the tax already paid to the supplier. Finally, the consumer buys it from the retailer for a value of $150 with $15 as the tax amount. So the retailer, as in the case with the previous dealers in the supply chain, will pay ($15 − $12 = $3) to the Government. In essence, for every dealer in GST,

$$Tax\,payable = (Output\,tax - Input\,tax) \qquad (1)$$

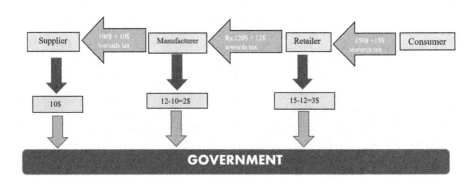

Fig. 1. Tax flow in GST

1.2   Tax Evasion

Tax evasion and taxation go hand in hand. Business dealers deliberately manipulate their monthly tax returns in order to maximize their profits. Tax enforcement officers design new rules and regulations involved in the payment of tax after studying the behavior of known evaders who exploit the loopholes in the existing taxation laws. In this never-ending cat and mouse game, evaders always try to stay one step ahead of the enforcement officers. Hence it is very important for the officials to track down the evasion as quickly as possible and close the loopholes before the techniques used in a particular evasion spreads to the other dealers. By doing so, the taxation officers will be able to limit the loss of state revenue due to tax evasion quite substantially.

In general, dealers commit tax evasion in the following ways:

1. The dealer will collect tax at a higher rate from the customer and remits to the Government at a lower rate.
2. The dealer does not report all the transactions made by her/him (suppressing the sales).
3. The dealer will arrive at lower taxable turnover by wrongly applying the prescribed calculations.
4. The dealer creates fictitious transactions where there is no movement of goods, but only the bills are floated in order to claim Input Tax Credit (ITC) and escape payment of tax. This is called bill trading [2].

1.3 Our Contribution

Goods and Services Taxation system, which came into effect in India in July 2017, unified the taxation laws in India. Under GST law, the dealers are expected to file tax-return statements every month by providing the details of sales and purchases that happened in the corresponding month. This new taxation system went through some teething problems, which were sensed and exploited by many business dealers. In fact, dealers who observed certain loopholes in the new system doctored their tax-return statements to reduce their tax liability. The objective of this work is to identify malicious dealers who manipulate their tax return statements to minimize their tax liability. For the same, clustering analysis is used on the dealers over specific sensitive correlation parameters and ratio parameters that are identified by the tax enforcement officers. We also used a parameter derived based on the link analysis algorithm, famously known as the *TrustRank algorithm* [8], in clustering. This idea can be applied in other nations, where multi-stage indirect taxation is followed.

The rest of the paper is organized as follows. In Sect. 2, we discuss the previous relevant works. In Sect. 4, we give a detailed description of the methodology used in this paper. Results obtained and the validation of the results are discussed in Sect. 5.

2 Related Work

In [6], Chandola et al. presented several data mining techniques for anomaly detection. In [1], Shuhan Yuan et al. used a deep neural network approach for fraud detection, which has its limitation as it will work only for the labeled data set. In [10], Hussein Issa and Miklos Vasarhelyi described classification-based and clustering-based anomaly detection techniques and their applications. As an illustration, they applied K-Means, a clustering-based algorithm, to a refund transaction dataset from a telecommunication company, with the intent of identifying fraudulent refunds. In [9], the authors proposed an algorithm to identify colluding sets in the instrument of future markets. In [12], Pamela et al. showed that it is possible to characterize and detect those potential users of false invoices in a given year, depending on the information in their tax payment, their historical performance, and characteristics, using different types of data mining techniques. In [11], Daniel de Roux et al. presented a novel approach for the detection of potentially fraudulent taxpayers using only unsupervised learning techniques and allowing the future use of supervised learning techniques. They demonstrated the ability of their model to identify under-reporting taxpayers on real tax payment declarations, reducing the number of potentially fraudulent taxpayers to audit. The obtained results demonstrate that their model doesn't miss on marking declarations as suspicious and labels previously undetected tax declarations as suspicious, increasing the operational efficiency in the tax supervision process without needing historic labeled data. In another literature [8], Jan Pedersen et al. proposed techniques to semi-automatically separate reputable, useful pages from spam. They first selected a small set of seed pages to be evaluated by an expert. Once they manually identified the reputable seed pages, they used the link structure of the web to discover other pages that are likely to be good. In this paper, they discussed possible ways to implement the seed selection and the discovery of useful pages. They presented results of experiments ran on the World Wide Web indexed by AltaVista and evaluated the performance of their techniques. Their results showed that we could effectively filter out spam from a significant fraction of the web, based on a good seed set of fewer than 200 sites. In [5], J. Mathew et al. gave an approach to predict the amount of tax lost by the State Government due to illegal activities by a set of suspicious dealers. In [3], the authors showed that they have comparatively lower effective tax rates if clients are engaged in better-connected individual auditors. In [7], Yusuf Sahin et al. worked on a credit card fraud detection problem where they developed some classification models based on Artificial Neural Network (ANN) and Logistic Regression (LR). This study is one of the first in credit card fraud detection with a real data set to compare the performance of ANN and LR. In [4], Zhenisbek Assylbekov et al. presented a technique employing statistical methods for detecting VAT evasion by Kazakhstani business firms. Starting from features selection they performed an initial exploratory data analysis using Kohonen self-organizing maps. In another literature [14], M. S. Rad et al. presented a parallel tax fraud detection algorithm where Bayesian networks have been used for parallelization.

3   Description of Data Set

We used two types of data sets of 1199 iron and steel taxpayers. One is GSTR-1 data, and the other is the monthly GST returns data.

GSTIN/ UIN	Invoice details			Rate	Taxable value	Amount				Place of Supply (Name of State)
	No.	Date	Value			Integrated Tax	Central Tax	State / UT Tax	Cess	
1	2	3	4	5	6	7	8	9	10	11
4A. Supplies other than those (i) attracting reverse charge and (ii) supplies made through e-commerce operator										
4B. Supplies attracting tax on reverse charge basis										
4C. Supplies made through e-commerce operator attracting TCS (operator wise, rate wise)										
GSTIN of e-commerce operator										

Fig. 2. GSTR-1 format

3.1   GSTR-1

GSTR-1 Overview: GSTR-1 is a monthly financial statement that should be submitted by every taxpayer. In this statement, the taxpayer has to provide complete details about every sales transaction done in the corresponding month. GSTR-1 is the monthly financial statement upon which the entire compliance structure in GST is based. Figure 2 gives a set of details to be provided by the taxpayer for every sales transaction.

GSTR-1 Data Set: Table 1 is a sample of this data set. Every row in Table 1 corresponds to one sales transaction. Every row contains information such as seller details, buyer details, invoice number, invoice value, rate of tax, the quantity of goods sold, *etc.* The data set we have taken contains fifteen million rows. The size of the data is 1.2 Terabytes. Figures 3, 4 show the distribution of values of sales transactions and tax on those transactions.

Table 1. GSTR-1 DATA

S. No	Month	Seller	Buyer	Invoice number	Amount (Rs)
1	Jan 2018	A	B	XY123	12000
2	Feb 2018	B	D	ZU342	18000
3	Jan 2018	B	C	UX5434	14000
4	July 2018	C	D	YS8779	15000
5	Mar 2018	D	A	ZX7744	12000

3.2   GSTR-3B

GSTR-3B Overview: GSTR-3B is a monthly self-declaration that has to be filed by the taxpayer. It is a simple return in which a summary of outward

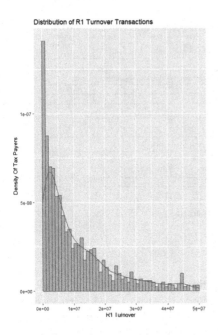

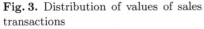

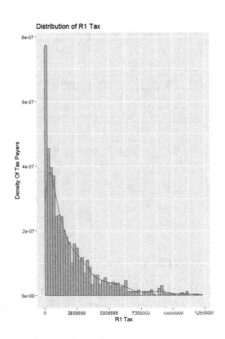

Fig. 3. Distribution of values of sales transactions

Fig. 4. Distribution of tax on sales transactions

supplies along with Input Tax Credit is declared, and payment of tax is affected by the taxpayer. The taxpayer need not provide invoice level details in this form. The taxpayer provides only a summary of inward supplies and outward supplies of the corresponding month. Figure 5 gives the details to be provided by the taxpayer.

GSTR-3B Data Set: Table 2 is a sample of GST returns data. Each row in this table corresponds to a monthly return by a taxpayer. *ITC (Input tax credit)* is the amount of tax the taxpayer paid during purchases of services and goods. The *output tax* is the amount of tax the taxpayer collected during the sales of services and goods. The taxpayer has to pay the Government the gap between the *ITC* and *output tax*, i.e., output tax - ITC. The actual database consists of much more information, like, tax payment method, return filing data, international exports, exempted sales, and sales on RCM (reverse charge mechanism). Figures 6, 7, 8, and 9 show the distribution of turnover, liability, input tax credit, and cash payments.

4   Methodology

The objective is to identify malicious dealers who manipulate their tax return statements to minimize the amount of tax they have to pay to the Government.

FORM GSTR-3B

Year
Month

1. GSTIN
2. Legal name of the registered person

3.1 Details of Outward Supplies and inward supplies liable to reverse charge

Nature of Supplies	Total Taxable value	Integrated Tax	Central Tax	State/UT Tax	Cess
1	2	3	4	5	6
(a) Outward taxable supplies (other than zero rated, nil rated and exempted)					
(b) Outward taxable supplies (zero rated)					
(c) Other outward supplies (Nil rated, exempted)					
(d) Inward supplies (liable to reverse charge)					
(e) Non-GST outward supplies					

3.2 Of the supplies shown in 3.1 (a) above, details of inter-State supplies made to unregistered persons, composition taxable persons and UIN holders

	Place of Supply (State/UT)	Total Taxable value	Amount of Integrated Tax
1	2	3	4
Supplies made to Unregistered Persons			
Supplies made to Composition Taxable Persons			
Supplies made to UIN holders			

4. Eligible ITC

Details	Integrated Tax	Central Tax	State/UT Tax	Cess
1	2	3	4	5
(A) ITC Available (whether in full or part)				
(1) Import of goods				
(2) Import of services				
(3) Inward supplies liable to reverse charge (other than 1 & 2 above)				
(4) Inward supplies from ISD				
(5) All other ITC				
(B) ITC Reversed				
(1) As per rules 42 & 43 of CGST Rules				
(2) Others				
(C) Net ITC Available (A) - (B)				
(D) Ineligible ITC				
(1) As per section 17(5)				
(2) Others				

5. Values of exempt, nil-rated and non-GST inward supplies

Nature of supplies	Inter-State supplies	Intra-State supplies
1	2	3
From a supplier under composition scheme, Exempt and Nil rated		

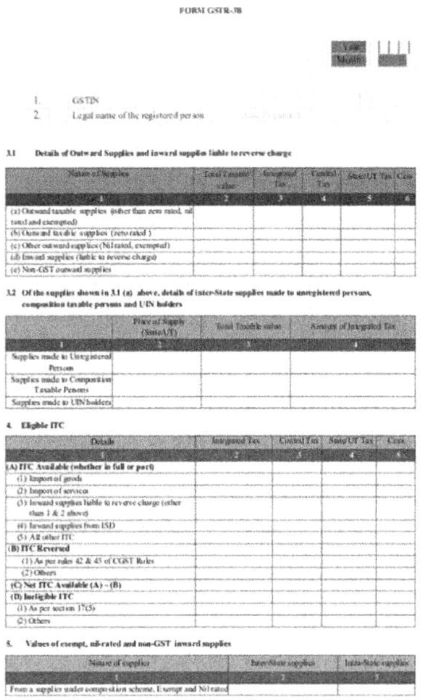

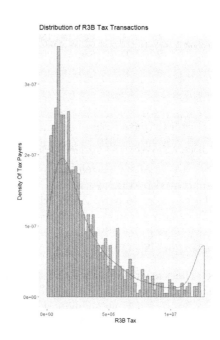

Fig. 5. GSTR-3B format

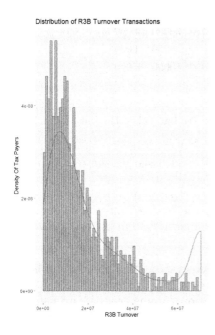

Fig. 6. Distribution of turnover

Fig. 7. Distribution of liability

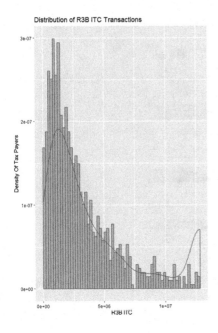

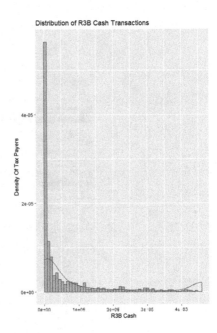

Fig. 8. Distribution of input tax credit   **Fig. 9.** Distribution of cash payments

Table 2. GST returns data

S.No	Firm	Month	Purchases	Sales	ITC	Output Tax
1	A	Feb-18	180000	220000	20000	26000
2	D	Sep-18	200000	280000	5000	9000
3	E	Oct-17	400000	480000	40000	48000

We cluster taxpayers based on ten parameters. We had taken the data from July 2017 to November 2019 to compute these.

- Six are sensitive correlation parameters derived based on the returns data submitted by taxpayers.
- Three are ratio parameters, which are also derived from returns data.
- The tenth parameter is derived based on the business interactions among taxpayers using a link analysis algorithm famously known as the *TrustRank algorithm* [8].

Once clusters are identified, we had taken taxpayers at the boundary of each cluster for further analysis towards identifying malicious taxpayers.

In Subsect. 4.1, we explain the six correlation parameters that are used in clustering. In Subsect. 4.2, we describe the three ratio parameters that are used in clustering. In Subsect. 4.3, the tenth parameter based on the business interactions among taxpayers derived using a link analysis algorithm famously known as

TrustRank is discussed. In Subsect. 4.4, we brief on the Spectral clustering [11] algorithm, which gave us the best clustering results when compared to other clustering techniques.

4.1 Correlation Parameters

In the Indian GST system, three types of taxes are collected, *viz.*, CGST, SGST, and IGST.

- *CGST:* Central Goods and Services Tax is levied on intrastate transactions and collected by the Central Government of India.
- *SGST:* State/Union Territory Goods and Services Tax, which is also levied on intrastate transactions and collected by the state or union territory Government.
- *IGST:* Integrated Goods and Services Tax is levied on interstate sales. Central Government takes half of this amount and passes the rest of the amount to the state, where corresponding goods or services are consumed.

Table 3 gives the six correlation parameters used in clustering. These are derived from month-wise data in Table 2. Total GST liability is the sum of CGST, SGST, and IGST liabilities. Total ITC is equal to the sum of SGST, CGST, and IGST ITCs.

Table 3. Correlation parameters

S. no	Correlation parameters
1	Correlation of Total Sales Amount and Total GST Liability
2	Correlation of Total GST Liability and SGST Liability
3	Correlation of SGST Liability and SGST paid in cash
4	Correlation of Total Sales Amount and SGST paid in cash
5	Correlation of Total Tax Liability and Total ITC
6	Correlation of Total ITC and IGST ITC

4.2 Ratio Parameters

These three ratio parameters used in clustering are derived from month-wise data in Table 2.

1. Ratio of *Total Sales* VS. *Total Purchases*: This ratio captures the value addition.
2. Ratio of *IGST ITC* VS. *Total ITC*: This ratio captures how much purchase is shown as interstate or imports compared to total purchases.
3. Ratio of *Total Tax Liability* VS. *IGST ITC*.

4.3   TrustRank

A Network of Taxpayers: One of the independent variables in clustering is the *TrustRank*. To compute this, we created an edge-weighted directed graph (social network). Each vertex (node) in this graph corresponds to a taxpayer. We placed a weighted directed edge from taxpayer a to taxpayer b, where edge weight is the amount of sales done by the taxpayer a to taxpayer b during the period July 2017 to November 2019. Then the min-max normalization of edge weights is performed. For the same, we used the sales data explained in Table 1. This graph will capture the scale of interaction and (or) the exchange of money between taxpayers.

Computing Trust Rank: The TrustRank algorithm is a procedure designed to assign a rating to web pages based on the quality and trustworthiness of the content [8]. This method takes the linking structure among web pages, just like the PageRank algorithm, to generate a measure for the quality of a page. TrustRank places a core vote of trust on a seed set of manually reviewed web pages. This trust is passed from the seed set to other pages through links from the seed pages. TrustRank is based upon the following idea:

- Good pages provide links only to good ones. Bad pages often provide links to good ones in an attempt to look good.
- The care with which people provide links in a page is often inversely proportional to the number of links provided in the page.
- The Trust score is attenuated as it passes from page to site.

Selecting Seed Set: Generally pages that contain a lot of out going links to other pages are selected as seed set. Then pages like DMOZ clones and pages which were not listed in any of the major directories can be excluded. From the remaining pages, select only pages that were backed by the government, educational, and corporate bodies.

Algorithm: We need to manually select some genuine web pages (seed set) as sources of trust initially, as explained earlier. This method then propagates the trust from seed pages to other pages based on the linking structure between pages. In the TrustRank Equation given below, T is the transition matrix derived from the linking structure among web pages. Vector d is a static vector of nonnegative real numbers summing up to one. Vector d can be used to assign a score manually to a set of seed pages; the score of such seed pages is then spread during the iterations to the pages they point to.

$$r = \alpha T * r + (1 - \alpha) * d \qquad (2)$$

We use the TrustRank [8] algorithm to assign weights to the dealers such that higher weights are assigned to the plausible genuine ones and lesser weights to the plausible fraudulent ones. We used the graph defined above to this.

Seed set is selected by domain experts with prior knowledge about taxpayers. This score is the 10th and the last parameter used to performing cluster analysis.

4.4 Clustering Dealers

Clustering is the most widely used data mining method. In every scientific field dealing with data, people attempt to understand their data by trying to find clusters (groups) of "similar behavior" in their data. Given a set of data points $x_1, \ldots, x_n$ and some notion of similarity measure $s_{ij} \geq 0$ between all pairs of data points x_i and x_j, the objective of clustering is to segregate the data points into several clusters (groups), such that points in the same group are similar, and points in different groups are dissimilar to each other.

Spectral Clustering. Spectral clustering [11,15,19] is one of the most popular clustering algorithms. It is easy to implement using standard linear algebra software. It outperforms traditional clustering algorithms, such as the k-means algorithm. Following is a brief sketch of spectral clustering.

Define a similarity graph $G = (V, E)$, where vertex $v_i \in V$ in this graph represents a data point x_i. Two vertices are connected by a weighted undirected edge with edge weight s_{ij}, if and only if the similarity s_{ij} between the corresponding data points x_i and x_j is more than a certain threshold. Now clustering problem can be restated using the similarity graph as follows: find a partition of the graph such that the edges between different clusters have very low edge weights (data points in different clusters are dissimilar from each other), and the edges within a cluster have high weights (points within the same cluster are highly similar to each other). Use eigenvectors of Laplacian of this similarity graph to perform clustering in fewer dimensions after performing dimensionality reduction. In [15], Andrew Ng *et al.* discuss the scenarios and reasons for the better working of spectral clustering in spite of the fact that it uses the K-means clustering algorithm. Following is the brief sketch of spectral clustering.

- Step 1: Create a similarity graph between taxpayers based on the ten parameters explained before.
- Step 2: Compute the eigenvalues and eigenvectors of the Laplacian matrix of this graph. Let k be the number of eigenvalues, whose magnitude is almost equal to zero.
- Step 3: Run k-means clustering on matrix, whose columns are k eigenvectors corresponding to these k eigenvalues, to separate vertices into k clusters.

4.5 Identifying Suspicious Taxpayers

We used kernel density estimation [16], which is a common technique in nonparametric statistics for the approximation of the unknown probability distribution of taxpayers within a cluster. It is a method to estimate the unknown probability distribution of a random variable, based on a sample of points taken from that distribution. The formula to make a prediction for any point x is: $1/n \sum_{i=1}^{n} K(x - x_i)$, where $x_1 \ldots x_n$ are a sample of points taken from the distribution, and K is the kernel function. We had taken the Gaussian kernel. Then we further investigate those taxpayers who are further from the center of each cluster [6].

5   Experimentation and Results Obtained

We had taken 1199 iron and steel dealers for clustering. We computed *TrustRank* for each of these dealers by constructing the graph defined in Subsect. 4.3. We took twenty genuine dealers and ten fraudulent dealers as the seed set for computing the *TrustRank*. We used prior knowledge of domain experts while selecting these dealers. We computed correlation and ratio parameters defined in Subsect. 4.1 and 4.2 for each dealer. Table 4 gives a snapshot of the parameters created for each dealer for performing cluster analysis. Figures 10 and 11 show the distribution of these parameters after normalization. We used *Gaussian kernel function* for computing the similarity matrix. Figure 12 shows the first fifteen eigenvalues of normalized Laplacian and unnormalized Laplacian of the similarity graph for different *k-nearest neighbors* (k = 2, 3, 4, 5, 6).

Figure 14 shows the eigenvalues of normalized Laplacian of the similarity graph with k-nearest neighbors is equal to six. There are four eigenvalues that are almost equal to zero. From this, we can conclude that there are four clusters in the data. Figure 15 gives the number of taxpayers in each cluster. Out of four clusters, two are small. All taxpayers in these two clusters are suspicious. The first cluster and third cluster contain 588 and 581 dealers, respectively. Out of these taxpayers we selected twenty seven taxpayers in cluster one and seventeen taxpayers in cluster three, who are at a distance more than three standard deviations from respective centers. These twenty-seven plus seventeen taxpayers are suspicious taxpayers.

Table 4. Snapshot of parameters

Unique ID	Corr1	Corr2	Corr3	Corr4	Corr5	Corr6	Total Sales /Total Purchases	IGST ITC /Total ITC	Total tax liability /IGST ITC	TrustRank score
1	0.9977	0.9998	0.2159	0.1967	0.9556	0.9988	1.0465	0.8717	1.3272	0.0091
2	0.9940	0.9799	-0.3371	-0.2486	0.6408	0.5539	1.1992	0.1347	7.4129	0.0077
3	0.9476	0.4556	0.0017	0.1286	-0.1620	0.9606	1.6991	0.8020	2.1824	0.0075

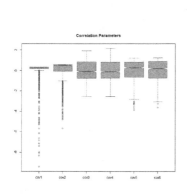

Fig. 10. Boxplot of correlation

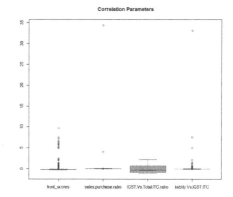

Fig. 11. Boxplot of TrustRank and ratios

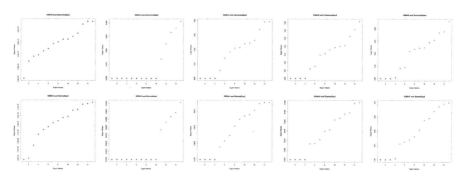

Fig. 12. Fifteen smallest eigen values

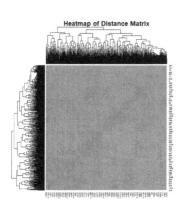

Fig. 13. Heat map of distance matrix

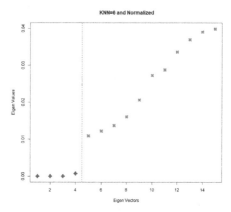

Fig. 14. Optimal number of clusters

cluster 1	cluster 2	cluster 3	cluster 4
588	29	581	1

Fig. 15. Sizes of four clusters identified

6 Conclusion

In this work, we analyzed the tax returns dataset of a set of business dealers in the state of Telangana, India, to identify dealers who perform extensive tax evasion. Dealers are clustered using two different types of features. One feature in clustering is the *TrustRank* value that is derived for each dealer based on his/her business interactions with other dealers using the TrustRank algorithm. The second type of feature is derived for each dealer based on their monthly returns. Dealers at the boundary of each cluster are further analyzed to identify malicious dealers.

Acknowledgment. We express our sincere gratitude to the Telangana state Government, India, for sharing the commercial tax data set, which is used in this work. This work has been supported by Visvesvaraya Ph.D. Scheme for Electronics and IT, Media Lab Asia, grant number EE/2015-16/023/MLB/MZAK/0176.

References

1. Yuan, S., Wu, X., Li, J., Lu, A.: Spectrum-based deep neural networks for fraud detection. In: Proceedings of the 2017 ACM on Conference on Information and Knowledge Management (CIKM 2017), pp. 2419–2422. ACM, New York (2017). https://doi.org/10.1145/3132847.3133139
2. Godbole Committee: Report on Economic Reforms of Jammu and Kashmir. Ministry of Finance, Government of Jammu and Kashmir (1998)
3. Bianchi, P.A., et al.: Professional Networks and Client Tax Avoidance: Evidence from the Italian Statutory Audit Regime, SSRN (2016). https://ssrn.com/abstract=2601570
4. Assylbekov, Z., Melnykov, I., Bekishev, R., Baltabayeva, A., Bissengaliyeva, D., Mamlin, E.: Detecting value-added tax evasion by business entities of Kazakhstan. In: Czarnowski, I., Caballero, A.M., Howlett, R.J., Jain, L.C. (eds.) Intelligent Decision Technologies 2016. SIST, vol. 56, pp. 37–49. Springer, Cham (2016). https://doi.org/10.1007/978-3-319-39630-9_4
5. Mathews, J., Mehta, P., Kuchibhotla, S., Bisht, D., Chintapalli, S.B., Visweswara Rao, S.V.K.: Regression analysis towards estimating tax evasion in goods and services tax. In: 2018 IEEE/WIC/ACM International Conference on Web Intelligence (WI), pp. 758–761, Santiago (2018)
6. Chandola, V., Banerjee, A., Kumar, V.: Anomaly detection: a survey. ACM Comput. Surv. **41**, 3, Article 15, 58 (2009). https://doi.org/10.1145/1541880.1541882
7. Sahin, Y., Duman, E.: Detecting credit card fraud by ANN and logistic regression. In: 2011 International Symposium on Innovations in Intelligent Systems and Applications. IEEE, June 2011. ISBN 978-1-61284-919-5

8. Gyöngyi, Z., Garcia-Molina, H., Pedersen, J.: Combating web spam with trustrank. In: Nascimento, M.A., Özsu, M.T., Kossmann, D., Miller, R.J., Blakeley, J.A., Schiefer, K.B., (eds.) Proceedings of the Thirtieth International Conference on Very Large Data Bases, VLDB Endowment (VLDB 2004), vol. 30, pp. 576–587 (2004)

9. Wang, J., Zhou, S., Guan, J.: Detecting potential collusive cliques in futures markets based on trading behaviors from real data. Neurocomputing **92**, 44–53 (2012)

10. Issa, H., Vasarhelyi, M.A.: Application of anomaly detection techniques to identify fraudulent refunds (2011). https://doi.org/10.2139/ssrn.1910468

11. de Roux, D., Perez, B., Moreno, A., del Pilar Villamil, M., Figueroa, C.: Tax fraud detection for under-reporting declarations using an unsupervised machine learning approach. In: Proceedings of the 24th ACM SIGKDD International Conference on Knowledge Discovery and Data Mining (KDD 2018), pp. 215–222. ACM, New York (2018)

12. González, P.C., Velásquez, J.C.: Characterization and detection of taxpayers with false invoices using data mining techniques. Expert Syst. Appl. **40**(5), 1427–1436 (2013)

13. Golub, G., Pereyra, V.: Separable nonlinear least squares: the variable projection method and its applications. Inverse Problems (IOP) **19**(2), R1–R26 (2003)

14. Rad, M.S., Shahbahrami, A.: High performance implementation of tax fraud detection algorithm. In: Signal Processing and Intelligent Systems Conference (SPIS), pp. 6–9, Tehran (2015)

15. Ng, A.Y., Jordan, M.I., Weiss, Y.: On spectral clustering: analysis and an algorithm. In: Advances in Neural Information Processing Systems, pp. 849–856 (2002)

16. Tran, L.T.: The L_1 convergence of kernel density estimates under dependence. Can. J. Stat./La Revue Canadienne de Statistique **17**, 197–208 (1989). http://www.jstor.org/stable/3314848

17. Rousseeuw, P.: Silhouettes: a graphical aid to the interpretation and validation of cluster analysis. J. Comput. Appl. Math. **20**(1), 53–65 (1987). https://doi.org/10.1016/0377-0427(87)90125-7

18. Ketchen, D.J., Shook, C.L.: The application of cluster analysis in strategic management research: an analysis and critique. Strateg. Manag. J. **17**(6), 441–458 (1996)

19. Luxburg, U.: A tutorial on spectral clustering. Stat. Comput. **17**(4), 395–416 (2007)

20. Dani, S.: A research paper on an impact of goods and services tax on indian economy. Bus. Econ. J. **7**(4), 1–2 (2016)

An Architecture for Multi-chain Business Process Choreographies

Jan Ladleif[(✉)], Christian Friedow, and Mathias Weske

Hasso Plattner Institute, University of Potsdam, Potsdam, Germany
{jan.ladleif,mathias.weske}@hpi.de,
christian.friedow@student.hpi.de

Abstract. An increasing number of organizations employ blockchain technology in their business process landscapes, especially when dealing with inter-organizational choreographies. Due to complex requirements with regard to data security and privacy in practice, however, no singular blockchain captures all use cases. Blockchains optimized for various levels of risk tolerance and confidentiality coexist in multi-chain environments, posing severe architectural challenges for blockchain-based Business Process Management Systems (BPMSs). Current state-of-the-art approaches lack the global perspective necessary, and focus on single-blockchain environments. In this paper, we alleviate these issues by developing a general architecture for multi-chain BPMSs for choreographies. We show the feasibility of our architecture by a prototypical implementation, and discuss future challenges using a concrete case study.

Keywords: Process choreographies · Business Process Management System · Blockchain-based process execution · Smart contracts

1 Introduction

Its immutability and non-repudiation properties make blockchain a widely applicable technology to securely manage data, logic, and digital assets among mutually distrustful entities [18]. Business Process Management (BPM) is no exception, and blockchain technology has proven to be a major force in the domain [13]. Blockchains have been shown to be advantageous in executing [11,14,16], monitoring [4], and mining [1] business processes. Especially when organizations interact towards a common business goal within a choreography, blockchains provide novel ways of thinking about their modeling and enactment [10].

For blockchains to be useful in practice, they need to be a core component of an organization's Business Process Management System (BPMS) on a conceptual as well as a technological level [13], breaking up the monolithic and centralized architecture of traditional BPMSs [3, Ch. 9.1.2]. This is especially necessary since it is becoming clear that no single blockchain will prevail in the mid-term. Rather, an entire landscape of purpose-built blockchains will coexist with different applications in mind, for reasons of data security and protection,

© Springer Nature Switzerland AG 2020
W. Abramowicz and G. Klein (Eds.): BIS 2020, LNBIP 389, pp. 184–196, 2020.
https://doi.org/10.1007/978-3-030-53337-3_14

scalability, as well as compliance with local legislation [18]. That means that an organization might need to interact with a variety of different blockchain technologies, sometimes even in the scope of a single choreography.

Current research, however, has only just started to appreciate the challenges involved in this. Most approaches are focused on a singular blockchain as a simple execution engine without potential for generalization. In this paper, we work towards understanding and specifying the impact of multi-chain environments on current state-of-the-art BPMS architectures. The paper analyzes typical components of BPMSs, rearranges them according to the requirements of multi-chain environments, and arrives at a proposed architecture for a multi-chain BPMS. Finally it shows its applicability in practice by introducing Mantichor, a prototypical implementation of the architecture. We also give an intuition into which challenges lie ahead in pursuit of a fully-featured multi-chain BPMS.

The paper is structured as follows. In Sect. 2 we introduce the reader to necessary background knowledge and related work in blockchain technology and BPM. We continue with a conceptual model of blockchain-based BPMS in Sect. 3, before introducing our prototypical implementation in Sect. 4. We discuss our results and future challenges in Sect. 5, and conclude the paper in Sect. 6.

2 Background

In this section, we introduce the reader to the required background knowledge in multi-chain environments as well as the connection of BPM and blockchain. We will then lay out the goals of the paper in more detail.

2.1 Choreographies in Multi-chain Environments

A multi-chain environment is characterized by the presence of a set of coexisting blockchains. These blockchains may be used in conjunction to achieve certain goals regarding data security, throughput, or other metrics. Especially in areas in which loosely coupled organizations work together in a dynamic market, possibly across different jurisdictions, those environments emerge organically [15].

Supply chain management is one of those areas, and has already been heavily affected by blockchain [9]. Organizations within a supply chain tend to have a rather distrustful relationship, which is shaped by constant supervision, tracking and logging of important events. Figure 1 shows a scenario which we will explore in the course of this paper. The basic premise is that a wholesaler in Australia sells goods to a buyer in Estonia, while shipping is provided by a third-party multinational shipping company. For tax reasons, the Australian Taxation Office (ATO) is involved as well.

From a technological perspective, the scenario exhibits a complex multi-chain environment. Australia[1] and Estonia[2] are some of the first countries to push

---

[1] https://www.australiannationalblockchain.com/, accessed 2019-11-27.
[2] https://e-estonia.com/, accessed 2019-11-27.

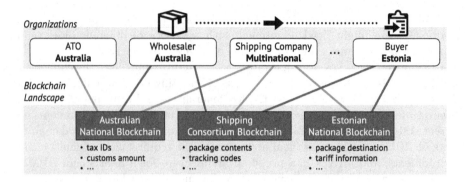

Fig. 1. Multi-chain ecosystem being used within one choreography

national blockchains, which are used to digitize, store, and execute contracts within their respective jurisdictions. Furthermore, the shipping company is part of an international industry consortium using a consortium blockchain to synchronize and secure their business. In practice, such initiatives already exist and are backed by major industry players, one example being TradeLens[3].

As a consequence, the supply chain choreography spans at least three different blockchains, with some participants having to access more than one of them to audit the parts of the choreography relevant to them. It is important to note that this scenario is not only due to short-term discrepancies in technology until every actor will converge to one single blockchain. Rather, blockchains are configurable and applications as well as deployment environments—e.g., different jurisdictions—may still call for purpose-built solutions in the future. For example, a blockchain platform will have to comply with European data security standards when running within the European market.

2.2   BPM and Blockchain

The first major work in connecting business processes with blockchain technology was done by Weber et al. [16]. One of their main insights was that organizations in a choreography may not directly interact with each other, but do so through transactions to a smart contract. That way, interactions can be tracked and monitored, and smart contract logic can be used to enforce the process. Being an early work in the field, the focus of the paper is on a concrete technology rather than an architecture level. Similarly, our own extension of said original approach is mainly concerned with modeling and enforcement issues rather than BPMS architectures [10]. Other approaches using artifact-centric [5] and declarative (DCR Graphs) [12] process specifications range on the same level of detail.

The main trade-off in blockchain-based BPM seems to concern the balance between off-chain and on-chain components. While the aforementioned approaches assume a situation where the choreography model is created and

---

[3] https://www.tradelens.com/, accessed 2019-12-04.

managed entirely outside the scope of the blockchain, for example, there are others which make much more radical changes. Klinger and Bodendorf [7] add further blockchain mechanisms to their blockchain-based collaboration execution architecture, e.g., a voting contract to democratize instance deployments [7]. Sturm et al. [14] do not only relay the execution engine to the blockchain, but also the process modeling tools themselves [14]. Participants collaborate building a model on-chain using a smart contract providing simple change operations.

With Caterpillar, López-Pintado et al. [11] provide the most complete vision of a blockchain-based BPMS to date [11]. They discuss several aspects of traditional BPMS architecture in face of blockchain technology, and move the entire execution engine to the blockchain. We further discuss Caterpillar's architecture as a baseline for our work in Sect. 3.1.

The approaches at blockchain-based process and choreography execution presented above share a significant restriction: they are conceptually limited to single-blockchain environments, and their architecture can not easily be adapted to fit a multi-chain environment. This circumstance poses a gap in research, and there are major challenges for the architecture of multi-chain BPMSs.

2.3   Goals and Requirements

In this paper, we will explore the limitations in current blockchain-based BPMSs and provide a more global view on the future architecture of such systems in light of multi-chain environments. We focus on two specific research questions:

(RQ1) *How can a BPMS architecture embedded in multi-chain environments be structured?* We investigate which components make up a typical BPMS, and how they can be distributed to on-chain and off-chain mechanisms in the blockchain context. Data security is an important factor, in that organizations should be able to use permissioned/consortium blockchains to control the spread of their business knowledge, and that sensitive information may stay local.

(RQ2) *How can crucial system properties be maintained in a multi-chain BPMS architecture?* In this context, we place special emphasis on flexibility in that various different blockchains can be managed at the same time without requiring to maintain entirely separate systems. A further goal is interoperability, in that choreographies can make use of multiple blockchains simultaneously. In Fig. 1, for example, the choreography is distributed along three blockchains.

Note that it is not the goal of the paper to solve the inherent problems associated with actual realizations of cross-chain communication. While methods like atomic swaps, relays or merged consensus are being developed to allow blockchains to interact with each other, they are—like blockchains themselves—still in their infancy. Our architecture should, however, not prohibit future use of such protocols, but rather reflect their eventual operation in practice.

3   Architecture of Blockchain-Based BPMSs

In practice, most BPMSs rely on one central assumption: that all process instances are executed on a single platform within the confines of one organization [3, Ch. 9.1.2]. This is not necessarily the case anymore with blockchain-based choreographies [10]. In this section, we will first discuss the state-of-the-art in blockchain-based BPMS architecture by introducing the single-chain architecture of the Caterpillar framework [11]. We will then generalize and develop a new architecture which is able to deal with multi-chain environments natively.

3.1   State-of-the-Art Architecture

Figure 2 shows an adapted overview of Caterpillar's architecture [11]. Note that we abstracted from the full architecture provided in the publication to only include relevant elements, omitting some components tied to the concrete technology used. We also relabeled some components to equivalent terms in our architecture for comparability, e.g., "Modeling Panel" to "Process Modeling Tool".

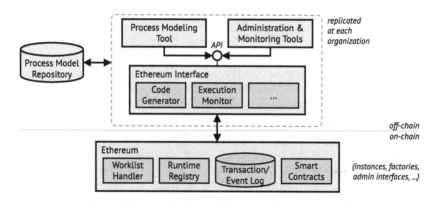

Fig. 2. Architecture of Caterpillar (adapted from López-Pintado et al. [11])

Caterpillar moves the *execution engine* of the traditional BPMS architecture to the blockchain, specifically Ethereum [17]. The execution engine is the central part of most BPMSs and handles the execution of processes, keeps track of instances, distributes tasks to worklist handlers, and manages correlated data [3].

In the architecture, this arrangement is represented by the respective components on the Ethereum blockchain—i.e., all components in the lower on-chain half of the figure. Each component is realized by one or more purpose-built smart contracts. The *runtime registry* keeps track of running instances. Using different interface methods, instances can be created, registered, and interlinked. Thus, there is an on-chain record of instances and their relationship. The *worklist handler* is the entry point to choreography instances, which are enforced via smart contracts generated off-chain within a code generator. The worklist handler queries these instances to find enabled tasks, which can then be executed within transactions to advance the choreography state.

Additional data is stored in two places. The *process model repository* is used to persist process models and is moved to distributed storage solutions such as InterPlanetary File System (IPFS). Ethereum's *transaction/event log* is utilized for logging purposes, both through its inherent ledger containing all past transactions and through custom logging of events in the smart contract logic. Since each task of the choreography instance is executed in the scope of some transaction, the entire execution can be reconstructed from the logs. Due to this, logs also benefit from the blockchain's inherent properties, namely (i) immutability, ensuring that logs cannot be forged, (ii) non-repudiation, making it impossible for participants to deny their actions, and (iii) transparency, allowing each participant to monitor the entire choreography execution.

Lastly, there are two further components. A *process modeling tool* allows the creation of process models within a web-based modeling framework. *Administration and monitoring tools* can be used to keep track of instances, and configure certain parameters for new and existing contracts.

An obvious distinction that will also be relevant in the generalized architecture later on is that between on-chain and off-chain components, as indicated by the dotted horizontal line in Fig. 2. Off-chain components are mostly local to a single organization, or possibly shared via some other mechanism than blockchain. This gives organizations more control over these specific components. A component is susceptible to an off-chain placement if it contains sensible business knowledge, or does not benefit from the blockchain's inherent properties.

3.2   Multi-chain Architecture

The state-of-the-art architecture as presented in the previous section includes relevant blockchain aspects, but lacks provisions for multi-chain landscapes. In this section, we fill this gap by proposing a new general architecture for BPMSs that incorporates multi-chain environments as an integral part of its design.

Figure 3 shows an overview of our architecture. There are several significant changes to the architecture introduced previously. These include (i) moving components from the blockchain back into the off-chain part of the BPMS, (ii) generalizing from one concrete blockchain technology to multi-chain adapters, and (iii) rearranging the process model and instance repositories.

On-Chain and Off-Chain Components. While the first major change of pulling components back from the blockchain may appear counter-intuitive, it is a direct consequence of multi-chain environments. The previous architecture assumes a situation in which each choreography instance resides on the Ethereum blockchain. Thus, the runtime registry, being an Ethereum smart contract itself, can keep track of all instances. This is not true anymore in multi-chain environments. The runtime registry needs to keep track of instances on several

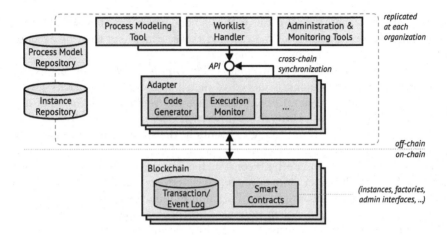

Fig. 3. Multi-chain BPMS architecture

blockchains in the simplest case, and even instances that are spread across multiple blockchains themselves. As a consequence, the worklist handler and runtime registry cannot be placed on-chain anymore, since the blockchain's closed-world nature limits the global perspective needed.

Rather, the worklist handler becomes part of each organization's local infrastructure, allowing an organization to autonomously decide on how to handle their worklist without disclosing internal operational details. The runtime registry is moved to a newly introduced instance repository, which allows for a more secure handling and interlinking of instances between organizations, which is discussed in more detail below.

Blockchain Interface. Because blockchains exhibit fundamental conceptual and technical differences, they need specific wrappers so other BPMS components have a common interface to interact with. To this end, we use the *adapter* pattern to generalize from Ethereum to arbitrary blockchains. An adapter, for example, contains a code generator which transforms a process model into a valid smart contract specification for that particular blockchain—e.g., a Solidity smart contract for Ethereum, or a Michelson smart contract for Tezos. The adapter is aware of its blockchain's concrete interfaces and knows how to access and monitor smart contracts. Thus, the worklist handler queries different adapters for changes in instances and resource management can proceed locally.

Adapters also have to care for including proper provisions to make smart contracts tractable, i.e., by adding logging, update, and migration mechanisms which can then be used by the administration and monitoring tools. Since blockchains differ heavily in their smart contract models, especially concerning immutability, these issues can not be solved at a higher level of the architecture. We also anticipate that adapters may need to interact and synchronize with other adapters, which they can do via the same interface the other components use.

Instance and Model Repositories. One inherent difficulty in choreography enforcement is the initial deployment and successive sharing of a process instance and its model. In essence, one or several participants in a process need to create an instance and then share appropriate correlation and identification information with the others. In practice, this could mean sharing an address to a deployed smart contract, or even have an on-chain voting system [7].

Since this issue is not directly part of the architecture itself, but more of a protocol issue, we decided not to specifically include different provisions for it. Instead, we envision two general components, a process model repository as well as an instance repository. These can be located within the organization, e.g., if models are especially sensitive, or shared using technologies like IPFS as Caterpillar suggests. In any case, concrete realizations of the architecture could implement different solutions through on-chain or off-chain channels. Mantichor, which we will present later in Sect. 4, is an example of a realization employing an off-chain negotiation and deployment system.

4 Prototypical Implementation

The goal of this paper is to not only have a conceptual architectural understanding of multi-chain BPMSs, but to also evaluate their feasibility. To this end, we are currently developing Mantichor, an open-source, web-based realization of our architecture. The entire source code of Mantichor is available online on GitHub[4]. In this section, we will introduce the current state of Mantichor and how it relates to the conceptual considerations devised in the previous sections.

4.1 Components

Mantichor's basic design is strongly centered around the general multi-chain architecture described in Sect. 3.2. In particular, we implemented four components: (i) a *frontend* bundles process modeling tool, worklist handler, as well as administration and modeling tools, (ii) a *share server* implements a basic process model and instance repository, and (iii) a Tezos[5] and (iv) a R3[6] adapter provide access to one blockchain, respectively.

Frontend. The frontend is based on current web frameworks that run in modern web browsers. It enables process experts to create choreography models and provides interfaces to track and interact with process instances—i.e., smart contracts—via the adapters. We decided to use (BPMN) 2.0 choreography diagrams as the underlying modeling formalism, which have been shown to provide adequate conceptual support for important aspects of blockchain-based smart contracts [10,16]. The frontend uses blockchain adapters that handle code generation and smart contract instantiation to deploy these models on the respective blockchain.

---

[4] https://github.com/bptlab, accessed 2019-12-04.
[5] https://tezos.com/, accessed 2019-11-26.
[6] https://www.r3.com/, accessed 2019-11-26.

Share Server. The share server implements both the process model as well as instance repository using a simple server connected to a database. Participants can upload process models or instance data, and share these using an automatically generated key. Our main requirement in designing the share server was not data security or access management, but rapid exchange of models and instance metadata between participants for testing. As such, it should not be seen as a general recommendation for practical uses.

Adapters. One of the central challenges of developing against several blockchains is the need for a single interface in spite of the considerable differences in features and architectures. These differences should be hidden to the frontend, which is supposed to have a unified view on the chains.

Table 1. Excerpt of the adapter interface

Verb	URI/description
POST	`/choreographies` deploy a new choreography instance
GET	`/choreographies/{choreographyId}/tasks`
	query the set of enabled tasks in the given choreography instance
POST	`/choreographies/{choreographyId}/tasks/execute`
	execute a task in the given choreography instance

Table 1 shows an excerpt of the interface functions we decided on for our specific use-case. As this specification illustrates, the adapters provide a unified interface to deploy choreography diagrams, check currently enabled tasks of instances, and execute tasks within an instance. Each adapter acts as a full-node for the respective blockchain. To deploy a diagram, the adapter converts it into smart contract code and instantiates a smart contract on the respective blockchain. Runtime monitoring capabilities are implemented on a per-adapter basis since each blockchain implementation works differently. Some may require standardized functions built into every process instance's smart contracts, while others may provide native concepts for state-machine like contracts.

4.2   Case Study

While Mantichor in its current state is not mature enough to properly support the supply chain use case introduced in Sect. 2.1, it is easy to visualize the desired end goal. Figure 4 shows how a possible future version of Mantichor could realize the entire choreography across multiple blockchains and organizations.

The four organizations each run Mantichor locally, using their individual sets of adapters connected to concrete blockchain deployments. In this case, these include Hyperledger Fabric for the Australian National Blockchain and KSI for the Estonian National Blockchain—both already in use by the respective country—, and Tezos as an example for a third blockchain used by the shipping consortium. The choreography is split along the blockchains in three smart

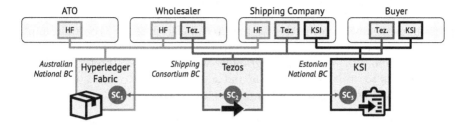

Fig. 4. Mantichor in practice with several adapters connecting to multiple blockchains

contracts SC_1, SC_2, and SC_3, which as a union enforce the entire choreography. Using cross-chain channels, either realized directly between the blockchains or through mediator capabilities within Mantichor, the fragments interact and exchange enforcement and runtime data. Each organization has a view on the choreography where relevant, e.g., the ATO can monitor tax-related issues on the Australian National Blockchain, but does not access the Estonian National Blockchain.

5   Discussion and Challenges

The architecture proposed in this paper was primarily guided by our initial research questions (see Sect. 2.3). We were able to partially answer them, in that (RQ1) our architecture proposes a structure for embedding BPMSs in multichain environments, allowing organizations to make use of purpose-built blockchains for security-relevant scenarios, and (RQ2) our prototypical implementation shows that blockchains can be connected via flexible adapters which are able to mediate between different blockchains. While we thus present an initial step towards a multi-chain BPMS, we do not solve all of the issues in the area.

Data Security and Privacy. Not only do organizations in a choreography own a wealth of business knowledge which needs to be protected, the content and nature of interactions themselves may be subject to data security and privacy considerations. These considerations were of special concern in our first research question and heavily influenced the design of our architecture:

For one, the architecture enables organizations to use a wide array of blockchains, which can be fine-tuned to different levels of confidentiality and risk tolerance [2]. While the concrete realization of data security measures is still the responsibility of the organizations themselves, they are empowered to make use of the whole spectrum of available technology. Secondly, there are further data artifacts that are not necessarily stored directly on any chain, notably the process model and instance repositories. While the architecture does not prescribe it, these repositories can and should be protected by various means, e.g., using an IPFS storing encrypted data which only relevant organizations can access.

Yet, there are certainly still challenges regarding data security and privacy. On an implementation level, the required technology has not been fully implemented yet. On a modeling level, standards like BPMN do not provide adequate modeling support to specify choreographies with added data protection. There are first works towards understanding and specifying those constraints [8], however, which will be of interest in the future.

Cross-Chain Communication. In our second research question we referred to support for interoperability between blockchain technologies as a core issue: In our running example discussed previously (see Fig. 4), an instance of one choreography actually spans multiple smart contracts SC_1, SC_2, and SC_3 on three blockchains. For the choreography to be correctly enforced by the union of all smart contracts, they need to communicate choreography data, but also runtime information such as lifecycle states or tokens.

We accounted for this in several places: For one, adapters for different blockchains can communicate among each other via a structured interface. Secondly, several components were moved off-chain as compared to the baseline BPMS architecture presented in Sect. 3.1 since they would otherwise be restricted to the closed-world perspective of a single blockchain. Still, there are unsolved challenges on a technological as well as a conceptual level.

Cross-chain communication protocols are starting to emerge [6], but there are currently no production-ready methods to transfer arbitrary data between blockchains. Techniques like atomic swaps are becoming more usable, but will probably need additional on-chain features. Similarly, smart contract migration is an issue. While early protocols can handle token transfer between blockchains, they can not transfer whole smart contracts yet. Switching choreography instances from one blockchain to another would thus be a non-trivial task to solve.

On a conceptual level, the area of choreography modeling was until now not concerned with multi-chain scenarios as discussed in our example. Deploying one choreography as a set of loosely coupled fragments across different implementation platforms poses new questions regarding enforceability and enactment, which will require significant conceptual work.

6   Conclusion

Blockchain technology offers great potential for BPM, but comes with various challenges for organizations aiming to introduce it into their BPMSs—especially when dealing with multi-chain environments. In this paper, we approached these challenges from a broader perspective and analyzed the core components of state-of-the-art blockchain-based BPMSs before rearranging them in a multi-chain choreography context. Core concepts of the resulting multi-chain BPMS architecture were then evaluated via a prototypical implementation. While the conceptual architecture and implementation are steps towards integrating multi-chain environments into BPM, numerous open challenges remain. In future work, we want to provide a more concrete architecture, as well as extend our implementation to show the full potential of multi-chain BPMSs.

References

1. Di Ciccio, C., et al.: Blockchain-based traceability of inter-organisational business processes. In: Shishkov, B. (ed.) BMSD 2018. LNBIP, vol. 319, pp. 56–68. Springer, Cham (2018). https://doi.org/10.1007/978-3-319-94214-8_4
2. Dinh, T.T.A., Wang, J., Chen, G., Liu, R., Ooi, B.C., Tan, K.L.: Blockbench: a framework for analyzing private blockchains. In: ACM International Conference on Management of Data, SIGMOD 2017, pp. 1085–1100. ACM (2017)
3. Dumas, M., Rosa, M.L., Mendling, J., Reijers, H.A.: Fundamentals of Business Process Management. Springer, Heidelberg (2013). https://doi.org/10.1007/978-3-642-33143-5
4. García-Bañuelos, L., Ponomarev, A., Dumas, M., Weber, I.: Optimized execution of business processes on blockchain. In: Carmona, J., Engels, G., Kumar, A. (eds.) BPM 2017. LNCS, vol. 10445, pp. 130–146. Springer, Cham (2017). https://doi.org/10.1007/978-3-319-65000-5_8
5. Hull, R., Batra, V.S., Chen, Y.-M., Deutsch, A., Heath III, F.F.T., Vianu, V.: Towards a shared ledger business collaboration language based on data-aware processes. In: Sheng, Q.Z., Stroulia, E., Tata, S., Bhiri, S. (eds.) ICSOC 2016. LNCS, vol. 9936, pp. 18–36. Springer, Cham (2016). https://doi.org/10.1007/978-3-319-46295-0_2
6. Jin, H., Dai, X., Xiao, J.: Towards a novel architecture for enabling interoperability amongst multiple blockchains. In: 38th International Conference on Distributed Computing Systems (ICDCS), pp. 1203–1211. IEEE (2018). https://doi.org/10.1109/ICDCS.2018.00120
7. Klinger, P., Bodendorf, F.: Blockchain-based cross-organizational execution framework for dynamic integration of process collaborations. In: 15th International Business Informatics Congress (2020)
8. Köpke, J., Franceschetti, M., Eder, J.: Balancing privity and enforceability of BPM-based smart contracts on blockchains. In: Di Ciccio, C., et al. (eds.) BPM 2019. LNBIP, vol. 361, pp. 87–102. Springer, Cham (2019). https://doi.org/10.1007/978-3-030-30429-4_7
9. Korpela, K., Hallikas, J., Dahlberg, T.: Digital supply chain transformation toward blockchain integration. In: Proceedings of the 50th Hawaii International Conference on System Sciences (2017)
10. Ladleif, J., Weske, M., Weber, I.: Modeling and enforcing blockchain-based choreographies. In: Hildebrandt, T., van Dongen, B.F., Röglinger, M., Mendling, J. (eds.) BPM 2019. LNCS, vol. 11675, pp. 69–85. Springer, Cham (2019). https://doi.org/10.1007/978-3-030-26619-6_7
11. López-Pintado, O., García-Bañuelos, L., Dumas, M., Weber, I., Ponomarev, A.: Caterpillar: a business process execution engine on the Ethereum blockchain. CoRR abs/1808.03517 (2018)
12. Madsen, M.F., Gaub, M., Høgnason, T., Kirkbro, M.E., Slaats, T., Debois, S.: Collaboration among adversaries: distributed workflow execution on a blockchain. In: Symposium on Foundations and Applications of Blockchain (2018)
13. Mendling, J., Weber, I., et al.: Blockchains for business process management - challenges and opportunities. ACM Trans. Manag. Inf. Syst. (TMIS) 9(1), 4:1–4:16 (2018). https://doi.org/10.1145/3183367
14. Sturm, C., Szalanczi, J., Schönig, S., Jablonski, S.: A lean architecture for blockchain based decentralized process execution. In: Daniel, F., Sheng, Q.Z., Motahari, H. (eds.) BPM 2018. LNBIP, vol. 342, pp. 361–373. Springer, Cham (2019). https://doi.org/10.1007/978-3-030-11641-5_29

15. Udokwu, C., Kormiltsyn, A., Thangalimodzi, K., Norta, A.: The state of the art for blockchain-enabled smart-contract applications in the organization. In: Ivannikov ISP RAS Open Conference, pp. 137–144. IEEE (2018)

16. Weber, I., Xu, X., Riveret, R., Governatori, G., Ponomarev, A., Mendling, J.: Untrusted business process monitoring and execution using blockchain. In: La Rosa, M., Loos, P., Pastor, O. (eds.) BPM 2016. LNCS, vol. 9850, pp. 329–347. Springer, Cham (2016). https://doi.org/10.1007/978-3-319-45348-4_19

17. Wood, G.: Ethereum: a secure decentralised generalised transaction ledger. Technical report EIP-150 (2014)

18. Xu, X., Weber, I., Staples, M.: Architecture for Blockchain Applications. Springer, Cham (2019). https://doi.org/10.1007/978-3-030-03035-3

Correlating Data Objects in Fragment-Based Case Management

Stephan Haarmann[(✉)] and Mathias Weske

Hasso Plattner Institute, University of Potsdam, Potsdam, Germany
{stephan.haarmann,mathias.weske}@hpi.de

Abstract. Business process management (BPM) supports organizations with their operational procedures. Traditional BPM focuses on structured processes but lacks support for flexible ones. Case management addresses this gap. The fragment-based case management (fCM) approach models processes as a set of repetitive, structured fragments. At run-time fragments are instantiated and composed to realize flexibility while data requirements synchronize their execution. So far, fCM does not consider data-to-data associations or object-to-fragment bindings. We investigate both by (i) extending fCM models and (ii) refining the execution semantics. For evaluation, we present a formal model based on colored Petri nets.

Keywords: Case management · Process modeling · Process management · Process execution

1 Introduction

Business process management (BPM) provides organizations with methods and tools to capture, enact, and analyze their processes [18]. Traditionally, BPM focused on well-defined and highly structured workflows. However, the interest broadened to model, execute, and analyze a wide range of processes spanning from structured ones to ad-hoc ones. Case management describes approaches that focus on knowledge-intensive processes [3,17].

The *fragment-based case management* (fCM) approach originates in the observation that the work of knowledge workers can be divided into multiple repetitive and structured fragments [8]. The knowledge workers execute and combine these fragments based on their expertise and experience. Only data requirements capture dependencies between fragments and enforce synchronization during their execution.

EXAMPLE 1: Physicians' and nurses' work depends highly on their training and the patient they treat. Many of their tasks, however, are structured: during diagnosis an X-ray can be performed requiring safety precautions, executing the X-Ray, and analyzing the image. For treatment, surgery may be conducted requiring the physicians to explain the risks to the patient, to apply a sedative,

© Springer Nature Switzerland AG 2020
W. Abramowicz and G. Klein (Eds.): BIS 2020, LNBIP 389, pp. 197–209, 2020.
https://doi.org/10.1007/978-3-030-53337-3_15

to perform the surgery, and to have a follow-up examination [7]. Although some activities are strictly ordered, traditional process models can be too restrictive or too incomprehensible (when incorporating flexibility).

EXAMPLE 2: An IoT system comprises a network of sensors and actuators. Expert knowledge is encapsulated in the business logic that defines how events are processed and how actuators act. Each node of the IoT system runs a fragment of a bigger process. Data of sensors is enriched and shared leading to synchronization between the distributively running fragments [4].

In data-centric case management approaches such as fCM, knowledge workers create and consult data objects (i.e., the diagnosis of a patient's disease or a prescription). A single task may require multiple data objects of different types, and there can be multiple objects of the same type. The knowledge workers are responsible for selecting the right data-objects and correlating them: A physician treats multiple patients. For a single patient, there might be multiple diagnoses and corresponding treatments. The physician must match and correlate the treatment to the diagnosis and vice versa. However, the case history often contains incidents of such correlations, which can be reused.

In this paper, we consider data associations in fCM and binding of data objects to fragment instances. Therefore, we extend the fCM modeling language, describe the adapted execution semantics, and evaluate our approach using colored Petri nets as a formal model.

The next section provides more details on fCM models and their execution semantics. In Sect. 3, we extend fCM on a model level and adapt the execution semantics to handle data-to-data correlation and binding objects to fragments. We evaluate our approach using colored Petri nets in Sect. 4. An overview of related work is given in Sect. 5. We conclude our contribution in Sect. 6.

2   Fragment-Based Case Management

The work of knowledge workers requires flexibility. This means, that their processes can have many variants, some of which are unknown at design-time [17]. In fCM, we model such processes as data-centric case models: fragments contain activities, which require, create, and alter data objects. Knowledge workers can run fragments and activities flexibly if all the requirements (data and control-flow) are satisfied. In case the knowledge workers encounter situations unknown at design-time, they can add new fragments at run-time. Eventually, a case reaches a predefined goal state, and the knowledge workers can close the case.

2.1   fCM Case Models

An fCM case model describes the behavior that can be performed as well as data that can be involved in a case. The data model describes the classes of data objects and valid state transitions for each class. In fCM, there is always one crucial class for each case—the *case class*. Each case has exactly one object of the case class. Figure 1a depicts the classes for an insurance claim handling process:

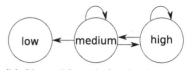

(a) Complete set of data classes of an insurance claim handling process

(b) Object life cycle for the class risk

Fig. 1. Excerpt from the data model of an insurance handling process.

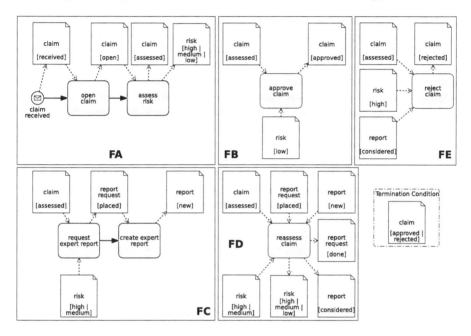

Fig. 2. Process fragments modeling a claim handling process at an insurance. A new case starts when a claim is received (fragment **FA**). Objects with a states separated by "|" can be in any of the specified states

claim, risk, report request, and *report.* The claim is the case class; there is no claim handling case without a claim. For each class, the case model contains an *object life cycle* defining the states and valid state transitions. Figure 1b depicts the object life cycle for the object risk.

Activities operate on data objects. Fragments contain the activities and define the variants of the process. In fCM, any fragment can be started as long as the data requirements for its first activity are satisfied. This allows multiple and concurrent executions of fragments. Figure 2 depicts the behavioral model of handling an insurance claim: first, the case is opened and the risk is assessed (**FA**). If the risk is low, the claim can be approved straight away (**FB**). Otherwise, expert reports must be requested and created (**FC**). Fragment **FC** can run multiple times sequentially and concurrently (as long as the data requirement of "request expert report" is satisfied). The claim's risk can be reassessed in respect to a report (**FD**). According to the object life cycle (cf. Fig. 1b), a high risk remains high or becomes medium, and a medium one can be reassessed as either low, high, or medium. If the new risk is low, the claim will be approved. If the risk is medium, further reports must be considered (and can be requested). If the risk is high, the knowledge workers may decide between rejecting it (**FE**) or considering/requesting more reports. The case can be closed once the claim is either rejected or approved (termination condition).

2.2   fCM Case Execution

The execution of a case is driven by knowledge workers and only constrained through data and control-flow requirements. The start event produces and instance of the case object *claim* in state *received*. Thus, only fragment **FA** can be executed. However, during execution, new data objects are created and existing ones are altered, which enables and disables activities and fragments.

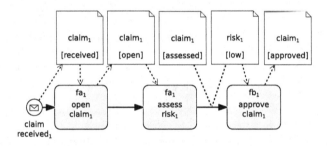

Fig. 3. "Happy" path: a claim with low risk is approved directly. Data objects that are connected to a control flow arc are created by the flow's source and read by its target

Figure 3 depicts a simple example case. We use BPMN [14] activity, event, and data object nodes to represent instances of the elements, respectively. The claim is opened, and the risk is considered to be low, which leads to the approval of the claim and the termination of the case. However, cases can be more complex including multiple instances of fragments, activities, and data classes. Figure 4 depicts another case. The risk is considered medium and two reports are requested. After the first report is created, the claim is reassessed, and the risk becomes high. The claim is rejected and the case is closed.

The second case (Fig. 4) contains two instances of fragment **FC**, fc_1 and fc_2. While instance fc_1 completes, "create expert report" of fc_2 is never executed. Furthermore, the instances run concurrently: first, two reports are requested and respective report request objects *report request*$_{1/2}$ are created. The activity instance *create expert report*$_1$ follows as part of fragment instance fc_1. However, it consumes *report request*$_2$, which was created in fc_2. Furthermore, the knowledge workers considered *report request*$_1$ and *report*$_1$ during the rejection: the

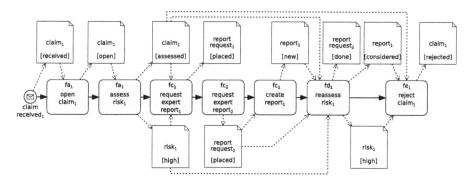

Fig. 4. A claim has a medium risk. After requesting two and considering one report the claim is reassessed as high and rejected

fragment and object instances mixed. It is the responsibility of the knowledge workers to avoid such situations. In the remainder, we show that this situation can be prevented automatically by adding associations between data objects and binding objects to fragment instances.

3   Design and Run-Time Support for Data Correlation

By nature, case management empowers knowledge workers by handing them responsibilities. One of the challenges is balancing between the degree of flexibility and the number of responsibilities. In fCM, the system orders activities within a fragment instance and enables tasks based on their requirements. However, the knowledge workers have to track associations among objects and use data in each fragment instance consistently. We extend fCM case models and their semantics so that the system enforces consistent data usage. To this end, we add data associations and bind objects to fragment instances.

3.1   Binding Object

During a case, multiple fragment instances can run concurrently. Knowledge workers can switch back and forth between them. Each fragment instance involves at least one but potentially many activity instances. Whenever knowledge workers start an activity, they can choose from all combinations of data objects that satisfy the activity's data requirements. By binding data objects to fragment instances, we limit these options.

Each fragment instance can involve a set of data objects, each of a specific type. In the example from Fig. 2, an instance of **FC** requires a *claim*, a *risk*, a *report request*, and a *report*. They are like free variables. While the knowledge workers execute the fragment instance, they choose from the objects that satisfy the requirements: when the workers execute an activity, we bind the involved objects, which are read or written, to the fragment instance. If the instance requires an object of the same type in the future, the knowledge worker must use the same object.

In the insurance example, an instance of fragment **FC** creates a *report request* object during "request expert report". Using the binding, we assert that "create expert report" reads the same report request object. This prevents traces in which objects of different fragments mix (see Fig. 4). A run which respects bindings is depicted in Fig. 5.

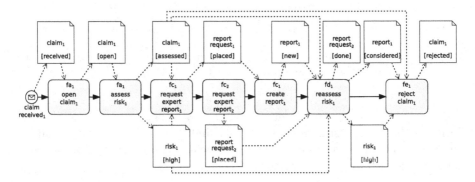

Fig. 5. An instance of the example that binds data objects to fragment instances. *reassess claim₁* can only access *report request₁* not *report request₂*

3.2   Correlating Data Objects

The example case in Fig. 5 still contains undesirable behavior: activity instance *reassess claim₁* reads *report request₂* and *report₁*, although *report₁* is a response to *report request₁*. So far, fCM neglects relationships among data objects. Naively two data objects are in a relationship if a single activity or fragment instance involves both, but this is not always desirable. We explicitly define possible relationships on the model level. Our extended fCM case model describes additionally binary relationships among data classes (cf. Sect. 2.1).

Figure 6 depicts the data classes of the insurance example including relationship types. The risk, as well as every report request, is associated to the claim. The report is only in a relationship with a report request. When knowledge workers execute tasks, they create data objects and instantiate the corresponding relationships. Relationships are created if

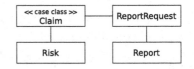

Fig. 6. The data model of the example extended with relationship types

- an activity instance reads a data object of type A, creates an object of type B, and the data model contains a relationship between A and B. EXAMPLE:

The activity "assess risk" reads an object of type claim and creates one of type risk, and a corresponding relationship has been modeled (Fig. 6); thus, the claim object and the newly created risk object are in a relationship.

- an activity creates an object of type A and an object of type B. Additionally, a relationship between A and B is in the data model.

The relationships among objects are considered during execution. In the simplest case, a single activity reads multiple data objects (i.e., "reassess claim" reads a claim, a risk, a report request, and a report). If the corresponding types are related in the data model (i.e., claim and risk, claim and report request, and report request and report), only associated data objects will be consumed by this activity. Figure 7 shows an example case. In contrast to Fig. 5, *reassess claim*$_1$ does not mix *report*$_1$ with *report request*$_2$.

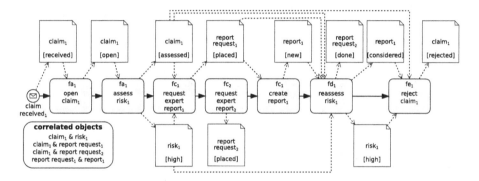

Fig. 7. An instance of the example respecting the binding of objects to fragments and relationships between objects (see box **correlated objects**)

Furthermore, our approach considers relationships between objects read by an activity and objects bound to the corresponding fragment instance. Let A and B be two associated classes. If an activity 1 reads an object of type A it is bound to the corresponding fragment instance. Assuming that activity 2 reads an object of type B and succeeds 1 within the same fragment instance then our approach requires that the objects are associated.

4   Evaluation

Our extensions enable an fCM system to enforce consistent data usage within a case. In order to build such a system, precise execution semantics are necessary. Formal execution semantics of fCM models have been defined in [9]. The authors map fCM models to classical Petri nets. This approach is insufficient for our extensions, since tokens are indistinguishable and incapable of representing identities, correlations, and bindings.

We extend the approach from [9] by mapping the execution semantics of the refined case models to colored Petri nets (CPNs)[1]. CPNs are formal models that support concurrency, typed data, as well as conditions and operations on data [10]. A rich set of tooling (e.g., CPNtools [11]) exists to model, simulate, and verify CPNs. To capture the semantics of refined fCM models, a mapping must

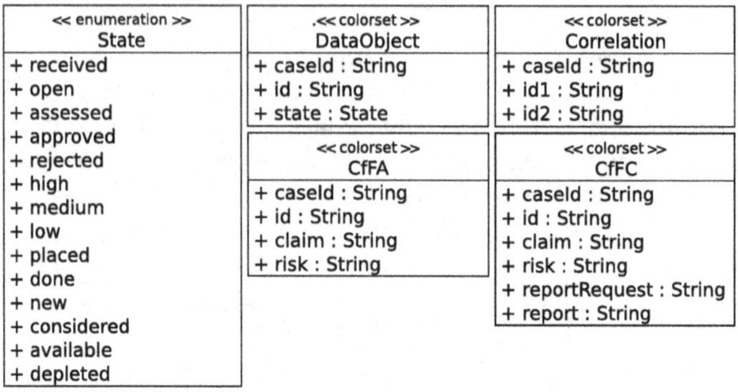

<< enumeration >> State	.<< colorset >> DataObject	<< colorset >> Correlation
+ received	+ caseId : String	+ caseId : String
+ open	+ id : String	+ id1 : String
+ assessed	+ state : State	+ id2 : String
+ approved		
+ rejected	<< colorset >> CfFA	<< colorset >> CfFC
+ high		
+ medium	+ caseId : String	+ caseId : String
+ low	+ id : String	+ id : String
+ placed	+ claim : String	+ claim : String
+ done	+ risk : String	+ risk : String
+ new		+ reportRequest : String
+ considered		+ report : String
+ available		
+ depleted		

Fig. 8. Color sets for the insurance example

– distinguish between entities to represent different data objects, fragment instances, and cases; and
– support structured types to represent the correlations among objects, and the binding of objects to fragment instances.

CPNs support both. We can implement ν-transitions. A ν-transition produces unique identifiers, which can be used to distinguish between different entities [15]. Furthermore, places and tokens are typed. We can define tokens, that store data objects (an identifier and a state), bindings (an identifier and references to data objects), and correlations (two identifiers). Figure 8 depicts the color sets for the example. We generalize the data objects into a single type `DataObject`. We have a color set supporting correlations and one for each non-atomic fragment representing the control flow and bindings. Each color set has an additional attribute referencing the case. On an abstract level, the CPN formalization has the same structure as the corresponding case model. Each activity is represented by a transition. It consumes and produces tokens for each data object that is read or created by the activity, respectively. If the activity has a successor or a predecessor, it produces and/or consumes a token for the control flow. The different fragments are only connected via data objects.

---

[1] Complementary files (CPNtools file, images, and screencast) are available at https:// owncloud.hpi.de/s/asYhqMnBFp73wMc.

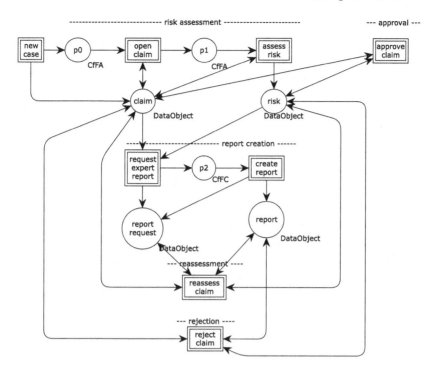

Fig. 9. The top-level CPN represents the insurance example formally. Correlation, binding and other details are hidden on this abstraction level. Each transition is a placeholder for a sub-Petri net

Figure 9 depicts an abstract CPN representing the structure of the insurance example. Places *p0*, *p1*, and *p2* store control flow tokens representing fragment instances (i.e., color sets CfFA and CfFC). Other places hold data objects corresponding to their label. Each transition hides a complete sub-net. The transition "new case" has no input places and represents the event "claim received". In the sub-nets, activities are mapped to transitions that consume and produce tokens representing data objects, fragment instances (including bindings), and correlations. Through guards and arc expressions, we assert bindings and correlations.

Figure 10 depicts two sub-nets Fig. 10a for the activity "request expert report" and Fig. 10b for "create report". The nets contain additional places for correlations (i.e., *request report* stores correlations between *report requests* and *reports*) and for counters, which are used to assign unique identifiers (e.g., *count request* holds the total number of *report request* objects). The activity "request expert report" marks the beginning of the fragment **FC**. The respective transition creates a control flow token. The token's attributes *claim*, *risk*, and *reportRequest* are set to the ids of the involved objects (e.g., attribute *claim* identifies the read claim). Attribute *report* has the value *NULL* since no *report* object is bound. The places *claim risk*, *request report*, and *claim request* have the type `Correlation`. Any token that they hold references two objects (according to the place's label) and indicates a corresponding correlation.

Whenever an activity uses a data object, the transitions in the CPN consume tokens according to the binding of objects to a fragment instance (only for fragments with more than one activity) and update the binding if necessary. The same holds for correlations. The transition "create report" consumes a control flow token of type CfFC. It furthermore consumes the report request that is referenced by the control flow token. The transition produces a new *report* token and a new token for the correlation of the *report request* and the *report*.

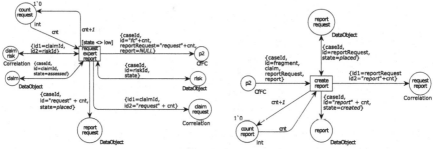

(a) Petri net for the activity "request expert report".

(b) Petri net for the activity "create report"

Fig. 10. Sub-Petri nets for the fragment **FC**. If an attribute's value is assigned to a variable with the same name (e.g, $CfFC.claim = claim$) we abbreviate it as the name (e.g., $\{caseId, id = riskId, state\}$ means $\{caseId = caseId, id = riskId, state = state\}$)

5   Related Work

Data plays an important role in business processes, and major modelling languages offer support for modelling their mutual relationship. The correlation of data objects to one another and to process instances is an important aspect in process choreographies. The WS-BPEL language supports property-based correlation of activities in so called *correlation sets* [13]; analogously, BPMN describes key-based correlation and context-aware correlation. The context-aware correlation uses data objects and properties of the process instance to dynamically correlate messages [14]. BPMN's and WS-BEPL's correlation mechanism for inter-organizational processes are content-based, while our correlation mechanism is action-based and used within a single case-model.

In business process modelling, a rich set of data-centric modeling languages have been proposed. Steinau et al. provide a framework and an overview of data-centric approaches in [16]. Out of the plethora of related work, we present some representative approaches.

Artifact-centric process models take a data-centric point of view: an artifact-centric model contains a state-transition system for each artifact. The transitions in different objects can depend on each other and be synchronized [5]. While artifact-centric approaches are data-oriented, they focus neither on flexibility per se nor on case-management.

Proclets are a Petri net-based notation for data-centric process models. Each Proclet is a process describing a "case" from a single object's point of view. The Proclet system consists of multiple proclets (for different objects) that exchange messages through channels [2]. Similar to fCM, Proclets support multiple objects of the same type and can express dependencies between objects [6]. While Proclets can define relationships between data more precisely than fCM, they break with the traditional notion of a case. In contrast to fCM, Proclets address data-driven processes instead of knowledge-driven and ad hoc ones.

Object centric behavioral constraints (OCBC) is another approach towards flexible and data-centric processes. OCBC is based on the observation, that real-world *event logs* do not always support a traditional notion of a process instance or case. The execution of activities depends on the available data as well as past activities. OCBC models define a data model (with associations and cardinalities) as well as constraints among activities. The data model and the behavioral model are connected: constraints between activities are quantified over data, and cardinality constraints between activities and relationships can be modeled. OCBC's main application area is process mining [1]. It is well suited to describe even complex event logs. In contrast to OCBC, fCM sustains a notion of a case as well as control flow. Also, fCM has mainly been used for process modeling, analysis, and execution and OCBC for process mining.

While many approaches address data-centric and flexible processes, fCM takes a unique standpoint. By splitting knowledge-driven processes into smaller repetitive fragments that share a common set of data objects, fCM captures multi-variant processes compactly [12].

6   Conclusion

BPM offers methods for use-cases ranging from the elicitation to the enactment and to the analysis of business processes. Case management approaches comprise such methods for flexible knowledge-driven processes. The fragment-based case management approach (fCM) captures multi-variant processes as a set of smaller process fragments that operate on a shared set of data objects. We extended fCM to support data-to-data relationships and object-to-fragment bindings.

Our extension to fCM case models captures relationship types between data classes and binds objects to fragments for consistent data usage. The adapted execution semantics allows managing data relationships through an automated system rather than manually by knowledge workers. Furthermore, we formalized the semantics of an extended case model using colored Petri nets. The formalization can be the foundation for verification and further implementation.

Acknowledgment. We acknowledge and thank Marcin Hewelt, the originator of fCM, for sharing his thoughts and work not only in his publications but also during presentations and in face-to-face interactions.

References

1. van der Aalst, W.M.P., Artale, A., Montali, M., Tritini, S.: Object-centric behavioral constraints: integrating data and declarative process modelling. In: Proceedings of the 30th International Workshop on Description Logics, Montpellier, France (2017). http://ceur-ws.org/Vol-1879/paper51.pdf
2. van der Aalst, W.M.P., Barthelmess, P., Ellis, C.A., Wainer, J.: Workflow modeling using proclets. In: Scheuermann, P., Etzion, O. (eds.) CoopIS 2000. LNCS, vol. 1901, pp. 198–209. Springer, Heidelberg (2000). https://doi.org/10.1007/10722620_20
3. van der Aalst, W.M.P., Weske, M., Grünbauer, D.: Case handling: a new paradigm for business process support. Data Knowl. Eng. **53**(2) (2005). https://doi.org/10.1016/j.datak.2004.07.003
4. Beyer, J., Kuhn, P., Hewelt, M., Mandal, S., Weske, M.: Unicorn meets chimera: integrating external events into case management. In: Proceedings of the BPM Demo Track Co-Located with the 14th International Conference on Business Process Management (BPM 2016), Rio de Janeiro, Brazil (2016). http://ceur-ws.org/Vol-1789/bpm-demo-2016-paper13.pdf
5. Cohn, D., Hull, R.: Business artifacts: a data-centric approach to modeling business operations and processes. IEEE Data Eng. Bull. **32**(3) (2009). http://sites.computer.org/debull/A09sept/david.pdf
6. Fahland, D.: Describing behavior of processes with many-to-many interactions. In: Donatelli, S., Haar, S. (eds.) PETRI NETS 2019. LNCS, vol. 11522, pp. 3–24. Springer, Cham (2019). https://doi.org/10.1007/978-3-030-21571-2_1
7. Gonzalez-Lopez, F., Pufahl, L.: A landscape for case models. In: Reinhartz-Berger, I., Zdravkovic, J., Gulden, J., Schmidt, R. (eds.) BPMDS/EMMSAD -2019. LNBIP, vol. 352, pp. 87–102. Springer, Cham (2019). https://doi.org/10.1007/978-3-030-20618-5_6
8. Hewelt, M., Weske, M.: A hybrid approach for flexible case modeling and execution. In: La Rosa, M., Loos, P., Pastor, O. (eds.) BPM 2016. LNBIP, vol. 260, pp. 38–54. Springer, Cham (2016). https://doi.org/10.1007/978-3-319-45468-9_3
9. Holfter, A., Haarmann, S., Pufahl, L., Weske, M.: Checking compliance in data-driven case management. In: Di Francescomarino, C., Dijkman, R., Zdun, U. (eds.) BPM 2019. LNBIP, vol. 362, pp. 400–411. Springer, Cham (2019). https://doi.org/10.1007/978-3-030-37453-2_33. https://surfdrive.surf.nl/files/index.php/s/1x7PMzTjsFRfw6y
10. Jensen, K., Kristensen, L.M.: Coloured Petri Nets - Modelling and Validation of Concurrent Systems. Springer, Heidelberg (2009). https://doi.org/10.1007/b95112
11. Jensen, K., Kristensen, L.M., Wells, L.: Coloured petri nets and CPN tools for modelling and validation of concurrent systems. STTT **9**(3–4) (2007). https://doi.org/10.1007/s10009-007-0038-x
12. Meyer, A., Herzberg, N., Puhlmann, F., Weske, M.: Implementation framework for production case management: modeling and execution. In: 18th IEEE International Enterprise Distributed Object Computing Conference, EDOC 2014, Ulm, Germany (2014). https://doi.org/10.1109/EDOC.2014.34

13. OASIS: Webservice business process execution language (2007). https://docs.oasis-open.org/wsbpel/2.0/OS/wsbpel-v2.0-OS.html
14. Object Management Group (OMG): Business process model and notation (BPMN) version 2.0 (2014). https://www.omg.org/spec/BPMN/
15. Rosa-Velardo, F., de Frutos-Escrig, D.: Name creation vs. replication in Petri net systems. Fundam. Inform. **88**(3) (2008). http://content.iospress.com/articles/fundamenta-informaticae/fi88-3-06
16. Steinau, S., Marrella, A., Andrews, K., Leotta, F., Mecella, M., Reichert, M.: DALEC: a framework for the systematic evaluation of data-centric approaches to process management software. Softw. Syst. Model. **18**(4) (2019). https://doi.org/10.1007/s10270-018-0695-0
17. Swenson, K., Palmer, N., Pucher, M.: How Knowledge Workers Get Things Done. BPM Books by Future Strategies Inc. (2012)
18. Weske, M.: Business Process Management - Concepts, Languages, Architectures, 3rd edn. Springer, Heidelberg (2019). https://doi.org/10.1007/978-3-662-59432-2

Countering Congestion: A White-Label Platform for the Last Mile Parcel Delivery

Luise Pufahl[1]([✉]), Sven Ihde[1], Michael Glöckner[2], Bogdan Franczyk[2], Björn Paulus[3], and Mathias Weske[1]

[1] Hasso Plattner Institut, University of Potsdam, Potsdam, Germany
{luise.pufahl,sven.ihde,mathias.weske}@hpi.de
[2] Information System Institute, Leipzig University, Leipzig, Germany
michael.gloeckner@uni-leipzig.de, franczyk@wifa.uni-leipzig.de
[3] Pickshare, Dortmund, Germany
bjoern@pickshare.de

Abstract. The success of online shopping combined with the convenience of home delivery leads to massive congestion in cities. CEP (Courier-Express-Parcels) service provider have increasing cost and service pressure, especially in the last mile parcel delivery. Therefore, we propose a process-based white-label last mile delivery platform as a smart city approach to counter congestion. It allows the consolidation of parcels on the last mile, considers customer preferences, and gives local carriers access to the parcel market. This platform is conceptualized based on insights from interviews and workshops with experts. Experiences from a pilot study are discussed.

Keywords: Inter-organizational process · Parcel delivery · Last mile logistics · Conceptual architecture · Smart cities

1 Introduction

An increasing revenue of e-commerce and the inherent increase of parcels being delivered can be recognized during recent years. This trend will continue resulting in an accelerated growth in the e-commerce sector and a forecast of the growth from 1.3 trillion global revenue per year in 2017 up to 2.1 trillion in 2022 is made [18]. Courier-express-parcel logistics service providers (CEP) are specialized in delivering those parcel ordered via the internet to the buyer. In general, the processes of big CEP, such as DHL or UPS, are highly optimized and integrated. One exception are the processes on the last mile (i.e., transportation from the final hub to the final receiver). Especially in the B2C business, the final recipient is a factor of uncertainty which results in a low drop rate, meaning parcels are not being delivered successfully on the first try. Instead, a second or third attempt is needed, or the parcels are being temporarily stored in a depot to be picked up

© Springer Nature Switzerland AG 2020
W. Abramowicz and G. Klein (Eds.): BIS 2020, LNBIP 389, pp. 210–223, 2020.
https://doi.org/10.1007/978-3-030-53337-3_16

by the receiver itself. The net of those depots is usually not very dense split up between the different CEP. This low drop rate leads to higher costs –up to 28 % of the total delivery costs [16]– and inefficiencies for the CEP, and a higher traffic in cities with have the negative effects, such as noise, air pollution, traffic jams etc.

Thus, the last mile delivery is one of the main points of interest in the context of smart cities. As the big players of CEP rely on the fact that their parcels are only to be handled by the CEPs' proprietary systems and infrastructure, one promising proposed solution is a white label approach,[1] using standards such as the Serial Shipping Container Code (SSCC) of GS1 [7]. If infrastructure and transport capacity could be shared with other CEP or even smaller CEP (denoted as 'local carriers' in this paper) or the crowd [12] due to such a white label approach, the last mile could be run more flexible and with higher efficiency, e.g. by shared depots (denoted as 'micro depots' in this paper) more widespread throughout the city or specialized local carriers enabling customer-oriented time slot delivery.

In order to establish such a white label approach and enable the afore-mentioned benefits of flexible collaboration of different stakeholders, the inter-organizational information and communication as well as flexible processes and services have to be ensured [13]. The technical challenge can be solved by a plat-form approach to flexibly connect all stakeholders, i.e. carriers, micro depots, couriers, and receivers, in order to deliver smart services based on flexible busi-ness process management [1]. The developed platform (1) allows the consolida-tion of parcels to be handled, operated and delivered by different CEP on the last mile, (2) considers dynamic receiver's preferences, and (3) involves local car-riers, usually more efficient in their neighbourhoods, and the concept of crowd logistics. Summarizing the contribution of this paper is a conceptual architecture for a smart service platform that enables internal (composition of processes) as well as external (connection to different and changing stakeholders) flexibility in order to enable customer-oriented and less traffic-oriented last mile logistics.

The Design Science Research approach in accordance to [9] is applied to cre-ate the conceptual architecture of such a smart service platform. The artifact is created with a prototyping method [10] and evaluated with the Framework for Evaluation in Design Science Research (FEDS) [20] in a first pilot study. Lead by a Design Science Research approach [6,9] the paper is structured as follows: in the first section the general problem and solution context are introduced as well as the methodological framework. Section 2 summarizes the status quo and the challenges of the current parcel delivery process, especially in Germany. Section 3 briefly introduces last mile solutions of the related work. Then, Sect. 4 presents the details of the developed artifact and shows the sub-systems of the archi-tecture as well as their interrelations and connections. Afterwards, the artifact is evaluated by a pilot study and lessons learnt presented in Sect. 5. Section 6 concludes the findings and gives an outlook on future research directions.

---

[1] http://smile-project.de/.

2   Current Parcel Delivery Process and Its Challenges

In this section, we present the current parcel delivery process in general and its current challenges, especially on the last mile of the delivery. In addition to the analysis of literature, interviews with experts from the CEP industry and related fields were conducted. Further, a survey[2] among receivers with an online-questionnaire was conducted with 318 participants.

2.1   Current Parcel Delivery Process

The process with its main actors and the messages exchanged between them is visualized in Fig. 1 as a BPMN collaboration diagram. When a sender prepares a parcel for sending ordered items to its customer, first, a CEP is selected and then the information about the parcel (e.g. size, weight) and the recipient (e.g. name, address) are provided. Then, the parcel is given to the CEP service. The CEP service tags this parcel with a proprietary ID and encrypted information about the sender and receiver.

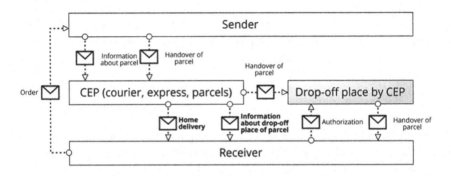

Fig. 1. Current parcel delivery process

In the next step, the CEP transports the parcel consolidated with other parcels of the same main direction to a hub close to the receiver. The distance from this hub to the receiver comprises the last mile where the CEP service provider brings the parcel from the hub directly to the address of the receiver.

However, receiver are often not at home during the standard (not-customized) delivery time windows, resulting in a failed delivery. Recently some CEP service providers try to increase their success rate in the first-attempt-delivery with different means e.g. private parcel box or a (static) preferred location, if the receiver has given an approval. But usually if the home delivery is not possible, then the parcel is dropped at a selected place by the CEP as near as possible to the receiver, e.g., in a post office, at a neighbor's apartment, or at other drop-off places such as, a public parcel box like DHL Packstation, parcel services

---

[2] http://smile-project.de/wp-content/uploads/2019/04/GS1_SMile_Broschuere.pdf.

integrated in a retail location. For this drop-off, each CEP has its own proprietary infrastructure. To save time and money the CEP service tries to drop all their parcels off at once, which in turn may lift the goal of choosing a place as near as possible to the customer. In order to enable the customer to get the parcel, the receiver is notified about the drop-off place. The receiver is then expected to visit this place and get the parcel. In addition, the receiver is bound to the opening hours of the drop-off places or the presence of the neighbor.

2.2   Challenges and Requirements

The following challenges arise from the current parcel delivery process for the CEP service providers, for the receivers, and for the communities:

High cost pressure for the parcel delivery: Home delivery is currently considered and advertised as a free service for customers. This leads to the CEP service provider optimizing the delivery process as much as possible to reduce its cost. However, as receivers are usually not available during the day, because they are working, parcels can often not be delivered on the first try. Thus, the CEP needs to either start a new delivery attempt, or find an alternative drop-off place, which results in multiple additional process steps and thus, increases the cost of the last mile.

Proprietary infrastructure of CEP: In order to stay competitive, CEP use multiple techniques (e.g., using proprietary parcel IDs and parcel information) to protect their volume. This leads to a high amount of redundancy. Each CEP manages, operates and maintains its own physical and informational infrastructure. In case that several parcels for one recipient are sent via different CEP services, they might end up at different drop-off locations, e.g. parcel 1 at the post office and parcel 2 at a neighbor. Thus, a high effort for the recipient to get all the shipments.

No service diversification in parcel delivery: No matter who is ordering in the process, each customer is handled the same. This is also observed to a greater extend independently of what kind of goods the order is containing. However, this leads to missed opportunities in customer satisfaction as, based on the combination of customer and good, different requirements of the receiver are expected to be met.

No consumer involvement during parcel delivery: As the success of a delivery is highly dependent on the availability of the receiver, CEP services miss the opportunity to actually involve the receiver for better results. Some ideas exist (e.g. personal drop-off places, time window selection, etc.) but they are not yet fully employed by CEP. Especially with the digitalization all necessary information are easily collectable, and thus it's possible to improve the flexibility and success of parcel delivery, enabling a service-oriented perspective involving the customer and/or its information as an important stakeholder for value creation.

Increased traffic and ecological consequences: As parcels from different CEP service provider are not consolidated on the last mile the cities have to face more and more problems due to an increased traffic from CEP trucks and parcel delivery vans of various companies serving the same location at the same time. Additionally with multiple attempts for the drop-off of a parcel, the communities and cities are facing an unnecessary CO_2-emission.

3   Last Mile Solutions in the Related Work

An overview on last mile solutions based on a literature survey is provided by Ranieri et al. [16]. In research, the problem solution of last mile logistics can be broadly summarized in two categories. The first category describes physical systems and infrastructures of the last mile, such as discussing different forms and locations of transshipment areas [4]. This will in turn lead to a reduced traffic between the depot and remote areas of distribution. Another solution is a modular box system in order to increase drop rate on the first attempt. Even though, an improved infrastructure can tremendously improve the situation by making every first drop attempt successful, this kind of infrastructure with modular parcel boxes is very expensive. Nevertheless, this approach shows the effectiveness of a dense net of drop locations. The idea of urban consolidation centers (UCC) is picked up by [8]. The UCC could be operated by governments' initiatives [14] or company alliances, and function as a cross docking point for re-ordering shipments provider-independent concerning destination area. The authors tried to optimize the profit of the UCCs by auction mechanism.

The second category focuses more on an advanced matching and synchronization of existing resources based on an increased flow of information. Liakos and Delis [11] propose a freight-pooling service in order to reduce traffic and increase occupancy rates. Petrovic et al. [15] suggest an advanced interactive end-to-end communication between service providers and customers in order to increase delivery quality. They also emphasize the need of integrating the information of all stakeholders, i.e. senders and recipients. Further, they interestingly outline a shift from location-oriented to person-oriented services in the last mile sector. Similarly, the paper [17] raises questions for future development effort. This comprises collaboration of multiple stakeholders (such as shippers, CEP, and customers) via a common platform, as well as the demand for a common framework and possibilities of visualization and real time data availability. Suh et al. [19] demonstrate the feasibility and increased efficiency of an intelligent last mile approach enabled by a mobile ICT platform providing real-time communication and thus an enhanced transparency. The authors state a main challenge is the amount and distribution of central pickup locations. Further research directions comprise the creation of individual receivers' networks in order to increase efficiency of the crowd approach as well as the integration of alternative transportation technologies. Approaches of mobile crowd sourcing are related to the topic of sustainable last mile logistics, e.g. see [21] or the participation of citizens in an urban context of smart cities in [2] and [3].

The result of the related work analysis shows several important points and challenges when tackling problems of the last mile logistics field. On the one hand, there is a need for improvement of the physical infrastructure, in terms of a *dense net of hubs* for pick-ups that are *carrier-independent*. On the other hand, there are several points to increase the efficiency in the use of existing resources by advanced information systems for a higher transparency. This comprises a collaborative approach with the *pooling of resources* as well as the *integration of information* and collaboration of several stakeholders. This can be realized via a *platform* that is ideally operated by an independent third party to avoid discrimination – a *white-label platform*. Especially, the integration of the *crowd and flexible recipients' networks* will foster sustainability and acceptance.

4   Design of a White-Label Last Mile Delivery Platform

In this section, we introduce our concept, combining existing ideas in one smart city approach, to improve the last mile with a *white-label last mile delivery platform*, which has the following goals:

- Provide a white-label pick-up infrastructure, from which receivers can select their desired place for delivery
- Provide a diversed customer-oriented last mile service portfolio from self-pick up to an individual home delivery
- Provide local carriers an access to the parcel market using a standard for parcel identification. Thus enabling the matching of receiver's demand and carrier's offers.

The conceptual architecture of the platform was developed in several work-shops with experts in different domains of the concept. The experts had background in the following domains: 1) standards and technologies in the CEP and sender market, 2) last mile logistics, 3) information technology in logistics, 4) tracking and tracing technology, and 4)business process-oriented technology. Additionally, we conducted a focus group discussion in which 15 CEP experts[3] of different sizes participated. Based on the results of this focus group discussion, we improved and adapted the concept.

In the first part of this section, the adapted delivery process with an improved last mile solution is described. Afterwards the conceptual architecture of the white-label last mile delivery platform is presented.

4.1   Improved Last Mile Delivery Process

The idea of our last mile delivery platform is that the customer can select a white-label pick-up place that is the most convenient for him/her. A pick-up place can be any place that can offer storage of parcels, for example, a regular

---

[3] They were invited from the German Federal Association of Courier-Express-Post (BdKEP) https://bdkep.de/.

visited fitness studio, a bakery, a retail shop, or even a private person that has the benefit of being available during the normal operating times of the CEP service providers. This pick-up place is *white-label* because it is not related to any specific CEP service provider in contrast to the general parcel-shops. Each receiver will choose his/her own desired pick-up depending on his/her trust in a place and the convenience to reach this place.

As visualized in Fig. 2, during the checkout of an order, the receiver will then substitute their own address with the address of the pick-up place combined with a user-identifying id. Then, after the sender has given the parcel to the CEP service provider, it will be delivered directly to the desired pick-up place in the first try as shown in Fig. 2.

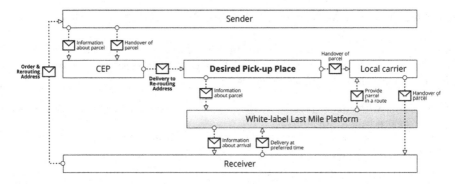

Fig. 2. Innovated parcel delivery process with a white-label last mile delivery platform.

Two advantages arise for the receiver from this: (1) The receiver can get the parcels any time in the open hours of the pick-up place, and (2) even, if the customer is expecting parcels from different companies, they will be consolidated at one pick-up place. Further, it reduces the failed delivery attempts of the CEP service providers.

As soon as the parcel arrives at the pick-up place, the customer has full control on how and when the parcel will be delivered to him/her. The parcel can be obtained by different means to the receiver. In the moment, we see the following three alternatives:

1. Customer pick-up the parcel on their own
2. Customer shares the parcel and allows another person to bring the parcel
3. Customer orders a local carrier to bring the parcel in a desired time frame

In Fig. 2 and more visually in Fig. 3, the third option is shown. After the parcel has arrived at the pick-up place, the receiver gets an offer for a possible home delivery at a time complying to the preferences of the customer. These time preferences are requested from the receiver by the platform beforehand. If he/she accepts the offer, a local carrier will deliver the parcel to the receiver.

Our online survey of receivers has shown that they are willing to pay some fees for the extra service of delivery at a desired time frame; in general 71% of the study participants are willing to pay for more service with almost 40% of study participants stating to at least pay 1 €. This number increases to 58% willing to pay 1.99 € if they expect an important delivery.

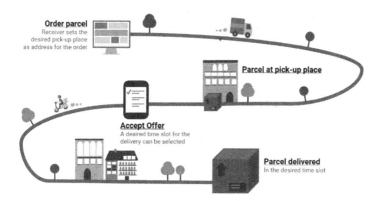

Fig. 3. Delivery of a parcel from the initial order via a pick-up place, finally delivered according to the receiver's preferences at a desired time.

On the other hand, the focus group has shown that local carriers, such as bike couriers working mainly in the B2B business today, are willing to take extra tours with parcels to consumers, if such a tour met their price conditions. It is the goal to involve these local carriers with their price conditions for a certain area in the process. The more parcels involved in a tour, the better the prices for the receivers. How exactly the local carriers are involved is further explained in the next subsection.

Still, when the receiver does not accept the offer, he/she has the flexibility to collect the parcel him/herself or to share it with a friend who will collect it then or just decide on a new time and place for future delivery. After having introduced the new last mile process connecting receivers and local carriers of a community, the next subsection introduces the conceptual architecture of the platform.

4.2   Last Mile Delivery Platform

A conceptual architecture of the envisioned last mile delivery platform is shown in Fig. 4. The platform will provide three user interfaces (UIs) to the different actors on the last mile, the receivers, the pick-places, and the local carriers:

- **Receiver UI:** Receivers can use the UI to enter a profile, to select a pick-up place, to find friends, whom they ask for taking a parcel with them, and to set time preferences, if a time delivery is preferred.

- **Pick-up Place UI:** Pick-up places can use the UI to create a profile with opening hours, to enter newly arrived parcels, and to verify the hand-over of a parcel to the receiver/friend/local carrier
- **Local Carrier UI:** Local carriers can use the UI to create profiles including their possible load, availability and their area of acting via postal codes, as well as their stop prices and prices per distance. Further, they can get their (pre-optimized) tours, and are guided through the tour. With a successful tour completion, the payment is also triggered via this UI.

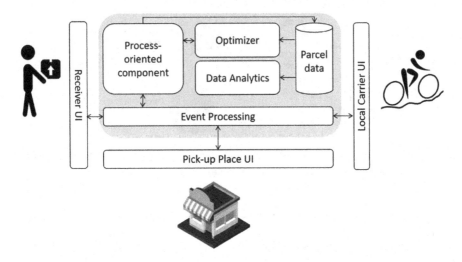

Fig. 4. Conceptual architecture of the last mile delivery platform

The last mile delivery platform in the middle between the UIs coordinates the communication between the different actors, distributes relevant information, and synchronizes the inter-organizational last mile delivery process. All three interfaces will communicate over an *event processing* [5] component with the core platform to react flexibly on events by the UIs and to distribute important events, such as the arrival of a parcel. In addition to the event processing component, the last mile delivery platform consists of three other internal components: a process-oriented component, an optimizer, and a data analytics component. All parcel-relevant information are stored in an internal database. The components are further explained in the following:

Process-oriented component. As soon as a parcel arrives in a pick-up place, the coordination and monitoring of the parcel is started in the last mile delivery platform. The process-oriented component is responsible that a receiver gets an offer for a delivery at the desired time, if a carrier is available in the delivery area. If the offer is not accepted, the component guides the self-pick-up or the sharing with friends. The goal is here to use a process model-driven implementation, which starts from a graphical representation of the delivery process. Verification techniques can be also used to check that the process is correctly implemented. During the process execution, the process-oriented component stores all relevant data in the internal data base which is used by the data analytic and optimizer component.

Event processing. Collaborative business processes, such as the last mile delivery process, require elaborated communication and transparency between the participants, which should be supported in an efficient manner by the platform. In a situation, where several receivers are connected with different carriers and pick-up places an asynchronous communication is the most suitable technique (i.e., compared to a synchronous communication). In practice, asynchronous communication is realized via sending and receiving relevant events. In a setting, such as the parcel delivery, a high amount of events regarding the status of the parcels has to be handled. Event processing platforms have the advantage of addressing this challenge and allow an efficient handling of multiple event streams. Moreover, such platforms allow for individual notification rules for each participant or business process case.

Optimizer. For generating tours involving several parcels, this component considers the parcels available, the preferred periods by the receivers, and the availability of carriers and their current workload. The optimization is a complex problem, which known as *vehicle routing problem with time windows*. It needs to be solved by an algorithm, which finds a good solution in a reasonable amount of time. For different regions, different algorithms might be relevant, such that the optimization is outsourced in an extra component. The optimization is planned to be started once a day in the early afternoon to create tours for the evening where most of the receivers would like to have their parcels. The price of a tour depends on the condition of the carrier of a specific region and will be split up equally among the parcel receivers included in the tour.

Data analytics. The data analytics component visualizes pre-defined key performance indicators of the last mile delivery process and allows a monitoring of the process based on the data produced by the process-oriented component.

5   Experiences from a Pilot Study

In this section, a pilot study is presented, in which especially the service of home delivery in a desired time frame by a local carrier was tested. Thereby, we wanted to study the experiences the receivers and local carriers have, when using the adapted last mile process with the white-label platform, which was prototypical implemented.[4]

---

[4] https://github.com/bptlab/chimera/tree/smile.

For the pilot, which was running for 6 months (February until July 2019), a pick-up place was opened in a mid-size city with 300.000 inhabitants in Germany. From this pick-up place, a bike carrier offered the service of home delivery of parcels in one hour time frames for three postal codes in a radius of around two kilometres. The carrier delivered with an electric cargo bike, highly frequently, where the customers could inform about their desired hour of delivery up to one hour before (i.,e. the *cut-off time*) the delivery. Different prices were provided for different times of delivery: delivery between 4–6pm for 2,49 €, delivery between 6–8pm for 3,49 € (due to the rush hour in the city), and delivery between 9–23pm for 2,99 € (due to being very late). The bike carrier was also running the pick-up place, which allowed the high-frequent delivery rates. A chatbot was selected as user interface in order to provide a more interactive experience for the receiver. With the help of the chatbot, the customers could actively reroute their parcels to the pickup place or ask for further explanations about the process. Usually the CEP delivered parcels between 10 am and 1 pm into the pick-up place. Arrived parcels were entered and their receivers were automatically informed about their parcels via the chatbot. Then, with the incoming wishes by the customers, the carrier arranged the parcels based on the times of delivery. From 4pm onward, the parcels were delivered to the customers. A customer payed the bike carrier as soon as the parcel arrived in cash or via credit card.

For finding potential customers, local marketing was done by flyers, in social media, and by a TV documentary in a local channel. The first delivery of a parcel was for free. Around 100 customers could be acquired for this small area and 562 parcel deliveries were conducted. In the following, we report on the experience of the customers, which were deduced by the chatbot (52 additional comments by the customers), and by the bike carrier, who was interviewed:

Customers. In general, they perceived the service as very comfortable and enjoyed to have a carrier who is familiar to them. However, they would not like to use the home delivery for each parcel, as the service was perceived as rather expensive. This feeling might have been intensified by paying the carrier directly after delivery. They would use it more, if several parcels could be also consolidated. This feature was possible but not very intuitively shown in the chatbot. The service was more used by people between 30 to 40 years, who did not enjoy so much the communication with the chatbot. They perceived it too manual and too complicated, especially the active rerouting where they had to use a different address when ordering products.

Bike Carrier. The bike carrier enjoyed driving in his neighbourhood. He was familiar with the area, and could drive very efficiently. Knowing the neighbourhood and also some people in this area, the carrier could feel a more personal contact in comparison to normal CEP drivers. This personal contact leaded also to a high valuation of his work by the customers. He also received tips from time to time. The highly flexible delivery rate was sometimes challenging to organize as the tour could be only planned one hour in advance. It was still possible because the driving radius was limited.

The following lessons learnt were deduced from this pilot:

- **Higher valuation of the home delivery:** The pilot study showed that a personal contact to the parcel carrier and also a price for home delivery leads to a higher valuation of this service again.
- **Integrated customer experience:** Customers of the pilot study described the chatbot as complicated to use because they needed to adapt to it. Additionally they had to manually provide their information for each checkout. Therefore, the idea is to provide an easier order process of e-commerce retailers by providing a checkout module that just automatically connects with our platform to provide the necessary information. Additionally, a smart phone app can be used, visualizing all ordered parcels and the different last mile services that can be selected.
- **Service diversification:** The pilot study showed that a flexible last mile delivery portfolio should be available as proposed in our last mile delivery platform. Customers do not like to pay for each type of parcel. Furthermore, they also like to have the option to consolidate different parcels for one home delivery, saving delivery costs.
- **Longer cut-off times:** In this special setting of a pilot, where the carrier was very familiar with the neighbourhood and also run the pick-up place, a cut-off time of one hour was possible. However, it will be not applicable for any carrier. Therefore, we strive for an approach, where the carrier can define its cut-off time individually.

6   Conclusion

The paper presented challenges of the established parcel delivery process, especially on the last mile, which are not only of concern by CEP service providers, but also by receivers due to decreasing service quality and by communities due to increasing traffic. Based on this, the paper presented a white-label last mile delivery platform adapting and innovating the last mile delivery process by consolidating parcels at pick-up places preferred by receivers and offering different means of the final delivery: (1) a self-pick-up, (2) sharing of the parcel pick-up with friends, or (3) the delivery at desired timeframes by local carriers. Opening the closed CEP infrastructure through white-label pick-up places, offers local carriers, often traveling with more economic-friendly solutions, e.g., bikes, to participate in the current parcel market of smart cities.

A first pilot study showed that customers valued the service, and the carrier preferred the delivery in an area being familiar with. However, an integrated customer experience beginning at the checkout of its products is important, as well as providing different pick-up and delivery options. After this initial pilot study, we are planning to conduct a bigger pilot study with several carriers and an improved user interface. Further, for the future, we are already in discussions with e-commerce retailers to provide a checkout module, enabling a better costumer journey.

Acknowledgement. The research leading to these results has been partly funded by the BMWi under grant agreement 01MD18012C, Project SMile. http:// smile-project.de.

References

1. Attaran, M.: Exploring the relationship between information technology and business process reengineering. Inf. Manag. **41**(5), 585–596 (2004)
2. Benouaret, K., Valliyur-Ramalingam, R., Charoy, F.: CrowdSC: building smart cities with large-scale citizen participation. IEEE Internet Comput. **17**(6), 57–63 (2013)
3. Chen, Z., et al.: gmission. Proc. VLDB Endow. **7**(13), 1629–1632 (2014)
4. Dell'Amico, M., Hadjidimitriou, S.: Innovative logistics model and containers solution for efficient last mile delivery. Procedia Soc. Behav. Sci. **48**, 1505–1514 (2012)
5. Etzion, O., Niblett, P., Luckham, D.C.: Event Processing in Action. Manning, Greenwich (2011)
6. Gregor, S., Hevner, A.: Positioning and presenting design science research for maximum impact. MIS Q. **37**(2), 337–355 (2013)
7. GS1: SSCC - serial shipping container code (2019). https://www.gs1.org/standards/id-keys/sscc
8. Handoko, S.D., Nguyen, D.T., Lau, H.C.: An auction mechanism for the last-mile deliveries via urban consolidation centre. In: 2014 IEEE International Conference on Automation Science and Engineering (CASE), pp. 607–612. IEEE (2014)
9. Hevner, A., March, S., Park, J., Ram, S.: Design science in information systems research. MIS Q. **28**(1), 75–105 (2004)
10. Lantz, K.E.: The Prototyping Methodology. Prentice-Hall, Englewood Cliffs (1986)
11. Liakos, P., Delis, A.: An interactive freight-pooling service for efficient last-mile delivery. In: 16th IEEE International Conference on Mobile Data Management, pp. 23–25. IEEE (2015)
12. Mladenow, A., Bauer, C., Strauss, C.: "crowd logistics": the contribution of social crowds in logistics activities. Int. J. Web Inf. Syst. **12**(3), 379–396 (2016)
13. Norta, A., Grefen, P., Narendra, N.C.: A reference architecture for managing dynamic inter-organizational business processes. Data Knowl. Eng. **91**, 52–89 (2014)
14. Park, H., Park, D., Jeong, I.J.: An effects analysis of logistics collaboration in last-mile networks for CEP delivery services. Transp. Policy **50**, 115–125 (2016)
15. Petrovic, O., Harnisch, M.J., Puchleitner, T.: Opportunities of mobile communication systems for applications in last-mile logistics. In: 2013 International Conference on Advanced Logistics and Transport, pp. 354–359. IEEE (2013)
16. Ranieri, L., Digiesi, S., Silvestri, B., Roccotelli, M.: A review of last mile logistics innovations in an externalities cost reduction vision. Sustainability **10**(3), 782 (2018)
17. de Souza, R., Goh, M., Lau, H.C., Ng, W.S., Tan, P.S.: Collaborative urban logistics - synchronizing the last mile a singapore research perspective. Procedia Soc. Behav. Sci. **125**, 422–431 (2014)
18. Statista: Revenue of e-commerce (B2C) in Germany (1999–2015), forecast for 2016 (2017). https://de.statista.com/statistik/daten/studie/3979/umfrage/e-commerce-umsatz-in-deutschland-seit-1999/

19. Suh, K., Smith, T., Linhoff, M.: Leveraging socially networked mobile ICT platforms for the last-mile delivery problem. Environ. Sci. Technol. **46**(17), 9481–9490 (2012)
20. Venable, J., Pries-Heje, J., Baskerville, R.: FEDS: a framework for evaluation in design science research. Eur. J. Inf. Syst. **25**(1), 77–89 (2016)
21. Wang, Y., Zhang, D., Liu, Q., Shen, F., Lee, L.H.: Towards enhancing the last-mile delivery: an effective crowd-tasking model with scalable solutions. Transp. Res. Part E Logist. Transp. Rev. **93**, 279–293 (2016)

Data-Driven Process Choreography Execution on the Blockchain: A Focus on Blockchain Data Reusability

Tom Lichtenstein[✉], Simon Siegert[✉], Adriatik Nikaj, and Mathias Weske

Hasso Plattner Institute, University of Potsdam, Potsdam, Germany
{Tom.Lichtenstein,Simon.Siegert}@student.hpi.de,
{Adriatik.Nikaj,Mathias.Weske}@hpi.de

Abstract. Process choreography diagrams are the standard way of representing interactions between different parties to reach a common business goal. In order to enact choreographies in a trust-less environment, blockchain-based implementations have been proposed. They support trustful interactions, i.e., information generated on the blockchain during execution is trustworthy. However, existing solutions employ blockchain data that are bound to a single choreography. This paper proposes a novel approach to implement choreographies on the blockchain in a way that the generated data can be reused by different choreographies leading to cost reduction without sacrificing data integrity. The approach is evaluated in terms of feasibility and costs by developing a prototype based on the Ethereum blockchain.

Keywords: Process choreography · Reusability · Blockchain · Ethereum

1 Introduction

Interaction with partners is one of the most essential aspects for running a successful business. With the growth of a company, the number of partners and the exchanged messages are likely to increase. The standard way to capture and track the interactions between them is the BPMN [10] process choreography model. Process choreographies abstract from the internals of the business processes involved and focus on the message flow between the participants. Choreographies, different from business processes, lack a central orchestration engine. Hence, the success of the execution depends entirely on the participants. In a trust-less environment this constitutes a problem.

The classical solution to this problem is solved by introducing a trusted third party, which orchestrates the choreography. As an alternative, blockchains solutions [6, 11, 12] have been proposed as implementation approaches for interacting processes recently. A blockchain allows trustworthy communication in a network of mutually non-trusting participants. Due to their decentralized architecture,

© Springer Nature Switzerland AG 2020
W. Abramowicz and G. Klein (Eds.): BIS 2020, LNBIP 389, pp. 224–235, 2020.
https://doi.org/10.1007/978-3-030-53337-3_17

blockchains are ideally suited to serve as a neutral execution engine for choreographies.

While there are blockchain solutions for choreographies enactment and monitoring, they are limited to creating an entire new smart contract for each business process instance. This means that the created process data is tightly coupled to the process instance and, therefore, hardly useful outside the instance. In this paper we address the problem of data reusability (or the lack thereof) on the blockchain. Reusing data already stored on the blockchain leads to several advantages: the costs of generating and re-storing the same information are eliminated; the users' manual error-prone inputs are reduced; and, there is the certainty that the integrated information has not been manipulated.

To tackle the reusability problem we propose an approach where the data exchanged between participants via the blockchain are considered as first-class citizens. The choreographies' control flows are implicitly enforced by the underlying blockchain data. This allows the insertion of existing data objects in other choreographies at run-time. Finally, the approach is evaluated in terms of feasibility and viability by creating a proof-of-concept prototype based on the Ethereum blockchain [4] and by comparing the execution costs against existing solutions.

This paper is organized as follows: Sect. 2 provides the prerequisite knowledge on business process choreographies and blockchain technologies; an overview on the state-of-the-art blockchain solutions for process choreographies enactment is provided in Sect. 3; Sect. 4 introduces the reader to our main approach on data-driven design and implementation of choreographies on the blockchain; Sect. 5 provides the details of a prototypical implementation of our approach on the Ethereum blockchain as well as a cost comparison of existing solutions; at last, Sect. 6 concludes the paper.

2   Foundation

This paper proposes a novel way of implementing choreographies on the blockchain. Therefore, we explain the concepts of process choreographies and blockchains below.

BPMN process choreography models can be considered as an abstraction of BPMN business process models. While the latter addresses internal business processes as well as their communication, process choreographies focus only on the communication between different business entities. Choreography tasks model the exchange of messages between participants (see Fig. 1). A task represents a sent message and optionally also the corresponding response message. Tasks are globally ordered based on their causality. This order is expressed via sequence flows similar to BPMN process models. Gateways allow decisions and parallel behavior within the communication flow [10]. Figure 1 shows an example of a choreography model capturing the interaction of the participants in a car rental process.

The successful enactment of choreographies is realized by the participants enacting their part independently of each other, yet, in alignment with the choreography model [1]. In a trust-less environment this constitutes a problem. In the example, it is important that the driver's license data comes from a trustworthy source to ensure that the customer actually has a driver's licence. The example choreography also shows the interaction of business parties who do not necessarily trust each other but nevertheless have to work together for a common business goal. To address these kind of problems, blockchain technologies are used to establish trusted communication between non-trusted participants.

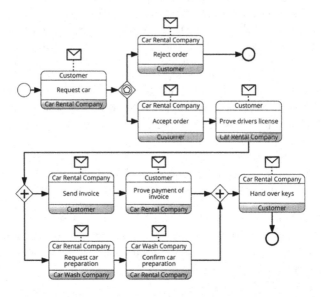

Fig. 1. Process choreography model of a car rental process

A blockchain is a distributed ledger that uses cryptographic algorithms to create a decentralized, immutable log of transactions [8]. Information is shared by transactions that get combined into blocks. The blocks form an unambiguous sequence by referring to their corresponding predecessor. The concatenation of the information leads to the immutability of the stored information. This is done with cryptographic signatures provided by hash algorithms. A consensus algorithm ensures that only valid and verified transactions are included in the blockchain [2]. On certain blockchain solutions, it is also possible to execute small programs, which are called smart contracts [3]. These enable a trustworthy execution environment in exchange for a certain fee. To implement our approach we focus on the blockchain Ethereum with the compiler language *Solidity 0.5.3*[1], but any blockchain technology that supports smart contracts can be used [4]. Blockchains are a suitable technology for implementing choreographies, as they

---

[1] https://solidity.readthedocs.io/en/v0.5.3/, 28.02.2019.

fill the trust gap that arises when several communicating parties have to transfer trust to an executing participant. Blockchains and choreographies have therefore already been in the interest of research together. There are already some approaches to implement process engines for blockchains, but these have certain disadvantages. For example, these methods are often costly for frequently used instances, are highly dependent on user input and often store redundant information.

3   Related Work

This paper addresses research areas in the context of BPMN 2.0 choreography processes and blockchain technologies as well as the implementation of tamper-proof collaborations on the blockchain. In this section we provide a brief overview of the ideas that are associated with our approach.

Because of their distributed storage technology, blockchains have already been discovered as tamper-proof alternatives to execute and monitor processes and collaborations [5]. In terms of business processes, Caterpillar is an open-source Business Process Management System using the Ethereum blockchain to create and execute instances of process models [6]. With regard to choreographies, there are already several approaches that use blockchains to address the lack of trust in collaborative processes. Weber et al. propose a way to model and execute choreographies on the blockchain using smart contracts [12]. They introduce two main ideas to implement choreographies on blockchains. The first one focuses on process monitoring, by storing the role assignment and the process execution status of each involved participant. Every data exchange between the parties is driven by one monitor in the form of a smart contract that ensures the correct execution order and conformance to the choreography. The second one extends the first approach by addressing the coordination of a choreography additionally to the other tasks. All choreography instances are generated by a factory smart contract. This approach thus uses a single smart contract as a process engine for an entire choreography instance. However, this leads to a new smart contract being deployed for each choreography instance, with each deployment causing additional costs. Furthermore, it is not possible to reuse information from older instances.

The consideration of semantic data objects as a central unit for data exchange has also already been investigated. [9] describes the representation and role of business artifacts in business process modeling.

4   Data-Driven Design and Implementation

This paper's main idea is to allow process choreographies to reuse existing data objects that are products or byproducts of already executed process choreographies on the blockchain. To this end, we need to accommodate the insertion of any choreography-produced data object into a new choreography, ideally at runtime. To realize this, we propose a data-driven choreography implementation on the blockchain. The term data-driven refers to the fact that in this approach the data exchanged in the choreography is the basis for the realisation of the control flow.

Figure 2 depicts the general idea of how choreography models are implemented as data object manipulations on the blockchain. Each data object has specific states, which participants can change by executing the choreography task, i.e., sending a message. Every message addresses exactly one data object. The overall choreography's execution state is implicitly captured in the corresponding data object states. The progression of the choreography execution is realized by changing the state of these data objects through messages as prescribed in the choreography diagram. Each data object is stored as a separate smart contract on the blockchain. This allows the data objects to be reused separately. The state is represented by variables with the respective data type on the corresponding data object smart contract.

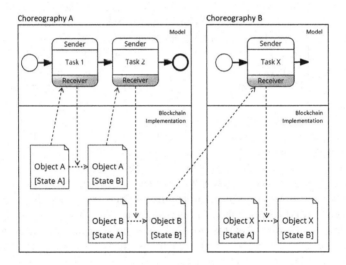

Fig. 2. Mapping of choreography behavior to data-driven blockchain implementation

The choreography control flow is enabled by the state of the data objects. Only when a data object is at a specific state the respective participant is allowed to execute the enabled choreography task. If several data objects are involved in the choreography, there is no smart contract that monitors the entire control flow. For this reason, the data objects and their states must be clearly defined alongside the choreography model. There are many ways to represent the additional information (e.g., using a UML class diagram). However, in order to remain compliant with the BPMN choreography specification and maintain only one model, we have decided to include this information in the choreography model. The information is added via graphical notations as Solidity source code like shown in Fig. 3. This also allows to formulate arbitrarily complex conditions and provides the basis for automatically generating smart contracts from a given choreography model. The annotations will be explained further in the next section.

4.1   Ensuring Process Choreography Correctness

To keep the choreography in a valid state, we extend the choreography task specification with conditions. State changes have to satisfy the modeled constraints. It is necessary to add those conditions to every task in order to enforce the flow of the choreography. To prevent incomplete state changes we use the pattern *Design by Contract*[7]. The pattern uses preconditions and postconditions to validate the changes in state by a software component in a program. If the software component receives an input that does not match the preconditions, the component will not be executed. Undesired state changes that do not match the postconditions will be reverted. This prevents state changes that should not be possible in the first place.

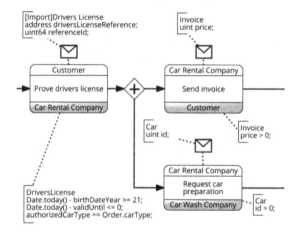

Fig. 3. Process section with solidity code annotations

To apply the design by contract pattern to our approach, we extend every choreography task with postconditions as shown in the annotations of the tasks in our car rental choreography excerpt in Fig. 3. To invoke a sequence flow, i.e., enforce the order between the tasks, we use the stated postconditions of the source task as the preconditions of the target task. In the choreography diagram, however, only the postconditions are modelled to avoid repetition. Once the source task of the sequence flow is executed and the postcondition is satisfied, the target task of the sequence flow is enabled and can be executed at any time by the respective participant.

In order to keep the execution order compliant with the choreography, it must be ensured that the conjunction of all postconditions of the predecessor can only be achieved by this particular task. This is essential for the entire choreography, because transitive postconditions such as $x < 10$ in one task and $x < 100$ in a sequential following task will lead to an illegal state where both tasks can be executed. If the previous element of a task is a gateway, we adjust the precondition accordingly to satisfy the flow conditions.

4.2   Decoupled Architecture for Data Injection at Runtime

Data objects are the central aspect of the data-driven choreography implementation. Since the data objects are stored on the blockchain, they serve as a tamper-proof source of information. To reuse data objects in another choreography, they must be decoupled from the choreography to which they originally belong. The decoupling is realized by creating an abstraction layer above the data objects by creating a smart contract for each participant as an interface for the choreography. Every message that can be sent by a participant is represented as a function in the corresponding participant interface smart contract. These functions implement the annotated choreography-specific state change rules according to the pattern Design by Contract, as described previously. The participants of the choreography, for example, the customer, the car wash company and the car rental company, exclusively use their own participant interface smart contract to change the state of the data objects via the provided functions and thus interact with the choreography. The decoupling of the process logic from the data layer enables a process independent reuse of the stored information in other choreographies.

To keep the state of the data object compliant to the state of the choreography that it originally belongs to, we have to limit the access rights of reused data objects to read only. Therefore we have to distinguish between choreography-specific (*internal*) and *external* data objects, which we also refer to as *imported* data objects. The former stores the choreography state and can be manipulated, whilst imported data objects are handled as additional sources of information for the choreography instance. We also want to support the import of data objects at run time to allow lazy imports of data objects. This is convenient when the data object might not exist at the deployment time of the choreography. To indicate an imported data object, we prefix the name of the data object in the choreography annotations with an *[Import]* tag as shown in Fig. 3 for the driver's license object. In this concrete case, the customer can insert his driver's licence at run time, given that the licence data object is provided by the respective authorities on the blockchain as part of driver's licence issuing process choreography.

In order to achieve this behaviour, we extend the initial contract architecture with another contract, the *data object store* (see Fig. 4). Every participant of the choreography has to store a reference to the corresponding data object store in their participant interface smart contract. It acts as a storage for all imported data object references. A data object can be referenced by the address of the contract that instantiated it. In the meta model, this flexibility is represented by the *DataObject* without fixed attributes or operations, since they vary depending on the specific instances.

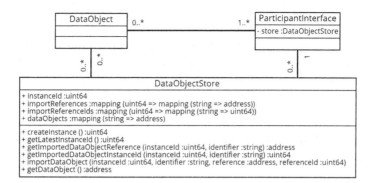

Fig. 4. Meta model of the data-driven approach

In Solidity, an address that refers to a smart contract can be casted into the according contract using a contract interface, which corresponds to a class interface in object-oriented programming. Therefore, every data object that can be imported must be modeled with the corresponding contract interface. The contract interfaces must only expose the view functions[2] of the original contract of the data object. This verifies that the imported data object can not be manipulated by a choreography which it does not originally belongs to. If the defined functions do not match the original smart contract, they will throw an error when they are executed. One advantage of storing references in the form of addresses instead of contract instances is that we do not need to know the specific data object types of the required imports in the data object store.

To store the data objects in the data object store, we use a mapping from an identifier to the data object contract address. We choose strings as identifiers for the data objects to improve the readability of the contract. With this data structure we can easily select already imported data objects from the data object store using the readable identifier. To import a data object, a participant has to invoke the *importDataObject* function of the data object store, which can be seen in Fig. 4, with the data object casted to the contract address and an unique identifier.

The data object store is also used to manage several instances of a choreography by accessing them using individual identifiers. This allows to deploy the same data objects only once and thus contributes to cost reduction.

5 Evaluation

In this section we introduce the prototype that implements our approach as well as cost analyses by comparing our approach with existing solutions.

---

[2] https://solidity.readthedocs.io/en/v0.4.21/contracts.html 25.02.2019.

5.1   Prototypical Implementation

To evaluate the feasibility of our approach, we built a prototype which can execute process choreographies on the Ethereum blockchain[3]. For visualization purposes we also created a screencast of the prototype[4]. Using the prototype we compare the varying costs of the data-driven approach with a reference approach that we designed based on the monitor approach by Weber et al. [12] described in Sect. 3. A direct comparison with the original monitor approach is not possible because the code is not provided. The reference approach is therefore interpreted as follows. Each choreography instance is implemented by a single smart contract that is shared by all participants. In this smart contract each choreography task of the choreography is represented by a function and a boolean variable. The boolean variable indicates whether the corresponding choreography task is enabled. The entirety of the boolean variable represents the state of the choreography instance. The participants use the functions to send messages and thus change the state of the choreography instance according to the implemented control flow rules. The content of the messages is stored in variables. For simplification reasons we do not include the factory contract described in the monitor approach.

The key aspect of this evaluation is the accruing cost of the data-driven approach compared to the reference approach. We consider the costs of deployment, instantiation and execution of the choreographies. In order to compare the two implementations, we define two scenarios that form the basis of the evaluation. To illustrate the impact of the contract overhead of the data-driven implementation we create a minimal choreography. The choreography consists of two choreography tasks that model the communication between two participants. There are no gateways involved. Each participant has to execute one task. To analyze the cost behavior in a real world use case scenario we modeled the choreography process of a car rental company[5] from Fig. 1. The scenario consists of nine choreography tasks and three gateways. Parallel executions are included. The scenario models the communication between three participants.

5.2   Costs Analysis

In the execution environment Ethereum, deployment and execution costs are measured in *gas*. A defined amount of gas is required for each atomic operation of the execution of a smart contract. The gas has to be payed in *Ether* which is the cryptocurrency of the Ethereum blockchain. The amount of Ether that has to be payed for one unit of gas is defined by the gas price [13].

---

[3] https://github.com/data-driven-choreographies/Car-Rental-Choreography-Visualization 28.07.2019.

[4] https://github.com/data-driven-choreographies/Car-Rental-Choreography-Visualization/blob/master/screencast/Screencast.mp4 28.07.2019.

[5] https://github.com/data-driven-choreographies/Smart-Contracts/blob/master/car-rental/car_rental_data_object_driven.sol 01.03.2020.

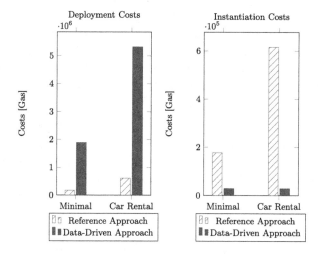

Fig. 5. Comparison of deployment and instantiation costs for both approaches

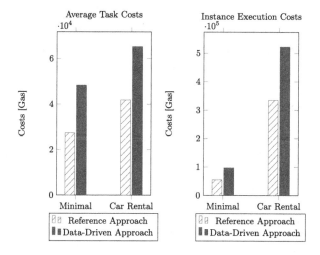

Fig. 6. Comparison of task and instance execution costs for both approaches

The diagrams shown in Fig. 5 compare the instantiation costs of both implementations regarding the first instantiation that includes the deployment of the contracts and all following instantiations. The diagrams shown in Fig. 6 compare the execution costs of both implementations regarding the average task cost and the execution cost of the whole choreography.

Splitting the choreography into multiple contracts and adding contracts for each participant as well as a storage system that dispatches all the requests adds certain costs to the deployment (c_{dpl}) of the data-driven approach. Despite being about three times more expensive to deploy, the data-driven approach also requires the resolution of references to other smart contracts in order to access the data objects, which increases the execution costs (c_{exec}) as well. The advantage of the data-driven approach in terms of costs is the creation of new instances (c_{inst}). While the reference approach has to deploy a new smart contract for each instance, the data object store can simply increase the instance number. This leads to lower overall costs of the data-driven approach in the long run, since the cost of a smart contract deployment is much higher than changing a single variable. According to our evaluation results, the data-driven approach gets cheaper than the reference approach after 20 executions of the minimal scenario as well as after 13 executions of the car rental company scenario. Therefore, the data-driven approach can be viable in financial terms when choreographies gets executed often and do not change. The following equation describes the calculation of the total costs (c_{total}) after the i-th executions, where $i > 0$.

$$c_{total} = c_{dpl} + c_{exec} + (i - 1) \times (c_{inst} + c_{exec})$$

In addition, we compare the cost of importing a data object with the cost of creating a final variable. In our approach, the cost of importing a data object is constant. A first import costs 67653 gas, overwriting the data object costs 32853 gas each. In comparison, the fixed storage of an additional 40 characters of character string at deployment time costs 67590 gas more. Importing data is therefore worthwhile even for small amounts of data. Furthermore, the import of a data object enables flexible handling of changing variables.

6   Conclusions

This paper introduces a new approach to model and execute choreographies on the blockchain. We provide a data-driven solution that allows the reuse of trustworthy data that are generated by executing choreographies on the blockchain.

The process choreography is implicitly represented as a set of data objects and their transitions. Each data object is manipulated through messages. The concrete manipulation is subject to the rules that are added as annotations in the choreography model. The correct process choreography flow is ensured by enforcing these rules on the blockchain.

The proposed implementation architecture decouples the data objects from the choreography execution in a way that data objects can be reused for multiple process choreographies. For additional flexibility, external data objects can be inserted at runtime. An instance management concept is introduced to reduce instantiation costs.

The approach is compared with an already existing implementation of choreographies on the blockchain by Weber et al. and evaluated in terms of costs [12]. The results show the feasibility and the viability of the data-driven implementation of choreographies on the blockchain. However, some aspects of process implementations on the blockchain have not been considered in this paper. The described approach with its decoupled data objects can be used to outsource some business process choreography parts in a trustworthy manner to an external participant. Access rights management, which plays an important role in a publicly accessible blockchain, is not considered. Furthermore, the data-driven approach does not offer the possibility to encrypt sensitive data. In the future, this approach can be enriched with further research in these areas.

References

1. van der Aalst, W.M.P., Weske, M.: The P2P approach to interorganizational workflows. In: Dittrich, K.R., Geppert, A., Norrie, M.C. (eds.) CAiSE 2001. LNCS, vol. 2068, pp. 140–156. Springer, Heidelberg (2001). https://doi.org/10.1007/3-540-45341-5_10. http://dl.acm.org/citation.cfm?id=646089.680214
2. Buterin, V.: Ethereum white paper: a next generation smart contract & decentralized application platform. First version (2014)
3. Cong, L.W., He, Z.: Blockchain disruption and smart contracts. Technical report, National Bureau of Economic Research (2018)
4. Dannen, C.: Introducing Ethereum and Solidity: Foundations of Cryptocurrency and Blockchain Programming for Beginners. Apress, Berkeley (2017). https://doi.org/10.1007/978-1-4842-2535-6_1
5. García-Bañuelos, L., Ponomarev, A., Dumas, M., Weber, I.: Optimized execution of business processes on blockchain. In: Carmona, J., Engels, G., Kumar, A. (eds.) BPM 2017. LNCS, vol. 10445, pp. 130–146. Springer, Cham (2017). https://doi.org/10.1007/978-3-319-65000-5_8
6. López-Pintado, O., García-Bañuelos, L., Dumas, M., Weber, I.: Caterpillar: a blockchain-based business process management system. In: Proceedings of the BPM Demo Track and BPM Dissertation Award Co-Located with 15th International Conference on Business Process Modeling (BPM 2017), Barcelona, Spain (2017)
7. Meyer, B.: Applying 'design by contract'. Computer 25(10), 40–51 (1992)
8. Nakamoto, S.: Bitcoin: a peer-to-peer electronic cash system (2008)
9. Nigam, A., Caswell, N.S.: Business artifacts: an approach to operational specification. IBM Syst. J. 42(3), 428–445 (2003)
10. OMG: Business Process Model and Notation (BPMN), version 2.0, January 2011. http://www.omg.org/spec/BPMN/2.0
11. Tran, A.B., Lu, Q., Weber, I.: Lorikeet: a model-driven engineering tool for blockchain-based business process execution and asset management. In: BPM (Dissertation/Demos/Industry), pp. 56–60 (2018)
12. Weber, I., Xu, X., Riveret, R., Governatori, G., Ponomarev, A., Mendling, J.: Untrusted business process monitoring and execution using blockchain. In: La Rosa, M., Loos, P., Pastor, O. (eds.) BPM 2016. LNCS, vol. 9850, pp. 329–347. Springer, Cham (2016). https://doi.org/10.1007/978-3-319-45348-4_19
13. Wood, G.: Ethereum: a secure decentralised generalised transaction ledger. Ethereum project yellow paper 151, 1–32 (2014)

Discovering Business Area Effects to Process Mining Analysis Using Clustering and Influence Analysis

Teemu Lehto[1,2](✉) ⓘ and Markku Hinkka[1,2] ⓘ

[1] QPR Software Plc, Helsinki, Finland
teemu.lehto@qpr.com
[2] School of Science, Department of Computer Science, Aalto University, Espoo, Finland

Abstract. A common challenge for improving business processes in large organizations is that business people in charge of the operations are lacking a fact-based understanding of the execution details, process variants, and exceptions taking place in business operations. While existing process mining methodologies can discover these details based on event logs, it is challenging to communicate the process mining findings to business people. In this paper, we present a novel methodology for discovering business areas that have a significant effect on the process execution details. Our method uses clustering to group similar cases based on process flow characteristics and then influence analysis for detecting those business areas that correlate most with the discovered clusters. Our analysis serves as a bridge between BPM people and business people, facilitating the knowledge sharing between these groups. We also present an example analysis based on publicly available real-life purchase order process data.

Keywords: Process mining · Clustering · Influence analysis · Contribution · Business area · Classification rule mining · Data mining

1 Introduction

Process mining helps organizations to improve their operations by providing valuable information about the business processes in easy to understand visual flowchart format based on transactional data in ERP systems. However, to provide these meaningful results, the data extracted from ERP systems may often contain different kinds of objects like 'apples and oranges' that should be analyzed separately. Using the procurement process as an example: the purchase order database tables may contain several different kinds of purchase orders items like services, equipment, raw materials, software licenses, high-cost items, free items, headquarter purchases, plant maintenance costs, manually approved items, and automatic replenishment purchases. Without appropriate tools the

W. Abramowicz and G. Klein (Eds.): BIS 2020, LNBIP 389, pp. 236–248, 2020.
https://doi.org/10.1007/978-3-030-53337-3_18

process analyst needs to either a) analyze all items separately - leading to potentially massive amount of work, b) analyze all items at the same time - leading to potentially meaningless results or c) rely on subjective information like asking business people which of the items should be analyzed separately or relying on own intuition. The techniques presenting in this paper help the analyst to discover those business areas (classification rules) that seem to have a major effect on the business process flow. These business areas are based on case attribute characteristics of the cases and thus easy to understand for the business people. Discovered business areas can be used to effectively guide the process mining analysis further in divide & conquer manner.

In this paper, we present methods to answer these three questions:

- How a business process can be analyzed based on the process flow of individual process instances in order to discover business-relevant clusters in such a way that a business analyst can easily understand the clustering results and use them for further analysis.
- How to find business areas that have a major effect on process flow behavior.
- How to further consolidate business area results to discover case attributes that have a significant effect on process flow behavior.

The rest of this paper is structured as follows: Sect. 2 is a summary of the latest developments. Section 3 present our methodology for Discovering Business Area Effects. Section 4 is a case study with real-life purchase order process data. Section 5 shows limitations and Sect. 6 draws the final conclusions.

2 Related Work

Process mining is an active research area that analyses business processes based on the event log data from IT systems in order to discover, monitor, and improve processes [16]. Process mining typically focuses on discovering the process flowchart as a control flow diagram, Petri net, or BPMN diagram. Other process mining types include conformance checking and enhancement. Root cause analysis as part of process mining has been studied in [14] as well as in our previous works [8] and [9].

One key challenge in process mining is that a single event log may often contain many different processes, in which case trying to discover a single process diagram for the whole log file is not a working solution. In the process mining context clustering has been studied a lot for with excellent results [3,5,12,15]. These previous work cover the usage of several distance measures like Euclid, Hamming, Jaccard, Cosine, Markov chain, Edit-Distance as well as several cluster approaches like partitioning, hierarchial, density-based and neuronal network. However, most of the previous research related to clustering within the process mining field has been directly focused on the process flowchart discovery with the prime objectives categorized as Process, Variant or Outlier Identification, Understandability of Complexity, Decomposition, or Hierarhization. In practice, this means that clustering has been used as a tool for improving the other process mining methods like control flow discovery to work better, i.e., clustering has divided the event log into smaller sub logs that have been directly used for further analysis. In this paper, we show how to use clustering for discovering those business areas that have a significant effect on process behavior. Yet another use case for clustering in the process mining field has been to perform structural feature selection in order to improve the prediction accuracy and performance [6].

Some recent research has started to address the challenge of how to explain the clustering results to business analyst [2]. It has been presented that when explaining the characteristics of clusters to business analysts, the role of case attributes becomes more important [11]. We show an easy-to-understand representation for showing cluster characteristics based on the difference of densities and case attribute information.

Substantial effort has also been spent in the process mining community to discover branching conditions from business process execution logs [4]. This has also lead to the introduction of decision models and decision mining [1] as well as a standard Decision Model and Notation (DMN) [10]. While the objective of the decision modeling is to provide additional details into individual branching conditions, our approach is to analyze the effect of any business area to the whole structure of the process flow, not just one decision branch at a time.

3   Discovering Business Area Effects

In this section, we present our methodology for Discovering Business Area Effects To Process Mining Analysis Using Clustering and Influence Analysis. Our approach is to do the clustering using process flow features and then use influence analysis to find those business areas that have the highest contribution for certain kinds of cases ending up in distinct clusters. If all process instance-specific business area values derived using any given case attribute are distributed randomly, then the contribution measure for each business area is zero, and the information for the analyst is that the particular case attribute does not correlate with the way how the clusters are formed. According to our methodology, it then means that the particular case attribute has no influence on the process flow behavior. In summary, our method finds those business areas and case attributes that have the highest contribution to the process flow behavior.

3.1   Clustering Cases

Feature Selection. To identify those business areas that have the strongest effect on the process execution, we first run clustering using relevant features representing the process execution characteristics. These features have been widely studied in Trace Clustering papers [6,7,12,15]. Clustering is a trade-off between quality and performance. As the amount of features is increased, the quality of the results potentially improves while performance gets slower.

- Activity profile: This profile contains one feature for each Event Type label in the data. The value of this feature is related to the number of occurrences of that particular event type within the case. If the number of occurrences is used as an exact value, then the clustering algorithm somehow needs to take into account the continuous values, ie. repeating activity A seven times is much more similar to repeating 6 or 8 times, compared to repeating the activity A only twice. One approach is to use value *zero* if the Event Log contains no occurrences of the Event Type for the given case and *one* if the log contains one or more occurrences. While this approach often works well, it may not be able to detect the repeating of a given Event Type multiple times with the log. For this reason, we recommend using value *zero* for no occurrences of the Event Type, *one* for only one occurrence and two for *two* or more occurrences.
- Transition profile: The transition profile captures all process flows from every activity to the next activity. In effect, it contains the process control flow information. Transition profile potentially provides a large number of features up to the square of the number of Event Types plus one for start and end transitions. For example, in the sample analysis presented in Sect. 4, we have 42 distinct event types, giving potentially $43^2 = 1849$ distinct transition. Luckily the control flow for 251.734 cases only contains 676 distinct transitions. Because the amount of transition features is high, we recommend using the coding *zero* if the transition does not occur in the case and *one* if it occurs once or more.

Clustering Algorithms. A comparative analysis of process instance cluster techniques has been presented in [15] and shows how various clustering techniques have been used to separate different process variants from a large set of cases as well as reducing the complexity by grouping similar cases into same clusters. Considering our method, the main functional requirement for the clustering algorithm is that it needs to put cases with similar process flow behavior into the same clusters, and all 20 approaches listed in [15] meet this requirement. If a particular clustering algorithm produces meaningful results and if there indeed is a correlation with a particular business area, then our method gives very high contribution values for that business area. If the clustering algorithm does not work perfectly but is still capable to some extent grouping similar cases together, then the contribution values are still likely to show the most significant business areas among the top contributors. The essential non-functional requirement for the clustering algorithm is performance, i.e., the ability to produce results fast with a small amount of memory. With these considerations, we have received good results with the algorithms and parameters below:

- *One-hot encoding.* Since our Activity and Transition feature profiles only include categorial values zero, one, and two, it is possible to use efficient one-hot encoding. This results in maximum of $(n(EventTypes) + 1)^2 + 2 * n(EventTypes)$ feature vectors.

- *Hamming distance* is the natural choice as the distance function with binary data like one-hot encoded features, because it completely avoids the floating-point distance calculations needed for common Euclid distance measure.
- *K-modes* clustering algorithm is suitable for categorical data. In our tests, k-modes produced well-balanced clusters and was fast to execute. The result of K-modes depends on the initial cluster center initialization. We also tested agglomerative clustering algorithms, but it produced highly unbalanced clusters.
- *Number of clusters* has a significant effect on the clustering. To discover the business areas, clustering should be done several times with different numbers of clusters. We found out that clustering four times with cluster sizes 2, 3, 5, and 10 clusters gave enough variation in the results providing meaningful results. When the number of clusters is less than five, the large business areas correlate more with the clustering. While clustering to 10 or more clusters, the smaller business areas like *Vendor, Customer, Product* having more distinct values correlate more with the clusters. Running the clustering several times is also an easy way to mitigate the random behavior of K-modes coming from initialization.

3.2   Influence Analysis

Business Areas. Examples of business area dimensions include: *company code, product line, sales unit, delivery team, geographical location, customer group, product group, branch offices, request category* and *diagnosis code*. All the case attributes that are relevant to business can be used as business area dimensions as such, for example, *product code*. However, a large organization may easily have thousands of low-level product codes in their ERP system, so it is beneficial to have access to product hierarchy and use each level as a separate business area dimension. Another example of a derived business area dimension is when a case attribute like *Logistics Manager Name* can be used to identify the *Delivery Team*. We again suggest having both the *Logistics Manager* and *Delivery Team* as business area dimensions; if one particular *Logistics Manager* has many cases and a major effect on process flow behavior then our method will show that person as the most significant business area in the *Logistics Manager* dimension. The third example of derived business areas is to utilize the event attributes. For example the *Logistics Manager Name* may be stored as an attribute value for the *Delivery Planning Done* activity. If there is always at most one *Delivery Planning Done* activity, then the attribute value can be used as such in the case level. If there are multiple *Delivery Planning Done* activities, then typical options include: use the first occurrence, use the last occurrence or use a concatenated comma-separated list of all distinct values from activities as the value on the case level. The outcome of forming business area dimensions is a list of case-level attributes that contain a specific (possibly empty) business area value for each case. To continue with our formal methodology, we now consider these business area dimensions as case attributes and the case attribute values as the corresponding business areas.

Interestingness Measures. We now present the definitions for interestingness measures used for finding the business areas that correlate with the clustering results. Let $C = \{c_1, \ldots, c_N\}$ be the set of cases in the process analysis. Each case represents a single business process execution instance. Let $P = \{p_1, \ldots, p_N\}$ be a set of clusters each formed by clustering the cases in C. $C_p = \{c_{p_1}, \ldots, c_{p_N}\}$ is the set of cases belonging to cluster p. $C_p \subseteq C$. Similarly $C_a = \{c_{a_1}, \ldots, c_{a_N}\}$ is the set of cases belonging to the same business area a, ie. they have the same value for the case attribute a.

Definition 1. *Let Density* $\rho(a, C) = \frac{n(C_a)}{n(C)}$ *where* $n(C_a)$ *is the total amount of cases belonging to the business area a and* $n(C)$ *is the total amount of all cases in the whole process analysis. Similarly, the Density* $\rho(a, C_p) = \frac{n(C_p \cap C_a)}{n(C_p)}$ *is the density of cases belonging to the business area a within the cluster P.*

Definition 2. *Let Contribution%* $(a \rightarrow p) = \rho(a, C_p) - \rho(a, C) = \frac{n(C_p \cap C_a)}{n(C_p)} - \frac{n(C_a)}{n(C)}$ *is the extra density of cases belonging to the business area a in the cluster p compared to average density.*

If business area a is equally distributed to all clusters, then the *Contribution%* $(a \rightarrow p)$ is close to zero in each cluster. If the business area a is a typical property in a particular cluster p_i and rare property in other clusters, then the *Contribution%* $(a \rightarrow p_i)$ is positive and other *Contribution%* $(a \rightarrow p_j, where j <> i)$ values are negative. Calculating the sum of all Contribution values for all clusters is always zero, so the extra density in some clusters is always balanced by the smaller than average density in other clusters.

We now want to find the business areas that have a high contribution in many clustering. We define:

Definition 3.
Let $BusinessAreaContribution(a) = \sum_{p_i \in P} \frac{n(C_{p_i})}{n(C)}(max\{Contribution\%(a \rightarrow p_i), 0\})^2$. *Here we sum the weighted squares of all positive contributions the business area a has with any clustering* p_i. *Positive values of Contribution%* $(a \rightarrow p_i)$ *indicate a positive correlation with the business are a and the particular cluster i, while negative values indicate that the business area a has smaller than average density in the cluster i. We found out that using only the positive correlations gives more meaningful results when consolidating to the business area level. Since a few high contributions are relatively more important than many small contributions, we use the Variance of the density differences, i.e., taking the square of the Contribution%* $(a \rightarrow p_i)$. *Since a contribution within a small cluster is less important than contribution in a large cluster, we also use the cluster size based weight* $\frac{n(C_p)}{n(C)}$.

Any particular business area a may have a substantial contribution in some clusters and small contribution in other, so the sum of all these clusterings is giving the overall correlation between business area a and all clusters $p_i \subseteq P$.

We use the term *Business area* in this paper for any combination of a process mining case attribute and a distinct value for that particular case attribute. *BusinessAreaContribution* thus identifies the individual case attribute-value combinations that have the highest effect on clustering results. It is then also possible to continue and consolidate the results further to Case Attribute level:

Definition 4. *Let* $AT = \{at_1, \ldots, at_N\}$ *be a set of case attributes in the process analysis. Each case* $c_i \in C$ *has a value* $at_{j\,c_i}$ *for each case attribute* $at_j \in AT$. $at_{j\,c_i}$ *is the value of case attribute* at_j *for case* c_i *and* $V_{at_j} = \{v_{at_{j_1}}, \ldots, v_{at_{j_N}}\}$ *is the set of distinct values that the case attribute* at_j *has in the process analysis.*

Definition 5. *Let* $CaseAttributeContribution(at)$ *be a sum of all BusinessAreaContributions from all the business areas corresponding to the given case attribute* at *as* $\sum_{v_{at_{j_i}} \in V_{at_j}} BusinessAreaContribution(at_{j_{v_{at_{j_i}}}})$

4   Case Study: Purchase Order Process

In this section, we apply our method to the real-life purchase order process data from a large Netherlands multinational company operating in the area of coatings and paints. The data is publicly available as the BPI Challenge 2019 [17] dataset. We made the following choices:

- **Source data.** We imported the data from the XES file as such without any modifications. To keep the execution times short, we experimented with the effect of running the analysis with a sample of the full dataset. Our experiments showed that the results remained consistent for sample size 10.000 cases and more. With the sample size of 1.000 cases, the results of the individual analysis runs started to change, so we decided to keep the sample size 10.000 cases.
- **Clustering algorithm.** We used the k-modes clustering as implemented in Accord.Net Machine Learning Framework [13] with one-hot encoding and hamming distance function. To take into account the different clustering sizes, we performed clustering four times, fixed to two, three, five, and ten clusters.
- **Activity profile features for clustering.** We used our default boolean activity profile, which creates one feature dimension for each activity and the value is *zero* if the activity does not occur in the case, value *one* if the activity occurs once and value *two* if it is repeated multiple times. There were 37 different activities in the sample, and the Top 20 activity profile is shown in Table 1.
- **Transition profile features for clustering.** Using a typical process mining analysis to discover the process flow diagram, we discovered 376 different direct transitions, including 13 starting activities, 22 ending activities, and 341 direct transitions between two unique activities. All of these 376 features were used as dimensions for clustering in a similar way as the activity profile, i.e., boolean value *zero* if transition did not occur in the case and *one* if it occurred once or multiple times.

- **Business area dimensions.** Since we did not have any additional information or hierarchy tables concerning possible business areas, we are using all available 15 distinct case attributes listed in Table 4 as business area dimensions. These case attributes have a total of 9901 distinct values, giving us 9901 business areas to consider when finding those business areas that have the most significant effect on process flow.

Table 1. Activity profile: Top 20 activities ordered by unique occurrence count

Name	Unique count	Count
Create purchase order item	10000	10000
Record goods receipt	9333	13264
Record invoice receipt	8370	9214
Vendor creates invoice	8310	8901
Clear invoice	7245	7704
Remove payment block	2223	2272
Create purchase requisition item	1901	1901
Receive order confirmation	1321	1321
Change quantity	707	853
Change price	443	498
Delete purchase order item	338	339
Cancel invoice receipt	251	271
Vendor creates debit memo	244	253
Record service entry sheet	232	10326
Change approval for purchase order	194	319
Change delivery indicator	112	128
Cancel goods receipt	109	136
SRM: in transfer to execution syst.	42	57
SRM: awaiting approval	42	50
SRM: complete	42	50

4.1 Clustering Results for Individual Clustering

Table 2 shows the results of clustering to fixed five clusters. We see that the first cluster contains 48% of cases, the second cluster 33%, third 17%, and both 4th and 5th one percent each. Here we show the five most important business areas based on the contribution%, which is calculated as the difference between Cluster specific density of that business area and Total Density. These results already give hints about the meaningful characteristics in the whole dataset, i.e.: Cluster

Table 2. Clustering results based on Contribution

Cluster	Business area a	Cluster density	Total density	Contribution
Cluster1 48% cases	Spend area text = Sales	0.36	0.26	0.11
	Sub spend area text = Products for Resale	0.34	0.24	0.11
	Spend classification text = NPR	0.41	0.32	0.10
	Item Type = Standard	0.96	0.87	0.09
	Item Category = 3-way match, invoice before GR	0.95	0.88	0.07
Cluster2 33% cases	Spend area text = Packaging	0.65	0.44	0.21
	Sub spend area text = Labels	0.39	0.24	0.16
	Spend classification text = PR	0.79	0.66	0.13
	Name = vendor_0119	0.14	0.05	0.08
	Vendor = vendorID_0120	0.14	0.05	0.08
Cluster3 17% cases	Item Category = Consignment	0.33	0.06	0.27
	Item Type = Consignment	0.33	0.06	0.27
	Name = vendor_0185	0.09	0.02	0.08
	Vendor = vendorID_0188	0.09	0.02	0.08
	Item = 10	0.33	0.26	0.07
Cluster4 1% cases	Sub spend area text = Metal Containers & Lids	0.19	0.08	0.11
	Name = vendor_0393	0.09	0.01	0.08
	Vendor = vendorID_0404	0.09	0.01	0.08
	Name = vendor_0104	0.11	0.04	0.07
	Vendor = vendorID_0104	0.11	0.04	0.07
Cluster5 1% cases	Spend classification text = NPR	0.59	0.32	0.27
	Spend area text = Sales	0.41	0.26	0.15
	GR-Based Inv. Verif. = TRUE	0.21	0.06	0.15
	Item Category = 3-way match, invoice after GR	0.21	0.06	0.15
	Sub spend area text = Products for Resale	0.38	0.24	0.14

one contains many *Standard* cases from spend areas related to *Sales*, *Products for Resale* and *NPR*. On the other hand cluster two contains more than average amount of cases from *spend area Packaging*, related to *Labels* and *PR*. *VendorID_0120* seems to be highly associated with the process flow characteristics of cluster 2. Cluster 3 is dominated by *Consignment* cases. Cluster 4 contains many *Metal Containers & Lids* cases as well as cases from *VendorIDs 0404* and *0104*. Further analysis of the top five business areas listed as characteristics for each cluster confirms that these business areas indeed give a good overall idea of the cases allocated into each cluster.

4.2   Discovering Business Areas

We clustered four times for fixed cluster amounts of 2,3,5 and 10 - yielding a total of 20 clusters, and then consolidating the results into business area level using Definition 3. The top 20 of all these 9901 business areas ordered by their respective Business Area Contribution is shown in Table 3. Clearly the business areas *Item Category = Consignment* and *Item Type = Consignment* have most significant effect on the process flow. Looking at the actual process model, we see that *Consignment* cases completely avoid three of the five most common

Table 3. Top 20 Business areas with major effect to process flow

Business area a	Contribution	nCases $n(C_a)$
Item Category = Consignment	0.051	576
Item Type = Consignment	0.051	576
Spend area text = Packaging	0.040	4382
Spend classification text = NPR	0.024	3175
Sub spend area text = Labels	0.022	2351
Spend area text = Sales	0.021	2574
Item Type = Standard	0.021	8740
Sub spend area text = Products for Resale	0.021	2390
Spend classification text = PR	0.019	6574
Item Category = 3-way match, invoice before GR	0.017	8760
Spend area text = Logistics	0.013	210
Item Type = Service	0.013	244
Item = 1	0.012	342
GR-Based Inv. Verif. = TRUE	0.012	623
Item Category = 3-way match, invoice after GR	0.012	625
Name = vendor_0119	0.007	549
Vendor = vendorID_0120	0.007	549
Sub spend area text = Road Packed	0.006	145
Name = vendor_0185	0.004	163
Vendor = vendorID_0188	0.004	163

activities in the process, namely *Record Invoice Receipt, Vendor creates invoice* and *Clear Invoice*. Similarly, the business area *Spend area text = Packaging* also has a high correlation with process flow characteristics. Analysis of the process model shows that, for example, 23% of *Packaging* cases contain activity *Receive Order Confirmation* compared to only 5% of the other cases. Further analysis of all the business areas listed in Table 3 shows that each of these areas has some distinctive process flow behavior that is more common in that area compared to the other business areas.

4.3 Clustering Summary for Case Attributes

Finally, Table 4 consolidates individual business areas into the Case Attribute level. *Item Type* with six distinct values and *Item Category* with four distinct values have most significant effects on process flow characteristics. To confirm the validity of these results we further analysis the materials provided in BPI Challenge 2019 website including the background information and submission reports [17]. It is clear that the *Item Type* and *Item Category* indeed can be regarded as the most important factors explaining the process flow behavior as

Table 4. Case Attributes ordered by effect on process flow

Case attribute at	Contribution	Distinct values $n(V_{at})$
Item Type	0.086	6
Item Category	0.080	4
Spend area text	0.077	19
Sub spend area text	0.056	115
Spend classification text	0.043	4
Name	0.025	798
Vendor	0.025	840
Item	0.016	167
GR-Based Inv. Verif	0.012	2
Purchasing Document	0.002	7937
Document Type	0.000	3
Goods Receipt	0.000	2
Company	0.000	2
Source	0.000	1
Purch. Doc. Category name	0.000	1

they are specifically mentioned to *roughly divide the cases into four types of flows in the data*. It is also interesting to see that the *Spend are text* and *Sub spend are text* have a significant effect on the process flow even though they have much higher number of distinct values (19 and 115) compared to *Spend classification text* which only has four distinct values.

5   Limitations

Forming business area dimensions is an essential step in our method. However, some relevant business areas may consist of several dimensions, for example, the process flow behavior could be very distinctive in a particular combination of business areas *SalesOffice = Spain* and *ProductGroup = Computers*. Automatically detecting this kind of significant combined business areas would be a useful feature. Another limitation is that the process flow behavior does not take into account the performance profile, i.e., the lead times between individual activities and the total case duration. Although the usage of this kind of numerical information would require a more advanced clustering technique, the influence analysis part of the method presented in this paper would already handle the discovery of related business areas.

6 Summary and Conclusions

In this paper, we have presented a method for discovering those business areas that have a significant effect on process flow behavior based on clustering and influence analysis. As a summary of our findings:

- Our presented method is capable of discovering those business areas that have the most significant effect on the process execution. Our method provides valuable information to business people who are very familiar with case attributes and attribute values but not so familiar with the often technical event type names extracted from transactional system log files.
- Our method supports any available trace clustering method. Our case study shows that using the k-modes clustering algorithm with activity and transition profiles provides good results.
- Clustering makes the analysts realize that not all the cases in the process model are similar. Using the *Contribution%* measure to explain clustering results works well for explaining the clustering results to business people.
- The case study presented in this paper confirms that the identified business areas indeed have distinctive process flow behavior, for example missing activities, higher than average amount of some special activities, or distinctive execution sequence for activities. Using our method, the business analyst may now divide the process model into smaller subsets and analyze them separately. It is a good idea to start the analysis of any process subset again by running the clustering to see if the cases are similar enough from both process flow point of view.
  Clustering reduces the need for external subject matter business experts. Naturally, it would be nice to have a person who can explain everything, but in real life, those persons are very busy, and some important details are always likely to be forgotten by busy business people.

Acknowledgements. We thank QPR Software Plc for the practical experiences from a wide variety of customer cases and for funding our research. The algorithms presented in this paper have been implemented in a commercial process mining tool QPR ProcessAnalyzer.

References

1. Bazhenova, E., Weske, M.: Deriving decision models from process models by enhanced decision mining. In: Reichert, M., Reijers, H.A. (eds.) BPM 2015. LNBIP, vol. 256, pp. 444–457. Springer, Cham (2016). https://doi.org/10.1007/978-3-319-42887-1_36
2. De Koninck, P., De Weerdt, J., vanden Broucke, S.K.L.M.: Explaining clusterings of process instances. Data Min. Knowl. Disc. **31**(3), 774–808 (2016). https://doi.org/10.1007/s10618-016-0488-4
3. De Leoni, M., Van Der Aalst, W.M., Dees, M.: A general process mining framework for correlating, predicting and clustering dynamic behavior based on event logs. Inf. Syst. **56**, 235–257 (2016)

4. de Leoni, M., Dumas, M., García-Bañuelos, L.: Discovering branching conditions from business process execution logs. In: Cortellessa, V., Varró, D. (eds.) FASE 2013. LNCS, vol. 7793, pp. 114–129. Springer, Heidelberg (2013). https://doi.org/10.1007/978-3-642-37057-1_9

5. de Medeiros, A.K.A., et al.: Process mining based on clustering: a quest for precision. In: ter Hofstede, A., Benatallah, B., Paik, H.-Y. (eds.) BPM 2007. LNCS, vol. 4928, pp. 17–29. Springer, Heidelberg (2008). https://doi.org/10.1007/978-3-540-78238-4_4

6. Hinkka, M., Lehto, T., Heljanko, K., Jung, A.: Structural feature selection for event logs. In: Teniente, E., Weidlich, M. (eds.) BPM 2017. LNBIP, vol. 308, pp. 20–35. Springer, Cham (2018). https://doi.org/10.1007/978-3-319-74030-0_2

7. Hinkka, M., Lehto, T., Heljanko, K., Jung, A.: Classifying process instances using recurrent neural networks. In: Daniel, F., Sheng, Q.Z., Motahari, H. (eds.) BPM 2018. LNBIP, vol. 342, pp. 313–324. Springer, Cham (2019). https://doi.org/10.1007/978-3-030-11641-5_25

8. Lehto, T., Hinkka, M., Hollmén, J.: Focusing business improvements using process mining based influence analysis. In: La Rosa, M., Loos, P., Pastor, O. (eds.) BPM 2016. LNBIP, vol. 260, pp. 177–192. Springer, Cham (2016). https://doi.org/10.1007/978-3-319-45468-9_11

9. Lehto, T., Hinkka, M., Hollmén, J.: Focusing business process lead time improvements using influence analysis. In: International Symposium on Data-Driven Process Discovery and Analysis (SIMPDA), pp. 54–67. Rheinisch-Westfaelische Technische Hochschule Aachen (2017)

10. OMG: Decision Model and Notation (DMN), vol. 1.2 (2019)

11. Seeliger, A., Nolle, T., Mühlhäuser, M.: Finding structure in the unstructured: hybrid feature set clustering for process discovery. In: Weske, M., Montali, M., Weber, I., vom Brocke, J. (eds.) BPM 2018. LNCS, vol. 11080, pp. 288–304. Springer, Cham (2018). https://doi.org/10.1007/978-3-319-98648-7_17

12. Song, M., Günther, C.W., van der Aalst, W.M.P.: Trace clustering in process mining. In: Ardagna, D., Mecella, M., Yang, J. (eds.) BPM 2008. LNBIP, vol. 17, pp. 109–120. Springer, Heidelberg (2009). https://doi.org/10.1007/978-3-642-00328-8_11

13. Souza, C.R.: The accord.NET framework. São Carlos, Brazil (2014). http://accord-framework.net

14. Suriadi, S., Ouyang, C., van der Aalst, W.M.P., ter Hofstede, A.H.M.: Root cause analysis with enriched process logs. In: La Rosa, M., Soffer, P. (eds.) BPM 2012. LNBIP, vol. 132, pp. 174–186. Springer, Heidelberg (2013). https://doi.org/10.1007/978-3-642-36285-9_18

15. Thaler, T., Ternis, S.F., Fettke, P., Loos, P.: A comparative analysis of process instance cluster techniques. Wirtschaftsinformatik **2015**, 423–437 (2015)

16. van der Aalst, W., et al.: Process mining manifesto. In: Daniel, F., Barkaoui, K., Dustdar, S. (eds.) BPM 2011. LNBIP, vol. 99, pp. 169–194. Springer, Heidelberg (2012). https://doi.org/10.1007/978-3-642-28108-2_19

17. Van Dongen, B.F.: Dataset BPI Challenge 2019. 4TU.Centre for Research Data (2019). https://doi.org/10.4121/uuid:d06aff4b-79f0-45e6-8ec8-e19730c248f1

Supporting Automatic System Dynamics Model Generation for Simulation in the Context of Process Mining

Mahsa Pourbafrani[1][✉], Sebastiaan J. van Zelst[1,2], and
Wil M. P. van der Aalst[1,2]

[1] Chair of Process and Data Science, RWTH Aachen University, Aachen, Germany
{mahsa.bafrani,s.j.v.zelst,wvdaalst}@pads.rwth-aachen.de
[2] Fraunhofer Institute for Applied Information Technology (FIT),
Sankt Augustin, Germany
{sebastiaan.van.zelst,wil.van.der.aalst}@fit.fraunhofer.de

Abstract. Using process mining actionable insights can be extracted from the event data stored in information systems. The analysis of event data may reveal many performance and compliance problems, and generate ideas for performance improvements. This is valuable, however, process mining techniques tend to be backward-looking and provide little support for forward-looking approaches since potential process interventions are not assessed. System dynamics complements process mining since it aims to capture the relationships between different factors at a higher abstraction level, and uses simulation to predict the effects of process improvement actions. In this paper, we propose a new approach to support the design of system dynamics models using event data. We extract a variety of performance parameters from the current state of the process using historical execution data and provide an interactive platform for modeling the performance metrics as system dynamics models. The generated models are able to answer "what-if" questions. Our experiments, using event logs including different relationships between parameters, show that our approach is able to generate valid models and uncover the underlying relations.

Keywords: Process mining · Scenario-based predictions · System dynamics · What-if analysis · Simulation

1 Introduction

Large amounts of event data are available in organizations, i.e., stored in information systems. Process mining provides the opportunity to exploit such data in a meaningful way, e.g., by discovering process models that describe the observed behavior in the organization, i.e., *process discovery* [2]. Furthermore, *conformance checking* [2] alongside process discovery assesses the level of similarity between the process model and the real executions of the process as captured

© Springer Nature Switzerland AG 2020
W. Abramowicz and G. Klein (Eds.): BIS 2020, LNBIP 389, pp. 249–263, 2020.
https://doi.org/10.1007/978-3-030-53337-3_19

in the event data. Moreover, *process enhancement* techniques improve the overall view of the process by extracting the information about the performance of the process [7,9]. For business owners, insight into their processes from different angles, especially from the performance view is highly valuable. These insights into their processes provide a platform to look forward and improve their processes. Therefore, these insights can be used to fill the gap between the current state of the process's performance and its desired future state. Business owners need to be supported in long-term decision making. Different approaches in process mining for the purpose of predicting a process's future behavior have been introduced. Most of these techniques are suitable for short-term prediction and they act at the *process instance level*, e.g., what is the next activity for a specific customer [22]. Others are highly dependent on explicit knowledge about the detailed processes such as [16]. Furthermore, hidden effects exist among the involved factors in the simulation models, e.g., the effects of increasing the workload of resources on their speed of performing the tasks, or the relationship between the level of difficulty of a task with the number of assigned resources.

Meanwhile, system dynamics techniques are able to cover different effects including human aspects and model the nonlinear relations at an aggregated level. Such techniques try to provide a holistic model of the system and include all possible effects in the system over time. However, most simulation-based approaches, including system dynamics, highly rely on the users and their understanding of the system. In [10,11], the idea of using process mining and system dynamics together at an aggregated level is presented which leads to designing the models including external factors. This approach generates system dynamics logs, i.e., a collection of measurable aspects from an event log. Then, the designed models are populated with the values of these measurable aspects referred to as SD-log. Hereafter, the validation step is performed to measure the similarity of the generated results by the model with the real values in the SD-log.

As shown in Fig. 1, the proposed approach depends on a modeling step which is based on the user insights into the system. In this paper, we propose a highly automated framework which supports businesses in an interactive manner for designing their simulation models. Our approach captures the influential factors in performance parameters and automatically generates the system dynamics models in order to predict the possible effects of future changes in the business processes. Afterward, these generated models can be populated with the values and the simulation and validation of the simulated results are possible.

The remainder of this paper is organized as follows. In Sect. 2, we introduce background concepts and basic notations used throughout the paper. In Sect. 3, we present related work. In Sect. 4, we present our main approach. We evaluate the proposed approach in Sect. 5. Section 6 concludes our work and discusses interesting directions for future work.

2   Preliminaries

In this section, we formalize the related concepts to our approach.

Fig. 1. Our proposed framework for using process mining and system dynamics together in order to design valid models to support scenario-based prediction of business processes in [10]. This paper focuses on the *automatic model generation*, i.e., the highlighted step.

Table 1. A simple event log. Each row refers to an event.

Case ID	Activity	Resource	Start timestamp	Complete timestamp
1	Register	Rose	10/1/2018 7:38:45	10/1/2018 7:42:30
2	Register	Max	10/1/2018 8:08:58	10/1/2018 8:18:58
1	Submit request	Eric	10/1/2018 7:42:30	10/1/2018 7:42:30
1	Accept request	Max	10/1/2018 8:45:26	10/1/2018 9:08:58
2	Change item	Eric	10/1/2018 9:45:37	10/1/2018 9:58:13
3	Register	Rose	10/1/2018 8:45:26	10/1/2018 9:02:05
...	...	...	...	...

Historic data, captured during the execution of a company's processes, provide the starting point for process mining [2]. Table 1, presents a simplified sample event log. It depicts the basic form of an event log in which each row represents an *event* and each *case ID* indicates an instance. An event log may include more data attributes, but, for simplicity, we abstract from these.

Definition 1 (Event Log). *Let $\mathcal{C}$, $\mathcal{A}$, $\mathcal{R}$ and $\mathcal{T}$ denote the universe of case identifiers, activities, resources, and the time universe, respectively. The universe of events is defined as $\xi = \mathcal{C} \times \mathcal{A} \times \mathcal{R} \times \mathcal{T} \times \mathcal{T}$. An event $e = (c, a, r, t_s, t_c) \in \xi$ refers to a case c, an activity a, a resource r, a start time t_s, and a complete time t_c. We define corresponding projection functions $\pi_\mathcal{C} : \xi \to \mathcal{C}$, $\pi_\mathcal{A} : \xi \to \mathcal{A}$, $\pi_\mathcal{R} : \xi \to \mathcal{R}$ and $\pi_\mathcal{T} : \xi \to \mathcal{T} \times \mathcal{T}$. Given $e = (c, a, r, t_s, t_e) \in \xi$, we have $\pi_\mathcal{C}(e) = c$, $\pi_\mathcal{A}(e) = a$, $\pi_\mathcal{R}(e) = r$, and $\pi_\mathcal{T}(e) = (t_s, t_c)$. An event log $L \subseteq \xi$ is a set of events.*

Consider the first event depicted in Table 1. In the context of Definition 1, the first row (which we denote as e_1), describes: $\pi_\mathcal{C}(e_1) = 1$, $\pi_\mathcal{A}(e_1) = Register$, $\pi_\mathcal{R}(e_1) = Rose$ and $\pi_\mathcal{T}(e_1) = (10/1/2018\ 7:38:45,\ 10/1/2018\ 7:42:30)$. Using such event data, process mining techniques can be used to discover process models, check conformance, uncover bottlenecks, predict process outcomes, and steer process improvement initiatives.

System dynamics provides a collection of techniques and tools to model and analyze capturing changes in complex systems over time [20]. Two main diagrams used within system dynamics are the *causal-loop diagram* and the *stock-flow*

$$Stock1 = Stock1_0 + \int_{t_0}^{t_n}(+Flow1 - Flow2)\mathrm{d}t \quad (1)$$

Fig. 2. A simple example stock-flow diagram and the underlying relation of $Stock1$ and its in/outflows ($Flow1$ and $Flow2$).

diagram which represent the conceptual relations between variables in the system and underlying equations respectively [13].

Definition 2 (The Causal-loop Diagrams). *A causal-loop diagram $CLD = (\mathcal{V}, \mathcal{L})$ is a set of nodes $\mathcal{V}$ and a set of directed links $\mathcal{L} \subseteq \mathcal{V} \times \mathcal{V}$. Directed link $l = (v_1, v_2) \in \mathcal{L}$ connects nodes v_1 and v_2 using a directed arc.*

The designed causal-loop diagram is a platform for designing a stock-flow diagram. Since all the elements and their relations are already modeled in the causal-loop diagram, a mapping between nodes in the causal-loop diagram and the elements in the stock-flow diagram should be made. A stock-flow diagram also is a diagram that indicates the same relationship in a causal-loop diagram for a system using three different basic elements, i.e., *stocks*, *flows* and *variables*[4]. Entities accumulated over time represented by numbers are usually mapped to a stock. Rate-based entities such as income per month can be considered as flows that can add to or remove from stocks.

Definition 3 (Stock-flow Diagram). *A stock-flow diagram is a tuple (S, F, A, M) with three disjoint set of elements, i.e., stocks S, flows F, and variables A, and $M \subseteq (S \cup F \cup A) \times (F \cup A)$ is a relation showing the flow of information between the elements. $\mathcal{V} = S \cup F \cup A$ and three subsets are pairwise disjoint. $\circlearrowright \in S$ is the boundary of the system.*

Each of the subsets introduced in Definition 3 is visualized with a specific shape the corresponding diagram, see Fig. 2. $Stock1 \in S$, $Flow1$ and $Flow2 \in F$ and $Variable1 \in A$ are the elements and the arcs between each two elements are derived from M. Also, there are two information flows from system boundaries to $Stock1$ and vice versa. Stock-flow diagrams are used for simulation using the specified underlying equations. The equation depicted on the right-hand side of Fig. 2 describes the underlying relation for the diagram. Consider t as time, $Stock1$ is equal to the amount in $Stock1$ at time t_0 plus the integral over the difference of the $Flow1$ and $Flow2$ over the time interval $[t_0, t_n]$. In each step, values of stock-flow elements get updated based on the previous values of the other elements that influence them.

3 Related Work

In this work, we propose a framework using process mining to provide insights that support the modeling techniques in system dynamics. The resulting models are used for the purpose of scenario-based prediction. We refer to [2] and [20] for an overview of process mining and system dynamics, respectively. Different approaches and techniques have addressed forward-looking in process mining. Among these approaches, we divide the ones w.r.t. performance into two main categories including simulation techniques and prediction techniques. Both categories of approaches aim to predict the future state of a process. In the first category, work such as [17] introduces discrete event simulation on the basis of discovered process models. Moreover, workflow management and simulation are combined in [18]. The authors considered both workflow design and event data to provide a model for the current state of the workflow. As the author in [1] mentioned, the factor of human behavior is missing in the proposed techniques.

In the second category, the approaches focus on predicting the performance aspects of the processes. The authors in [5] aim to predict the remaining process time or outcome of specific cases. In [23] a survey on the approaches which use prediction techniques in process mining is provided. The proposed approaches are mainly focused on the short-term prediction, e.g., predicting the time in which a specific process instance, i.e., a customer process will be finished or what will be the next activity [21]. The importance of context and interaction with the factors outside the processes for the prediction and simulation has been shown extensively [3,8]. Yet, existing approaches tend to abstract from these.

A combination of system dynamics with the process management field is proposed in [6]. In the context of processes also in [15], a business process of SAP is introduced in the form of system dynamics models that covers the factor of employees' productivity. However, in system dynamics models such as most of the simulation models, it is difficult to assess the reliability of the prediction results [1]. Moreover to the provided techniques in process mining and system dynamics, a combination of both fields in order to perform scenario-based analyses in the business processes has been recently proposed [10]. In this approach, the freedom in choosing the level of detail in modeling using system dynamics modeling, and the possibility of extending the factors outside of the processes are provided. Using event logs in process mining the validity of the designed models based on the result of the simulation is assessed. In our approach, we extend the main framework presented in [10] and propose a standalone interactive framework that supports the designing step using the event logs and structure of the system dynamics diagrams.

4   Approach

In this section, we explain the main approach focusing on the automatic generation of system dynamics models. As Fig. 3 shows, we transform an event log into a sequence of measurable performance parameters of a process, i.e., SD-log. In the parameter extraction module, we get the performance questions in the context of scenario-based analysis, e.g., how does the increase in the number of arrival affect the average waiting time in the process? Then, we extract the possible measurable parameters related to the questions over time. The calculated values of these parameters on the specified window of time form the SD-log.

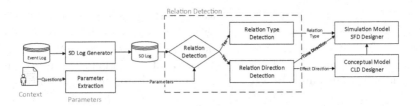

Fig. 3. The main approach including the SD-log generation, relation detection and the discovery of the type and direction of the relations. Our approach continues with the automatic generation of causal-loop diagrams (CLD) and Stock-flow diagrams (SFD). The type of relationship is used to form the underlying equations in SFD and the effect and time directions are automatically used to design the CLD as a backbone of SFD.

Our approach continues using the generated SD-log to detect any possible relationship between the parameters in which each relation has a type and a direction. The type of a relation can be linear or nonlinear and the direction of the relationship exists in two dimensions, time and effect. For instance, there is a relation between the arrival rate per day and the average waiting time per day. The type of this relation can be linear/nonlinear and negative/positive. The direction of the effect is from arrival rate to average waiting time that means arrival rate influences the average waiting time. At the same time, the effect of increases in the number of arrivals may only be visible with some delay, e.g., after two hours in the average waiting time, which shows the direction in time. These discovered relations and their types and directions are used to build the system dynamics simulation models.

4.1   SD-Log Generator

An event log is the starting point of any analyses in process mining, therefore, the possible parameters are highly dependent on the available data in the event log. In our approach, we consider the basic type of event logs in Definition 1. Hence, time-related performance parameters w.r.t events, cases, resources, and activities, e.g., service time of a case/event can be generated. Instead of extracting and computing the parameters at the instance level, in this work, we define aggregated parameters over a specific period of time such as δ. Reconsider Table 1. We extract the average duration of performing activity "Register" in each hour as δ in the log instead of extracting the values for each case separately. Having all the values in the possible steps considering the time window forms the SD-log, which we define as Definition 4. Performance parameters can be defined and extracted from an event log w.r.t. the scenario-based questions which we consider as a set $\mathcal{V}$.

Definition 4 (SD-log).   *Let $L \subseteq \xi$ be an event log, $\mathcal{V}$ be a set of process parameters, and δ be the selected time window. Assume e_1 is the first event in the event log starting at time t_S and assume e_n is the last event in the event log completing at time t_C. Given the time window δ, there are $k = \lceil (t_C - t_S)/\delta \rceil$*

subsequent time windows to go from t_S to t_C. An SD-log is a function $SD: \mathcal{V} \rightarrow \mathbb{R}^k$, where $\mathbb{R}^k$ is a sequence of real numbers of length k. Furthermore, for any $v \in \mathcal{V}$ and $0 \leq i < k$, we use $\pi_i(SD(v))$ to denote the $(i+1)^{th}$ value for parameter v, i.e., if $SD(v) = \langle x_0, x_1, \ldots x_{k-1} \rangle$ is the sequence for parameter v, then $\pi_i(SD(v)) = x_i$.

Assume event log $L \subseteq \xi$ with a total duration of $10\,h$, $\delta = 1\,h$ implies $k = 10$. An example SD-log including one variable *Arrival rate* is: $SD(Arrival\ rate) = \langle 11, 13, 10, 10, 12, 9, 10, 13, 8, 11 \rangle$, i.e., $\pi_2(SD(Arrival\ rate)) = 10$, representing that in the third hour, 10 cases arrived in the process.

It is important to note that each parameter is mapped onto a sequence of real numbers. Each real number is computed over the event log and focuses on the combination of a parameter and a time window, e.g., number of customers handled, number of customers queuing, average waiting time, percentage rejected, etc. Selection of the parameters is highly dependent on the "what-if" questions. Using these types of questions, the parameters are extracted and are being used in the next steps. We refer to [10] for detail of calculation method and dealing with overlapping features in multiple time windows.

4.2 Relation Detection

System dynamics models are based on the effects of the system's parameters on each other. In designing the system dynamics models such as causal-loop diagrams and stock-flow diagrams, the relations between the elements are crucial.

Knowing these relations and their directions makes it possible to create causal-loop diagrams and eventually stock-flow diagrams that can be used to simulate different scenarios. In the relation detection part, we discover all the possible relations and use them to automatically design the simulation models.

Considering the values of the parameters from the SD-log in the specified time window, we calculate both linear and nonlinear correlation using *Pearson correlation* and *Distance correlation* techniques [19]. In addition to the relation between two parameters regarding one influencing another one, this influence can happen in different time windows. Therefore, a relation has two aspects, i.e., the direction of

Fig. 4. A sample SD-log with k values for two parameters v^1 and v^2. The black arrows show that by investigating relations for the next time window (one step shift), one value for each parameter gets ignored, i.e., v_k^1 and v_0^2.

the effect that shows which parameter causes changes and also the time window in which these changes would influence the second parameter. One parameter may affect another parameter at a later time, hence it is not sufficient to calculate correlations between values in the same time window in this situation. In our previous example, if we look for the relations only at the same time window the effect of changes in the arrival rate on the average waiting time which appears after two hours would not be captured for the time window of one hour.

Consider Fig. 4 showing a possible time-shifted relation between two parameters. Due to the shift, we lose some values at the beginning and end of the SD-log. Therefore, to compare the values of parameters in different steps of the time window, an indicator is needed in order to preserve a sufficient number of values. Assume $s \in \mathbb{N}$ as the maximum possible shift in the time windows to look for the cause and effect between parameters where $s \leq k(1 - \theta_{sd})$ and θ_{sd} is the minimum percentage of values of the parameters that we are willing to use and $k \in \mathbb{N}$ is the number of values presented in the SD-log. Accordingly, prior to defining the relation detection algorithm, we need to define a shift function that provides the values of parameters with the given shift for detecting their underlying relations. We define the shift function in Definition 5 and use this function as an input of the relation detection algorithm in Algorithm 1.

Definition 5 (Shift Function). *Let $SD: \mathcal{V} \to \mathbb{R}^k$ be an SD-log with parameters $\mathcal{V}$ and s the maximum possible shift. For any two parameters $v^1, v^2 \in \mathcal{V}$, and a shift i with $0 \leq i \leq s$: $ShiftFun_i : \mathbb{R}^k \times \mathbb{R}^k \to \mathbb{R}^{k-i} \times \mathbb{R}^{k-i}$ relates values for v^1 with later values for v^2, i.e., $ShiftFun_i(SD(v^1), SD(v^2)) = (\langle \pi_0(v^1), ..., \pi_{k-(i+1)}(v^1) \rangle, \langle \pi_i(v^2), ..., \pi_{k-1}(v^2) \rangle)$ (note that $k = |SD(v^1)| = |SD(v^2)|$)*

Consider v^1 and v^2 as the arrival rate per day and the number of waiting cases per day and $SD(v^1) = \langle 11, 13, 10, 10, 12, 9, 10, 13, 8, 11 \rangle$, $SD(v^2) = \langle 2, 3, 0, 1, 4, 0, 1, 3, 0, 2 \rangle$ are the values of two parameters, applying the *ShiftFun$_2$* will result in $SD(v^1) = \langle 11, 13, 10, 10, 12, 9, 10, 13 \rangle$, $SD(v^2) = \langle 0, 1, 4, 0, 1, 3, 0, 2 \rangle$.

In the relation detection algorithm for each pair of parameters in the SD-log, the shift function is applied repeatedly bounded by the maximal possible shift s. The maximum value of the correlation is compared with the threshold θ_rel to assess how strong the relationship is. The comparison with the threshold shows whether the relationship exists or not. Each pair of parameters as an output of the algorithm will define the relations between parameters and be the root of the automatic causal-loop diagram designer in our approach.

Conceptual Model: Causal-Loop Diagram Designer. Using Algorithm 1, the relations are extracted and automatically transformed into a causal-loop diagram. In the transformation step, the domain knowledge of the user is also considered in indicating and selecting the desired relations in the output causal-loop diagram. Consider, for example, the effect of an increase in the arrival rate per time window on the average waiting time of cases. In this example, the output of Algorithm 1 is $(arrival\ rate, average\ waiting\ time)$, showing that changes in the values of arrival rate over time would cause changes in the average waiting time. The parameters in $\mathcal{V}$ from the SD-log are mapped to the nodes in the CLD diagram and the relations are represented as the links L. The extracted relations such as the example, form the causal-loop diagram automatically, e.g., in the causal-loop diagram, there is a link from *arrival rate* to the node labeled as *average waiting time*.

Algorithm 1: Relation Detection Algorithm

Input: SD_Log
Input: Maximum possible shift s and threshold of accepting a relation θ_rel
Output: All relations between pair of parameters $\in \mathcal{V}$

1 **foreach** v^m and $v^n \in \mathcal{V}$ **do**
2 | **foreach** $0 \leq i \leq s$ **do**
3 | | Generate $score = correlation(ShiftFun_i(v^m, v^n))$;
4 | | Add $score$ to the set Set_scores;
5 | **end**
6 | return $Max(Set\_scores)$ as max_score;
7 | **if** $max\_score \geq \theta\_rel$ **then**
8 | | return (v^m, v^n) as a relation;
9 | **else**
10 | | return null;
11 | **end**
12 **end**

Simulation Model: Stock-Flow Diagram Designer. Having a causal-loop diagram, the platform for designing the stock-flow diagram is provided. The parameters as mentioned in Sect. 4.1 are divided into three types namely, rate, number, and duration based which automatically are mapped into the stock-flow diagram as flows, stocks or variables, respectively. For a generated CLD, all nodes $v \in \mathcal{V}$ are mapped to a $s \in S, f \in F$ or $a \in A$ and for each link $l \in L$ such as (v^1, v^2) the corresponding notation in stock-flow diagram is replaced. Therefore, information flow M in Definition 3 is corresponding to the links in CLD with the replaced notations. The constraints in relations in stock-flow diagrams definition in Definition 3, automatically preserved in the mapping. A stock and a flow can influence a variable but a variable can only influence a flow or a variable, also a flow is able to influence a stock. Also, the option of including the user's domain knowledge regarding the simulation scenarios is provided.

Domain knowledge can be inserted interactively to the mapping step, e.g., the number of cases waiting in the system as a parameter can be treated as both stock or a variable which based on the scenario are exchangeable. For all the generated models, the system dynamics files (i.e., *mdl* files) are generated. These *mdl* files can be used for the scenario-based analyses presented in [10], but also analyzed using system dynamics simulation software such as *Vensim*.[1] The ability to generate system dynamics models from event logs provides an integrated approach that is both forward-looking and backward-looking.

5 Evaluation

We use synthetic event logs including different types of relationships between process' performance parameters to evaluate our approach. Moreover, we design

---

[1] https://vensim.com.

scenarios in which the input of the approach is the output of the predefined simulation model using which we can measure the similarity between the generated model and the original one.

Evaluating the Feasibility of the Approach. First, we evaluate our model with an event log which is intentionally designed with multiple linear and nonlinear relations. The hidden cause and effects inside the parameters of the process are specified in the process model. The process designed using CPN tools [14] simulates the process inside the call center of a car rental agency in which two types of requests are handled, requests for cars and requests for cars with a driver. The requests are randomly generated using the ratio of 60% for cars and 40% for cars with driver. For this reason, three resources are assigned for car requests and 2 resources are assigned for a car with a driver requests. The working hours of the call center are between 8:00 in the morning until 17:00 in the afternoon for 7 days per week with a higher number of requests around 10:00 and 15:00. Also, if the number of customers in line for getting the service is above 30 customers, the call will be rejected automatically.

We designed the model in a way that the operators perform the process of the calls faster if the number of calls in the line is higher. This effect of length of queues of the inline calls on the time of the processing calls is modeled as nonlinear relation, using an exponential function. Furthermore, an increase in the arrival rate influences the number of finish rates and the number of calls waiting inline to get the service. The generated event log using the presented model including 2000 cases used as an input of the approach and the detected relation using a *Daily* time window is shown as Fig. 5. For calculating the maximum shift, we set the minimum of data to 90%.

All the underlying effects at the instance level are captured. Then the generated model can be used as a basis of the simulation and scenario-based analysis w.r.t the changes in the process parameters. As the designed model by the approach indicates, only for the parameter *Number of unique resources* per day which we expect to have an effect on the number of handled requests, our approach could not find any strong relations. The reason is that we consider a fixed number of the resources in our CPN model.

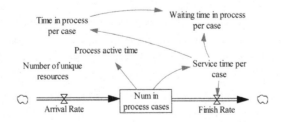

Fig. 5. The automatically generated output of the approach using the event log of the call center (the result is available as an *mdl* file that can be used in various system dynamics tools, e.g., Vensim).

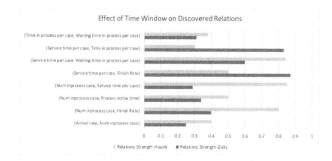

Fig. 6. A comparison between the strength of discovered relations for different SD-logs generated using *Hourly* and *Daily* time windows. As expected, the relations in the *Hourly* time window are stronger. However, the strength of the relations in the *Daily* time window are also above the threshold.

We extended our experiments by choosing different time windows to see if the approach is able to capture the relations in different time windows. Figure 6 illustrates the comparison of the discovered relations in the time window of *Daily* and *Hourly*. The strength of the relations as a result of Algorithm 1 are scaled between 0 and 1 which 1 shows the strongest relationship. The relations in *Hourly* time windows are stronger than the *Daily* manner, however, in both time windows, the expected relations are discovered. For instance, consider the relation between the number of cases in the process and the service time of cases, the relation can be seen strongly in *Hourly* time window, 0.84 rather than in *Daily* time window 0.29. As we expected, the approach is able to discover the hidden relations at an aggregated level, however, choosing the appropriate time window regarding the context of the process and modeling is important.

Evaluating the Accuracy of the Approach. We also designed a system dynamics model manually and generated a set of values over a defined *Daily* time window and used the generated data to feed to our approach to compare the discovered system dynamics model with the original one. Figure 7 shows the designed evaluation scenario.

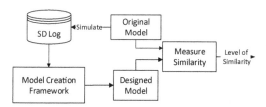

Fig. 7. The designed evaluation scenario in order to measure the similarity of the designed model using the proposed approach and the original model.

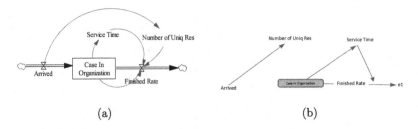

(a)   (b)

Fig. 8. Evaluation results in the similarity measurement scenario. The model presented in (a) is designed and the simulation results are used to feed the proposed approach. Model (b) is the designed model by the approach.

Assume R is the universe of all relations between parameters and $R_o, R_d \subseteq R$ represent the set of relations in the original diagram which is manually designed and the set of relations in the designed diagram using the proposed approach, respectively. We define two similarity measures regarding the accuracy and precision of the designed model [12]: Eq. 2 and Eq. 3. *Accuracy* considers the similarity between the existing relations in the original and in the discovered model and *Precision* considers the relations which in the original model do not exist but are discovered wrongly in the designed model.

$$Accuracy = \frac{|R_o| \cap |R_d|}{|R_o|} \tag{2}$$

$$Precision = \frac{|R_o| \cap |R_d|}{(|R_o| \cap |R_d|) + (|R_d| \cap |R_o|/R)} \tag{3}$$

Figure 8a shows the original model which is simulated for 30 weeks and generates an SD-log out of it. This model includes the relationship between the number of resources and the arrival rate using the equations based on the root equation. If the arrival rate increases, the number of resources will increase up to a specific point. Also, the finish rate is based on the number of cases in the process and the average service time, whereas service time itself gets influence from the number of cases in the process, i.e., more cases waiting in the process, resources work faster hence the service time will decrease. We give the generated SD-log as an input to our framework and Fig. 8b shows the discovered model. Relations in the original model and discovered ones in the designed model result in an *Accuracy* of 0.72 and a *Precision* of 1. It should be considered the calculated measures are for the designed model without the interference of any domain knowledge from users.

The results of both simulation scenarios indicate the effectiveness of our approach. More importantly, all the relations were at the instance level, i.e., for each specific case and our approach is able to catch the aggregated effects without considering performance metrics and details of the process's steps which happened for each case.

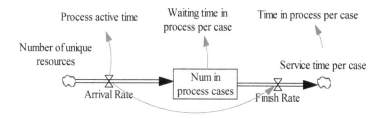

Fig. 9. Generated model for the real event log using the proposed approach.

Evaluating the Approach Using Real Event Log. In order to test our approach in practice we applied our framework on the real event log *BPIChallenge2017* [24]. The event log includes different executions of processes for taking a loan by customers. The number of cases is 31500 and the number of recorded events is 382454. The discovered relations between process parameters and the automatically generated model considering the *daily* behavior of the process is shown in 9. As expected, there is a strong relationship between the number of served people in the process per day and the number of people arrived per day. However, since not all the relations inside the event log are known, therefore the evaluation can only be performed based on the hypothesis and background knowledge of the processes. As indicated in Fig. 9, the number of people in the process of taking service is also directly influencing the process active time and the average waiting time per case. The reason for not indicating any strong relationship between the service time and the finish rate can be that the requests need a specific amount of time and not related to any other factors. The experiment using the real event log shows that the proposed approach is able to capture the most expected relations inside a real process.

6 Conclusion

In this paper, we proposed a novel approach to support designing system dynamics models for simulation in the context of operational processes. Using our approach, the underlying effects and relations at the instance level can be detected and modeled in an aggregated manner. For instance, as we showed in the evaluation, the effects of the amount of workload on the speed of resources are of high importance in modeling the number of people waiting to be served per day. In the second scenario, we focused on assessing the accuracy and precision of our approach in designing a simulation model. As the evaluations show, our approach is capable of discovering hidden relations and automatically generates the valid simulation models in which applying the domain knowledge is also possible. By extending the framework, we are looking to find the underlying equations between the parameters. The discovered equations help to obtain accurate simulation results in an automated fashion without user involvement. Moreover, we aim to apply the framework in case studies where we not only have the event data, but can also influence the process.

Acknowledgments. Funded by the Deutsche Forschungsgemeinschaft (DFG, German Research Foundation) under Germany's Excellence Strategy – EXC 2023 Internet of Production- Project ID: 390621612. We also thank the Alexander von Humboldt (AvH) Stiftung for supporting our research.

References

1. Aalst, W.M.P.: Business process simulation survival guide. In: vom Brocke, J., Rosemann, M. (eds.) Handbook on Business Process Management 1. IHIS, pp. 337–370. Springer, Heidelberg (2015). https://doi.org/10.1007/978-3-642-45100-3_15

2. van der Aalst, W.M.P.: Process Mining - Data Science in Action, 2nd edn. Springer, Heidelberg (2016). https://doi.org/10.1007/978-3-662-49851-4

3. van der Aalst, W.M.P., Dustdar, S.: Process mining put into context. IEEE Internet Comput. **16**, 82–86 (2012)

4. Binder, T., Vox, A., Belyazid, S., Haraldsson, H., Svensson, M.: Developing system dynamics models from causal loop diagrams. In: Proceedings of the 22nd International Conference of the System Dynamics Society, Oxford, Great Britain, 25–29 July 2004 (2004)

5. van Dongen, B.F., Crooy, R.A., van der Aalst, W.M.P.: Cycle time prediction: when will this case finally be finished? In: Meersman, R., Tari, Z. (eds.) OTM 2008. LNCS, vol. 5331, pp. 319–336. Springer, Heidelberg (2008). https://doi.org/10.1007/978-3-540-88871-0_22

6. Duggan, J.: A comparison of Petri net and system dynamics approaches for modelling dynamic feedback systems. In: 24th International Conference of the Systems Dynamics Society (2006)

7. Leemans, S.J.J., Fahland, D., van der Aalst, W.M.P.: Process and deviation exploration with inductive visual miner. In: Proceedings of the BPM Demo Sessions 2014 Co-Located with the 12th International Conference on Business Process Management (BPM 2014), Eindhoven, The Netherlands, 10 September 2014, p. 46 (2014) http://ceur-ws.org/Vol-1295/paper19.pdf

8. de Leoni, M., van der Aalst, W.M.P., Dees, M.: A general process mining framework for correlating, predicting and clustering dynamic behavior based on event logs. Inf. Syst. **56**, 235–257 (2016). https://doi.org/10.1016/j.is.2015.07.003

9. Mannhardt, F., de Leoni, M., Reijers, H.A.: The multi-perspective process explorer. In: Proceedings of the BPM Demo Session 2015 Co-Located with the 13th International Conference on Business Process Management (BPM 2015), Innsbruck, Austria, 2 September 2015, pp. 130–134 (2015). http://ceur-ws.org/Vol-1418/paper27.pdf

10. Pourbafrani, M., van Zelst, S.J., van der Aalst, W.M.P.: Scenario-based prediction of business processes using system dynamics. In: Panetto, H., Debruyne, C., Hepp, M., Lewis, D., Ardagna, C.A., Meersman, R. (eds.) OTM 2019. LNCS, vol. 11877, pp. 422–439. Springer, Cham (2019). https://doi.org/10.1007/978-3-030-33246-4_27

11. Pourbafrani, M., van Zelst, S.J., van der Aalst, W.M.P.: Supporting decisions in production line processes by combining process mining and system dynamics. In: Ahram, T., Karowski, W., Vergnano, A., Leali, F., Taiar, R. (eds.) IHSI 2020. AISC, vol. 1131, pp. 461–467. Springer, Cham (2020). https://doi.org/10.1007/978-3-030-39512-4_72

12. Powers, D.M.: Evaluation: from precision, recall and F-measure to ROC, informedness, markedness and correlation. J. Mach. Learn. Technol. **2**, 37–63 (2011)
13. Pruyt, E.: Small system dynamics models for big issues: triple jump towards real-world complexity. TU Delft Library (2013)
14. Ratzer, A.V., et al.: CPN tools for editing, simulating, and analysing coloured Petri nets. In: van der Aalst, W.M.P., Best, E. (eds.) ICATPN 2003. LNCS, vol. 2679, pp. 450–462. Springer, Heidelberg (2003). https://doi.org/10.1007/3-540-44919-1_28
15. Rosenberg, Z., Riasanow, T., Krcmar, H.: A system dynamics model for business process change projects. In: International Conference of the System Dynamics Society, pp. 1–27 (2015)
16. Rozinat, A., Mans, R.S., Song, M., van der Aalst, W.M.P.: Discovering colored Petri nets from event logs. STTT **10**(1), 57–74 (2008). https://doi.org/10.1007/s10009-007-0051-0
17. Rozinat, A., Mans, R.S., Song, M., van der Aalst, W.M.P.: Discovering simulation models. Inf. Syst. **34**(3), 305–327 (2009). https://doi.org/10.1016/j.is.2008.09.002
18. Rozinat, A., Wynn, M.T., van der Aalst, W.M.P., ter Hofstede, A.H.M., Fidge, C.J.: Workflow simulation for operational decision support. Data Knowl. Eng. **68**(9), 834–850 (2009). https://doi.org/10.1016/j.datak.2009.02.014
19. Schober, P., Boer, C., Schwarte, L.A.: Correlation coefficients: appropriate use and interpretation. Anesth. Analg. **126**(5), 1763–1768 (2018)
20. Sterman, J.D.: Business Dynamics: Systems Thinking and Modeling for a Complex World. McGraw-Hill, New York (2000)
21. Tax, N., Teinemaa, I., van Zelst, S.J.: An interdisciplinary comparison of sequence modeling methods for next-element prediction. CoRR abs/1811.00062 (2018). http://arxiv.org/abs/1811.00062
22. Tax, N., Verenich, I., Rosa, M.L., Dumas, M.: Predictive business process monitoring with LSTM neural networks. CoRR abs/1612.02130 (2016). http://arxiv.org/abs/1612.02130
23. Teinemaa, I., Dumas, M., Rosa, M.L., Maggi, F.M.: Outcome-oriented predictive process monitoring: review and benchmark. TKDD **13**(2), 17:1–17:57 (2019). https://doi.org/10.1145/3301300
24. Van Dongen, B.F. (Boudewijn): BPI challenge 2017 (2017). https://doi.org/10.4121/UUID:5F3067DF-F10B-45DA-B98B-86AE4C7A310B

Supporting the Development and Realization of Data-Driven Business Models with Enterprise Architecture Modeling and Management

Faisal Rashed[(✉)] and Paul Drews

Institute of Information Systems, Leuphana University of Lüneburg, Universitätsallee 1, 21335 Lüneburg, Germany
faisal.rashed@outlook.de, paul.drews@leuphana.de

Abstract. Designing and realizing data-driven business models (DDBMs) are key challenges for many enterprises and are recent research topics. While enterprise architecture (EA) modeling and management proved their potential value for supporting information technology-related projects, EA's specific role in developing and realizing DDBMs is a new and rather unexplored research field. We conducted a systematic literature review on big data, business models, and EA to identify the potentials of EA support for developing and realizing DDBMs. We derived 42 EA concerns from the literature, structured along the dimensions of the business model canvas and the status of realization (as-is, to-be).

Keywords: Enterprise architecture · Data-driven · Business model · Concerns

1 Introduction

Advancements in information technology, especially in machine learning, big data, cloud, and Internet of Things (IoT) technologies, have continuously increased the importance of data for business development and innovation. Today, many practitioners in business, as well as academic communities, perceive data as the 'new fuel' of the economy [1]. Nevertheless, the failure rate of big data and artificial intelligence projects remains disturbingly high [2]. Especially, incumbent companies are expected to rest on huge unused data treasures, facing several challenges in monetizing their data and seizing new business opportunities [3].

Over the last years, a new research field has emerged, which investigates data-driven business models (DDBMs). DDBMs are characterized by data as a key resource, data processing as a key activity, or both [4]. They highly rely on information systems for their core operations of data capturing, processing, and distribution. Various DDBM representations have been proposed by scholars. The latest efforts in academia have focused on extending the Business Model Canvas (BMC) as a widely accepted modeling framework to the special needs of data-driven businesses [4, 5]. These models help practitioners to envision and document the design of DDBMs in the first step and to further detail and realize the design in the second step [6]. Research on DDBMs is still

© Springer Nature Switzerland AG 2020
W. Abramowicz and G. Klein (Eds.): BIS 2020, LNBIP 389, pp. 264–276, 2020.
https://doi.org/10.1007/978-3-030-53337-3_20

at an early stage and requires detailed knowledge of tool support for DDBM design [5]. Especially, the needs of incumbent companies, with their existing data resources and structures, are currently not addressed.

In this paper, we explore if and how enterprise architecture (EA) modeling and management can be a beneficial approach for developing and realizing DDBMs. Previous research already highlighted the potential of EA in this context as it helps to gain transparency across relevant social and technical elements and their interdependencies [6]. Especially in the design phase, where companies struggle to understand the existing data resources and capabilities, EA can provide answers to stakeholder concerns based on models and tool support. However, the support of EA modeling and management has only been briefly described in the past. Therefore, our study focuses on the following research questions: What application fields for EA modeling and management in DDBM design and realization exist in the literature? What DDBM-specific EA concerns can be derived from the literature? To answer these questions, we conducted a structured literature search and analysis.

2 Research Background

Our study is grounded on two research fields. First, recent papers have explored the challenges associated with the development of DDBMs while taking into account the general research on business models (BMs). Second, research on EA has shown the potential benefits of EA models and EA management for projects related to business transformation.

2.1 Data-Driven Business Models and Their Representations

Data is traditionally been perceived as a crucial component of business operations, strategic decision making, and new business development. The terms under which it has been investigated has varied in the past decades, ranging from business intelligence, business analytics, and big data to big data analytics [7]. The potential value contribution of data has been researched in three major areas, namely improved decision making, enhanced products and services, and new BMs [8]. In the third area, the latest technological advancements have contributed to the current enthusiasm for new DDBMs. Several definitions of DDBM have been proposed in the literature. All commonly state that data has to be an essential component. Accordingly, Hartmann, Zaki, Feldmann, and Neely define DDBM as "a business model that relies on data as a key resource" [4, p. 6]. Bulger, Taylor, and Schroeder [9] and Brownlow, Zaki, Neely, and Urmetzer [10] similarly highlight the fundamental role of data for DDBMs. Schüritz and Satzger argue that a clear threshold of required data for a DDBM is not defined and that companies alter from a traditional BM to a DDBM, with increased application of the data for the value proposition [11]. In the context of this study, we clearly distinguish between enhancements of existing BMs and new DDBMs that are centered on data (data as a key resource and/or data processing as a key activity).

The conceptual structure of any business can be represented with business modeling techniques. Several modeling frameworks have been proposed in the past, varying in characteristics and components. The most popular BM framework is the BMC [12], comprising nine components: partners, key activities, key resources, value proposition, customer relationships, channels, customer segments, cost structure, and revenue stream. The conceptualization and definition of DDBMs rely on the BMC, with data as a key resource and/or with key activities focusing on data processing. Research on DDBMs is still at an early stage. The latest efforts in academia have focused on extending the BMC to the special needs of data-driven businesses [4, 5]. The literature in this area is analyzed as part of our literature review.

2.2 Enterprise Architecture

The EA discipline is rooted in the information system research body of knowledge [13]. Research on EA goes back to the Zachman framework in the 1980s, which provides an ontology for modeling the fundamental structure of an organization and its information systems [14]. Today, EA is essential for many organizations to support technology-driven transformations as it helps maintain an overview of complex sociotechnical systems. The Federation of Enterprise Architecture Organizations defines EA as "a well-defined practice for conducting enterprise analysis, design, planning, and implementation, using a comprehensive approach at all times, for the successful development and execution of strategy" [15, p. 1]. The Open Group provides a narrower definition of EA, in line with the ISO/ICE/IEEE Standard 42010 of architecture definition, that is, "the structure of components, their inter-relationships, and the principles and guidelines governing their design and evolution over time" [16]. Hence, researchers and practitioners sometimes refer to EA as the practice and sometimes as the actual architecture of an organization. We use the term EA for the practice comprising the related modeling techniques, frameworks, and management function within an organization (EA management). The actual architecture of an organization is noted as as-is architecture, while planned future states are called to-be architecture [13].

EA aims to improve information system efficiency and effectiveness. For this purpose, it provides artifacts, such as meta-models, frameworks, tools, guiding principles, and management methods. Many organizations have established an EA management function concerned with the aforementioned aim. An organization's key components and their interdependencies are represented in EA models [17]. These models provide transparency and support for strategic planning, top management decision making, and project management [18]. The modeling concepts and case-specific models help in quickly addressing stakeholder concerns. A diversity of modeling frameworks has been proposed in the literature, with different layers, elements, and relations to represent the enterprise [13]. The models built based on these meta-models are concerned with either the current state (as-is) or the desired state (to-be) of the enterprise. The EA management function supports the transition from the as-is to the to-be state through several intermediate architecture stages.

3 Methodology

To identify the current state of the literature on the interplay of DDBMs and EA, we conducted a structured literature review, following a proposed methodology [19]. We queried the following databases with keyword searches conducted in August and September 2019: (1) AIS Electronic Library, (2) EBSCO Host Business Source Complete, (3) Google Scholar, (4) IEEE Xplore, (5) JSTOR, (6) Science Direct, and (7) Web of Science. Since the DDBM belongs to an interdisciplinary field, its research is reflected in the intersection of BM and big data [8]. Our search comprised keywords covering both areas. We added the research stream of EA to understand the interplay of these research fields. The keywords "data-driven," "business model," and "enterprise architecture" were selected based on the resulting four intersections (see Fig. 1: A, B, C, and D). To further extend the literature search, the terms "big data" and "analytics," which are s associated with "data-driven," were integrated into the search as well. This led to a total of 10 search strings. All hits were screened based on their titles and abstracts. The first 100 hits from Google Scholar were considered, acknowledging their decreasing relevance. Irrelevant, duplicate, and non-peer-reviewed results were excluded. The remaining 80 articles were reviewed based on their full texts. With the objective of identifying the literature on the interplay of EA and DDBMs, we looked at intersection A for articles with a central focus on this topic. To gain a broader understanding of potential application fields of EA for DDBMs, we examined intersection C for articles addressing EA support in big data ideation and realization, as well as intersection D for EA support for BM design and implementation. Additionally, we analyzed the literature in intersection B, whose articles have a central focus on DDBMs, to identify the requirements for EA. Based on these requirements, we derived DDBM-related EA concerns. If required, we reallocated relevant articles to a better-fitting intersection. For example, one article from intersection B [20] was transferred to A.

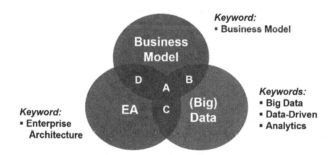

Fig. 1. Research fields and keywords.

Only four articles were identified in intersection A, of which two demonstrate contributions to the interplay of EA and DDBMs. We analyzed both articles in depth to gain insights into the state of the literature on the interplay of EA and DDBMs. From a total of 45 results, 17 articles are concerned with DDBMs. From this set, eight articles provide results that demonstrate the usefulness of our derivation of the concerns. These contributions were further analyzed.

The literature search on EA and big data resulted in a total of 16 articles, of which five were evaluated based on their full texts. In intersection D, we identified 16 articles concerned with EA and BMs. From this set, six were further analyzed. The relevant articles from intersections C and D were coded, allowing the derivation of EA support in the big data and the BM contexts.

To gain a deeper understanding of the potential application areas of EA for DDBMs, we analyzed the literature in the overlap of EA and big data, as well as that of EA and BMs. Based on the findings in intersection B, we derived EA concerns to support the viewpoints of the people involved in DDBM design. For this purpose, we drew on the top-down conceptual analysis proposed by Uzzle [21]. We analyzed the literature in depth and structured the derived concerns along the BMC fields, as the BMC is the most popular framework for BM representation [22]. We coded the results.

The literature in intersection B discusses DDBMs from various angles with regard to design requirements, representation requirements, and challenges. By detecting these characterizing requirements and challenges for DDBMs, we were able to derive EA concerns through deduction. For example, Kühne and Böhmann present design requirements for DDBMs, such as "customers and partners can provide data which can cause a connection in the representation. [...] data can be provided from customers or acquired by external sources" [5, p. 7]. Similarly, Hunke, Seebacher, Schuritz, and Illi mention the need for "an alignment of value propositions with involved partners and a definition of the ownership and access rights of data of different parties" [23, p. 4]. We formulated the related EA concern for the as-is architecture as follows: "Which data resources exist in the ecosystem (customers, partners, and data providers)?" The forward-looking concern for the to-be architecture was derived accordingly, as follows: "Which data could be generated in collaboration with actors from the ecosystem?" To cite another example, Zolnowski, Anke, and Gudat investigate costs and revenues from DDBMs and state, "revenues can be generated from [...] the sale of data to third parties" [24, p. 188]. We derived the EA concern accordingly, as follows: "Which data objects/sets can be offered to third parties?"

4 Results

4.1 EA Support for DDBM Modeling and Management

The DDBM literature overwhelmingly addresses the challenges and the requirements in DDBM conceptualization and design. All eight articles in intersection B address the urgent need for transparency of data and system components for successful DDBM design and realization. For example, Bulger, Taylor, and Schroeder highlight data quality, reliability, and availability as "obstacles" to DDBMs and emphasize the need for the transparency of the technology and science behind data sources [9, p. 24]. Zolnowski, Anke, and Gudat take the financial standpoint, stressing the need for transparency from a cost and revenue perspective [24]. They underline the cost related to data storage and with this, the importance of accurate tracking and management. Furthermore, sensors and information system channels are described as key for data capture and revenue generation. The gained transparency across relevant social and technical elements and their interdependencies infuse the DDBM design, which is primarily captured in the BMC [6]. The latest efforts in academia have focused on extending the BMC to the special needs of data-driven businesses [4, 5]. For example, Kühne and Böhmann propose nine key requirements for DDBM representation, concerned with data sources, security, quality, and data-processing capabilities [5]. We argue that EA can be beneficial in gaining transparency and infusing the design and realization of DDBMs.

EA has proven its potential in the context of big data project realization, as well as BM realization. EA supports the former with requirements and meta-models, specifically focusing on the data layer, as well as with management and modeling frameworks [18, 25–28]. In the BM context, EA modeling and management concepts are used for further detailing BMs and support their implementation. Specifically, the integration of the BMC into the popular EA modeling language ArchiMate and The Open Group Architecture Framework (TOGAF), as well as TOGAF ADM, has been investigated [17, 29].

The literature results of intersections A, C, and D discuss potential application fields of EA in the contexts of big data, BMs, and DDBMs. Based on these contributions, we have constructed an overview of the potential EA support for DDBMs. The relevant articles in intersection A propose two major phases for DDBM development, namely design and realization [6, 20]. Additionally, these articles distinguish between support from EA modeling and EA management. The contributions of both are described along the two phases. This structure has been used to map the potential application fields of EA in the DDBM context, as derived from the literature. Table 1 illustrates the results.

EA supports the DDBM design phase by providing transparency and the tools for stakeholders to envision future systems. EA management supports the realization phase by managing the architecture toward the desired target state and providing various tools and methods. EA offers blueprints, reference models, frameworks, and assessment tools to make the as-is state transparent and to develop the to-be state. A large set of contributions concentrates on EA for the realization of big data and with that, on DDBMs, while very little research has focused on EA for the design of DDBMs. This scarcity of studies may be due to the complex nature of EA as a tool for system developers [17]. Nevertheless, EA has the potential to drive innovation and the discovery of new DDBMs [30]. The design and realization of DDBMs require support from EA beyond the traditional project realization support. To justify huge investments in organizational transformation, as required for DDBMs, EA must first and foremost support the DDBM design phase. A good understanding of data assets and system components and their intertwinement, as well as the interlink to the ecosystem, is key for successful DDBM realization.

The extended literature provides insights on EA support for DDBM realization through various contributions on EA for big data and EA for BM realization. Furthermore, it highlights the scarcity of the contributions on EA for DDBM design. Our study supplements the research by depicting application fields of EA in the DDBM context using a structured literature-based approach and by contributing EA concerns to foster the intertwinement of EA in the DDBM design.

4.2 Derived EA Concerns

The literature results of intersection B are concerned with DDBMs. Based on the findings in this intersection, we have derived DDBM-related EA concerns. These concerns have been formulated from the viewpoints of practitioners in the DDBM design phase. They are related to either the current state of the architecture (as-is) or the desired state (to-be), which is envisioned in the design phase. Both have been derived deductively based on

Table 1. EA support for DDBMs.

		DDBM Design	DDBM Realization
EA modeling	As-is	• EA models for transparency [6, 30]	• EA models for transparency [6, 20] • EA meta-models for big data [18] • Big data-related EA concerns [18]
	To-be	• EA models for target state design [6, 20]	• Development of target architecture [6, 20] • Development of transition architectures [6, 20] • EA frameworks for big data realization [27, 28] • EA meta-models for big data [18] • Big data-related EA concerns [18]
EA management			• EA management for technical feasibility assessment [6] • EA management for enterprise transition [6] • Development of implementation roadmap [6] • Big data-related EA management requirements [26] • EA management method for big data [25]

the coded literature. The results are structured along the BMC fields and illustrated in Table 2. Each concern corresponds to an element of the BMC. Most findings relate to data, which is the key resource of DDBMs. In the rest of this subsection, we present the concerns and illustrate their usage and value contributions in an exemplary use case.

The BMC is a popular framework to capture the design of a BM by populating the nine key characterizing fields. Even DDBM research has adopted the BMC for BM representation [4, 22]. Since DDBMs have information systems as their engines, compared to traditional BMs, a set of profound EA concerns might be faced while populating the BMC fields. The complexity is further heightened because most incumbent organizations do not build up on a green field, such as startups do [3]. The existing resources and structures can be advantages or disadvantages, depending on how well they are understood and leveraged. Considering EA in DDBM design does not intend to assess feasibility but to provide a basic understanding of organizational structures. Designing DDBMs requires an understanding of EA in the same way that traditional BMs require basic business knowledge. For example, EA considerations can help prevent the proposition of a BM that generates €500,000 per year but requires tremendous investments

Table 2. DDBM-related EA concerns.

BMC field	As-is	To-be	Sources
Key activities	*C1:* What are the available data-processing capabilities at the application level?	*C1B:* What are the required data processing capabilities at the application level?	[5, 22, 31, 32]
	C2: What are the available data analytic capabilities at the business capability level?	*C2B:* What are the required data analytic capabilities at the business capability level?	[5, 32, 33]
	C3: How long does it take to process the data?	*C3B:* How can the data be processed within the time constraints of the new business model?	[32]
Key resources	*C4:* Which data resources exist across the organization?	*C4B:* Which additional data resources are required for the new business model?	[22, 23, 32, 33]
	C5: Which data resources exist in the ecosystem (customers, partners, and data providers)?	*C5B:* Which data could be generated in collaboration with actors from the ecosystem?	[5, 23]
	C6: How well are the data resources integrated?	*C6B:* How can the data resources be integrated to enable the new business model?	[22]
	C7: How often are collected data resources synchronized?	*C7B:* How often must the data be synchronized for the new business model?	[32]
	C8: Which measures are currently taken for realizing data security?	*C8B:* Which measures are necessary to ensure data security within the new business model?	[23]
	C9: What is the availability status of the data (e.g., company-owned existing data versus third-party-owned data that is not captured yet)?	*C9B:* What is the availability status of the data for the new business model?	[5, 22, 31]
	C10: Who is the owner of the data, legally and within the company?	*C10B:* Who owns the data processed for the new business model?	[5, 22, 23]
	C11: Where is the data stored (e.g., specific country, customer side)?	*C11B:* Where is the required data for the new business model stored?	[22, 23, 32]

(continued)

Table 2. (*continued*)

BMC field	As-is	To-be	Sources
	C12: What are the data privacy constraints for internal and external usage?	*C12B:* How is data privacy ensured within the new business model?	[5, 9, 22, 23]
	C13: What is the data quality in terms of consistency and completeness?	*C13B:* How is data quality ensured in terms of consistency and completeness?	[5, 9, 22, 31]
	C14: How long can the data be stored, and when must it be deleted?	*C14B:* How long must the data be stored for the new business model?	[22, 23]
Revenue model	*C15:* Which data objects/sets can be offered to third parties?	*C15B:* Which data objects/sets could be made available to third parties?	[24]
	C16: Which technical value-capturing mechanisms exist?	*C16B:* How can the value-capturing mechanism be realized from a technology perspective?	[23]
Cost structure	*C17:* What computing efforts are required to process the data (e.g., cloud facilities, analytic platforms)?	*C17B:* How can the computing efforts for data processing be reduced?	[22, 24]
	C18: How much does it cost to purchase/capture and store the existing data?	*C18B:* How can the cost of data purchase/capture and storage for the new business model be reduced?	[24]
Channels	*C19:* Where are the data import and export interfaces?	*C19B:* What are the required data import and export interfaces?	[22–24, 32]
	C20: How are the systems connected to customers and providers?	*C20B:* What are the required connections to customers and providers?	[22, 24]
	C21: What is the existing infrastructure to integrate data sources?	*C21B:* What are the required infrastructure components to enable the new business model?	[31, 32]

in the information system landscape and business capabilities amounting to over €20 million that would require 10 years for realization. Architects can infuse the DDBM design by providing information, models, guidance, and inspiration. The derived EA concerns equip architects to respond to the business demand while populating the BMC framework.

Günther, Rezazade Mehrizi, Huysman, and Feldberg [3] present the following example of how organizations may fail in realizing DDBMs. A European postal service organization aimed to sell addresses of potentially relevant households to business clients for targeted advertising. The company faced three major challenges. First, the company did not own the data, whose sale required the agreement of the owning party. Second, the company acquired a startup to obtain the data to implement its DDBM, which it failed to integrate into the organization model. "The decision to acquire appeared to be influenced largely by supra-organizational drivers, as the organization needed to access data from elsewhere and was pressured by a shrinking market" [3, p. 12]. The data resources' landscape and their availability within the organization and its ecosystem were not well understood. The existing structures had to be transformed in order to adopt the new resources. Lastly, due to the historical evolution of the company's BM, the sales team, who was used to selling contracts, struggled to interpret the data characteristics in their client conversations. The DDBM lacked the required business capabilities.

We assume that our derived EA concerns provide the equipment to prevent such a failure. The first challenge of data ownership could have been foreseen by utilizing EA concerns C10 (Who is the owner of the data, legally and within the company?) and C12 (What are the data privacy constraints for internal and external usage?) Raising these concerns leads to an early consideration of legal boundaries. EA models and the management function have to adopt in order to address these concerns effectively. The second challenge requires a good understanding of the existing data within the organization and throughout the ecosystem. Furthermore, it must be understood how well the data sources are integrated and what their availability status is. These concerns are reflected in C4 (Which data resources exist across the organization?), C5 (Which data resources exist in the ecosystem [customers, partners, and data providers]?), C6 (How well are the data resources integrated?), and C9 (What is the availability status of the data [e.g., company-owned existing data versus third-party-owned data that is not captured yet]?). The last challenge regarding the required business capabilities for the DDBM could have been addressed with C2B (What are the required data analytic capabilities at the business capability level?).

5   Discussion

Designing and realizing DDBMs require a creative yet analytic and structured set of activities. Both can benefit from the rich discipline of EA and its management. To shed light on the interplay of EA and DDBMs, we have investigated the literature using a structured search process. Our paper makes a twofold contribution to research. First, we have summarized the state of the knowledge about EA and DDBMs by conducting an extended literature search. We have analyzed the sources from the intersection of three research areas to provide a comprehensive view on previous research efforts. Our work has revealed the limited number of articles focusing on this highly demanded topic. The results of the literature related to this interdisciplinary research field have been used to construct an overview of potential application areas of EA in DDBM design and realization. With this overview, we have demonstrated the diverse application possibilities of EA in DDBM realization and have highlighted the current gap in EA support for DDBM design. Second, we have identified DDBM-related EA concerns to foster EA support for DDBM development and realization. By deriving 42 concerns from our literature-based approach, we have demonstrated the clear need for EA support in DDBM design and have laid the foundation for researchers to further investigate the contributions of EA for data-driven businesses.

Our study's results bear some limitations. First, the selection of keywords restricts the set of results. Though we have iteratively refined the search terms, some related work might have been overlooked. Although we have chosen an extended literature search approach, our research is not intended to be exhaustive. By using general databases, such as Google Scholar, we have also aimed at including work from other related research fields. Second, the results are limited in terms of validity as they are purely based on research articles and are not validated in an evaluation or an empirical study. Nevertheless, our work should help in guiding further studies in this area. We recognize the need for empirical research to extend and test the derived concerns.

Our study's results have implications for both academia and practice. For academia, we have gathered and analyzed the extended literature on the intersection of EA and DDBMs. Our findings have opened new research avenues. Additionally, the derived concerns lay the foundation for advanced support of EA in DDBM design. From a practitioner perspective, the overview of the current literature is beneficial for targeted knowledge development. Furthermore, the potential application areas of EA for DDBMs can be inspiring for organizational EA practice. Ultimately, the derived concerns provide starting points for the development of EA models to address DDBM-related concerns.

6   Conclusion and Future Work

Data has long been acknowledged as a key driver for business. Nevertheless, the recently emerging opportunities rooted in technological advancements, which allow BMs centered on data, are fairly new to research and practice. The design and realization of such DDBMs can benefit from the sophisticated EA practice, which not only enables strategic planning but also supports organizational transformation. We have conducted a systematic review to investigate the current state of the literature in the intersection of EA and DDBMs. Considering the related literature on big data and BMs, we have presented potential application areas of EA for DDBMs. Our findings reveal a gap in EA support in the DDBM design phase. To foster EA support in this phase, we have derived DDBM-related EA concerns.

Additional research is required to enrich our findings with empirical evidence. We plan to conduct expert interviews to advance the research in the intersection of EA and DDBMs. The rich literature on big data, EA, and BMs shows that DDBM implementation is not hindered by technological or methodical restrictions but more by creative design concepts that justify tremendous investments in DDBM realization.

References

1. Brynjolfsson, E., McAfee, A.: Big data: the management revolution. Harvard Bus. Rev. **10**, 1–12 (2012)
2. Redman, T.C.: Do your data scientists know the "why" behind their work? (2019). https://hbr.org/2019/05/do-your-data-scientists-know-the-why-behind-their-work
3. Günther, W.A., Rezazade Mehrizi, M.H., Huysman, M., Feldberg, F.: Debating big data: a literature review on realizing value from big data. J. Strategic Inf. Syst. **26**(3), 191–209 (2017)
4. Hartmann, P.M., Zaki, M., Feldmann, N., Neely, A.: Big data for big business? A taxonomy of data-driven business models used by start-up firms. Cambridge Service Alliance (2014)
5. Kühne, B., Böhmann, T.: Requirements for representing data-driven business models – towards extending the Business Model Canvas. In: Twenty-Fourth Americas Conference on Information Systems, pp. 1–10. AIS, New Orleans (2018)
6. Vanauer, M., Bohle, C., Hellingrath, B.: Guiding the introduction of big data in organizations: a methodology with business- and data-driven ideation and enterprise architecture management-based implementation. In: 48th Hawaii International Conference on System Science, pp. 908–917. IEEE, Hawaii (2015)
7. Chen, H., Chiang, R.H.L., Storey, V.C.: Business intelligence and analytics: from big data to big impact. MIS Q. **36**(4), 1165–1188 (2012)
8. Engelbrecht, A., Gerlach, J., Widjaja, T.: Understanding the anatomy of data-driven business models – towards an empirical taxonomy. In: Twenty-Fourth European Conference on Information Systems, pp. 1–15. ECIS, Turkey (2016)
9. Bulger, M., Taylor, G., Schroeder, R.: Data-driven business models: challenges and opportunities of big data. Oxford Internet Institute (2014)
10. Brownlow, J., Zaki, M., Neely, A., Urmetzer, F.: Data and analytics – data-driven business models: a blueprint for innovation. Cambridge Service Alliance (2015)
11. Schuritz, R., Satzger, G.: Patterns of data-infused business model innovation. In: 18th IEEE Conference on Business Informatics, vol. 1, pp. 133–142. IEEE, Paris (2016)
12. Osterwalder, A., Pigneur, Y.: Business Model Generation: A Handbook for Visionaries, Game Changers and Challengers. Wiley, Hoboken (2010)
13. Winter, R., Fischer, R.: Essential layers, artifacts, and dependencies of enterprise architecture. J. Enterp. Archit. **3**(2), 7–18 (2007)
14. Zachman, J.A.: Zachman International (2008). https://zachman.com/about-the-zachman-framework. Accessed 12 Nov 2019
15. Federation of EA Professional Organizations: a common perspective on enterprise architecture. Architecture and Governance Magazine, pp. 1–12 (2013)
16. The Open Group: TOGAF. https://www.opengroup.org/togaf. Accessed 06 Oct 2019
17. Musulin, J., Strahonja, V.: Business model grounds and links: towards enterprise architecture perspective. J. Inf. Organ. Sci. **42**(2), 241–269 (2018)
18. Burmeister, F., Drews, P., Schirmer, I.: Towards an extended enterprise architecture meta-model for big data – a literature-based approach. In: Twenty-Fourth Americas Conference on Information Systems (AMCIS), pp. 1–10. AIS, New Orleans (2018)
19. vom Brocke, J., Simons, A., Niehaves, B., Reimer, K., Plattfaut, R., Cleven, A.: Reconstructing the Giant: on the importance of rigour in documenting the literature search process. In: European Conference on Information Systems, pp. 2206–2217. ECIS, Verona (2009)
20. Chen, H.-M., Kazman, R., Garbajosa, J., Gonzalez, E.: Big data value engineering for business model innovation. In: 50th Hawaii International Conference on System Sciences, pp. 5921–5930. IEEE, Hawaii (2017)
21. Uzzle, L.: Using metamodels to improve enterprise architecture. J. Enterp. Archit. **5**(1), 49–61 (2009)

22. Kühne, B., Zolnowski, A., Böhmann, T.: Making data tangible for data-driven innovations in a business model context DSR methodology view project service dominant architecture view project. In: Twenty-Fifth Americas Conference on Information Systems, pp. 1–10. AIS, Cancun (2019)

23. Hunke, F., Seebacher, S., Schuritz, R., Illi, A.: Towards a process model for data-driven business model innovation. In: 19th Conference on Business Informatics, CBI, vol. 1, pp. 150–157. IEEE, Thessaloniki (2017)

24. Zolnowski, A., Anke, J., Gudat, J.: Towards a cost-benefit-analysis of data-driven business models. In: 13th International Conference on Wirtschaftsinformatik, pp. 181–195. WI, St. Gallen (2017)

25. Kearny, C., Gerber, A., Van Der Merwe, A.: Data-driven enterprise architecture and the TOGAF ADM phases. International Conference on Systems. Man, and Cybernetics, pp. 4603–4608. IEEE, Hungary (2017)

26. Kehrer, S., Jugel, D., Zimmermann, A.: Categorizing requirements for enterprise architecture management in big data literature. In: 20th International Enterprise Distributed Object Computing Workshop, pp. 98–105. IEEE, Vienna (2016)

27. Lněnička, M., Máchová, R., Komárková, J., Čermáková, I.: Components of big data analytics for strategic management of enterprise architecture. In: 12th International Conference on Strategic Management and Its Support by Information Systems, pp. 398–406. Curran Associates, Inc., Ostrava (2017)

28. Lnenicka, M., Komarkova, J.: Developing a government enterprise architecture framework to support the requirements of big and open linked data with the use of cloud computing. Int. J. Inform. Manag. **46**, 124–141 (2019)

29. Bouwman, H., De Reuver, M., Solaimani, S., Daas, D., Haaker, T., Janssen, W., Iske, P., Walenkamp, B.: Business models tooling and a research agenda. In: 25th Bled eConference – The First 25 Years of the Bled eConference, pp. 235–257. AIS, Bled (2012)

30. Petrikina, J., Drews, P., Schirmer, I., Zimmermann, K.: Integrating business models and enterprise architecture. In: 18th International Enterprise Distributed Object Computing Conference Workshops and Demonstrations Integrating, pp. 47–56. IEEE, Washington (2014)

31. Kühne, B., Böhmann, T.: Data-driven business models – building the bridge between data and value. In: 27th European Conference on Information Systems, pp. 1–16. ECIS, Stockholm & Uppsala (2019)

32. Exner, K., Stark, R., Kim, J.Y.: Data-driven business model: A methodology to develop smart services. International Conference on Engineering. Technology and Innovation, vol. 2018, pp. 146–154. IEEE, Madeira Island (2018)

33. Dremel, C., Wulf, J.: Towards a capability model for big data analytics. In: 13th International Conference on Wirtschaftsinformatik, pp. 1141–1155. WI, St. Gallen (2017)

Social Media

Customer Interaction Networks Based on Multiple Instance Similarities

Ivett Fuentes[1,2(✉)], Gonzalo Nápoles[2], Leticia Arco[3], and Koen Vanhoof[2]

[1] Computer Science Department, Central University of Las Villas,
Santa Clara, Cuba
ivett@uclv.cu
[2] Faculty of Business Economics, Hasselt University, Hasselt, Belgium
[3] Artificial Intelligence Lab, Vrije Universiteit Brussel, Brussels, Belgium

Abstract. Understanding customer behaviors is deemed crucial to improve customers' satisfaction and loyalty, which eventually is materialized in increased revenue. This paper tackles this challenge by using complex networks and multiple instance reasoning to examine the network structure of Customer Purchasing Behaviors. Our main contributions rely on a new multiple instance similarity to measure the interaction among customers based on the mutual information theory focuses on the customers' bags, a new network construction approach involving customers, orders and products, and a new measure for evaluating its internal consistency. The simulations using 12 real-world problems support the effectiveness of our proposal.

Keywords: Customer networks · Complex network construction · Multiple instance similarities · Customer purchasing behavior · Community detection

1 Introduction

Online store companies maintain large information systems for capturing records about customer purchasing transactions in a cost-effective manner. Understanding customer behaviors from this huge data is deemed crucial to improve customers' satisfaction and loyalty, which eventually is materialized in increased revenue [8,12,16].

Customer clustering techniques applied with proper pre-processing of the data are well suited to help address the challenge of transforming massive transactional information into knowledge and improving the business revenue [16]. Customer Purchasing Behavior (CPB) from discovered segments can give important insights not only in promotion and marketing, but also in deriving more efficient strategies for customer management and business profit [2]. Consequently,

The authors would like to thanks the anonymous commercial partners for providing the data sources and other resources used in this research. We are also grateful to Hasselt University for supporting this research with the special fund for incoming mobility.

W. Abramowicz and G. Klein (Eds.): BIS 2020, LNBIP 389, pp. 279–290, 2020.
https://doi.org/10.1007/978-3-030-53337-3_21

CPB analysis is considered a cornerstone when designing Decision Support Systems (DSS) [2,20] in commercial scenarios, due to the customer behavior understanding can help design more accurate models by considering common behaviors from customer segments [16]. Although CPB analysis is necessary in almost every company, a deep understanding is difficult since many entities come to play, such as products, customers, transactions and revenues [2].

The literature includes several customer clustering approaches that analyze different aspect of the Information Technology (IT) [16,18,20]. Some of them are heavily influenced by the extraction of additional features (i.e., recency, frequency, monetary value and volume of purchases) from the purchase history of customers before applying a classical clustering algorithm [16,18]. Other approaches conduct the exploratory market basket analysis under the transformation of each customer's order as a multi-category vector before the application of the segmentation process [16]. Although most of these models allow characterizing the customer behaviors, they do not explicitly consider a strategy for analyzing all entities involved in the problem and discovering which products and customers are related to each other in a more accurate way.

To cope with this issue, researchers have relied on different customer models for quantifying the similarity between them looking at the entirety basket market history. The authors in [10,11] proposed a recommendation systems based on a customer model in which, each customer is represented as a single vector of purchased products, ignoring the interaction between products in a specific order. In an attempt to get more accurate results, the authors in [6] present a new customer model by considering customers as bags of orders. In the last approach, customers are vectors of the multiple instance space and the interactions between them are computed by through Multiple Instance (MI) distances. Although this approach shows to be a suitable way to derive segments, it leads to some issues related with the loss of information for the similarity process aggregation by using Hausdorff multiple instance distances [7]. Therefore, the consistency relative to the coverage of common characteristics on segments are affected.

In this paper, we extended the customer purchasing behavior analysis into a complex network level, and propose a methodology to discover communities of customers with the same behavior, which can be used to design more effective marketing campaigns without spamming the customers. Our research comprises three main theoretical contributions, namely: (i) a new MI similarity function based on the mutual information theory that focuses on the customers' bags, (ii) a new network construction approach involving customers, orders and products and (iii) a new measure for evaluating how homogeneous is the behavior of customers who share the same community.

The remaining of this paper is organized as follows. Section 2 revises the methods reported in the literature. Section 3 introduces our customer complex network approach for CPB analysis. The issues related with the network construction are discussed in Sect. 4, while the experimental results are unveiled in Sect. 5. Towards the end, Sect. 6 outlines conclusions.

2   Customer Purchasing Representations

The CPB analysis from customer clustering, has been attaining more and more attention in recent years [6,8,10,11], due to clustering algorithms can analyze behavioral data and suggest a solution founded on observed data patterns. Customer clustering is the process of segment customer into distinct and internally consistent clusters with similar characteristics [6,10,11].

The transactions are the main aspect of the IT for the customer purchasing clustering. The transaction data usually has a multitude of dimensions that can be used for characterizing the CPB. For that reason, it is possible to discover different customer segments depending on the features used. However, most approaches comprise information about customers, products and customer-product purchases. Likewise, information about customer-product purchases (i.e., transactions) is represented by time, sales amount and place, etc. [9]. For the sake of simplicity, we assume that products that are bought in the same visit to the store, have the same timing component and form an order. Also, our approach can be used from both store information systems: online and offline.

The customer purchasing representation behind the recommendation systems proposed in [10,11] describes each customer by a single feature vector. Formally, a customer X corresponds to a vector in the space $\Re^d$, such that d denoted the cardinality of the product set P. In [6] the authors present a new customer model by considering customers as bags of orders. Formally, a customer is characterized by a bag $X = \{o_1, ..., o_n\}$, where n is the number of orders and o_i represents the instance that describes the i-th order of customer X. Each bag is allowed to have a different size, which means that the value n can vary among the bags in the dataset. In the above approach, due to customers are vectors of the multiple instance space, the authors aggregated the interactions between the orders of customers by using Hausdorff multiple instance similarities: the minimal ($MinH$), the maximal ($MaxH$) and the average ($AveH$) Hausdorff similarities [6]. Although, the Hausdorff distances (or similarities) are suitable techniques to determine to which extent one bag differs (or equals) from another, under certain restrictions they actually not work.

The minimal Hausdorff distance [7] uses the overall minimum pairwise distance between instances. This variant might not correctly express the semantics between bags due to the fact that a single instance determines the distance between bags. When the number of instances per bag is low and there is a very dense concept (i.e., it covers a small area in the feature space), this measure may actually work. The maximal Hausdorff distance [7] initially identifies for each instance in one bag the closest instance in the other bag. The maximal distance between instances is used to define the bag distance from all the closest matches. Both the minimal and maximal Hausdorff distances are sensitive to outliers that can dominate the distance value. The average Hausdorff distance [7] is not as noise-sensitive as the overall minimum. Nevertheless, the $AveH$ similarity does not consider the frequency, as illustrated in the next example. Henceforth, we use the lower case letters to represent orders (e.g., o) and upper case letters to represent customers (e.g., X). The notation o_i^k is used to denote the i-th order of customer k. However, when establishing a distinction between customers and orders is not required, we represent orders by only using a sub-index to keep the notation simple. Let us suppose that we apply the $AveH$ similarity over two costumers with bags $X_1 = \{o_1^1, o_2^1, o_3^1\}$ and $X_2 = \{o_1^2, o_2^2, o_3^2, o_4^2, o_5^2, o_6^2\}$ such that $o_1^1 = o_1^2 = o_2^2$, $o_2^1 = o_3^2 = o_4^2$ and $o_5^2 = o_6^2$, then $AveH(X_1, X_2) = 1$ although they do not have the same CPB. The results in [6] have confirmed the need for a function capable of capturing the structural differences between the bags associated to customers.

Several types of MI distances can be defined over any set as pointed out in [3] based on two main approaches: (I) *bag as a point set*, which aggregates the distances between instances (i.e., points) and (II) *bag is a distribution of points*, which compares the distributions represented by each point set. It should be stated that the second approach is not taken into account due to the following reasons: (1) it is difficult to estimate a probability density function in a high-dimensional feature space, (2) it is computationally demanding to estimate the difference, or overlap, between two probability distributions, and (3) in our study we need to aggregate the similarity between order pairs. Due to each order is a multi-set, the similarity between two orders o_i and o_j that belong to different customers is quantifying by using the $MSJaccard$ similarity, which is based on Jaccard index [19]. $MSJaccard$ measures the ratio between the intersection to the union of orders, where $multi_{p_k}(o_i)$ is the multiplicity of the product p_k in the ith order.

Aiming at overcoming limitation concerning to the compact structure within the segments discovered from these MI similarities, we exploit the potentiality of complex networks for modeling and analyzing CPB in real-world business scenarios. Besides, we explore the potentiality of discovering communities into the customer network constructions based on the threshold estimation for reducing weak interactions. In Sect. 4, we introduce a new multiple instance similarity function considering each bag as a point set of a subset of a high-dimensional space to cope with the challenge of compact customer segmentations.

3   Complex Network and MI Similarities for CPB Analysis

In this Section, we will introduce a new methodology for supporting customers' purchase decision-making process. Figure 1 portrays the blueprint of our methodology and the main contributions addressed by the Knowledge Discovery in Database (KDD) process model. Notice that our approach can provide different points of view depending on the marketing objectives to be addressed.

Aiming at providing more insights into our methodology, next we provide further details about the above-mentioned stages:

– **Stage 1:** (*Transformation*). In this stage we capture the information attached to customers (i.e., customer, product and order information). A bag space dataset from the original records is generated. The data is preprocessed for discarding irrelevant transaction records, through the data gathering, stratification and cleaning steps.

- **Stage 2:** (*Complex network construction*). This stage is devoted to determining the semantics of edges and nodes in the complex network, while quantifying the interactions between each pair of customers.
- **Stage 3:** (*Analysis*). This stage studies on each developed network model the best network topology in term of more compact customer community structure.
- **Stage 4:** (*Marketing*). The main idea of this stage is to show how the CPB study and characterization of each community can contribute to a personalized marketing strategy for supporting business decision making and increasing the loyalty and revenue. This stage is not addressed in this paper since it depends on the application problem to be modeled.

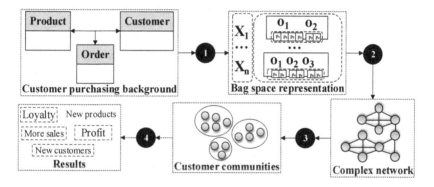

Fig. 1. General approach for supporting customers' purchase decision-making process addressed by the KDD process model.

Our methodology starts by mapping the purchasing information of customers in an MI bag space. Each customer is regarded as a set of orders, called the bag of orders that describe their behavior. This way of modeling customers distinguishes our proposal from other approaches that usually omit much information by simplifying the modeling of the customers and their orders. Customer interactions are measured by using MI similarities, thus allowing them to form a weighted complex network. The stage devoted to network construction is deemed the most important one since it comprises the main theoretical contributions of our research. Different network topologies are obtained by using a combination of an MI similarity and the threshold approaches selected to remove the weak edges. Aiming at investigating the quality of the community structure on each developed network topology different community detection (CD) algorithms are employed. The community detection performance discovering in these customer network topologies are evaluated in terms of more compact purchasing behaviors.

4   Complex Network Construction

According to the graph theory a customer complex network G can be understood as a pairwise of two information sets where the first component C represents the set of customers ($|C| = N$) and the second E denotes a set of interactions between them. Classic approaches often assume that objects are connected to each other by a single type of static edges, which encapsulates all connections between them [5]. Similarly to other proposals [9,14], our solution considers that each node in the network represents the object under study. Unlike the previous models, each node is a customer in our approach. In addition, an edge between two customers is considered if they have a similar purchase behavior. It is worth mentioning that compositional variables or calculated attributes (e.g., recency, frequency, monetization [2]) are not explicitly considered when building the network.

To tackle the issues discussed in Sect. 2, we introduce a new multiple instance similarity involving both the order similarities and frequencies. The new measure $MInteraction$ considers each bag as a point set or a subset of a high-dimensional space, while the similarity between two bags is calculated from the aggregation of the similarities among the instances comprised into each bag. Equations 1–3 show how to compute the $MInteraction$ similarity measure, which aggregates the $MSJaccard$ similarities [19], where $BagSim(X_i, X_j)$ denotes the similarity between customer bags X_i and X_j based on the membership of the order o to the bag X, represented by $\mu_X(o)$. This membership value is obtained via the maximum instance similarity of the order o to the instances in the bag X.

$$MInteraction(X_i, X_j) = \frac{min\{|X_i|, |X_j|\}}{max\{|X_i|, |X_j|\}} \cdot \frac{BagSim(X_i, X_j) + BagSim(X_j, X_i)}{2} \tag{1}$$

$$BagSim(X_i, X_j) = \frac{\sum_{k=1}^{|X_i|} \mu_{X_j}(o_k^i)}{|X_i|} \tag{2}$$

$$\mu_X(o) = max_{x \in X}\{MSJaccard(x, o)\} \tag{3}$$

Multiple copies of the same instance can be included into the same bag and different bags are also allowed to be overlapped, as shown in the following example. In the case that two customers have exactly the same composition and number of orders (e.g., $X_1 = \{o_1, o_2, o_3\}$ and $X_2 = \{o_3, o_1, o_2\}$) we can expect that our measure and the Hausdorff-based similarities perform identically. Let us suppose that we apply the Hausdorff similarities over two costumers with bags $X_1 = \{o_1^1, o_2^1, o_3^1, o_4^1\}$ and $X_2 = \{o_1^2, o_2^2, o_3^2, o_4^2, o_5^2\}$ such that $o_4^2 = o_5^2$, $o_j^1 = o_j^2$ and $j \in [1, 4]$ then $MinH(X_1, X_2) = MaxH(X_1, X_2) = AveH(X_1, X_2) = 1$ although they do not have the same CPB. However, $MInteraction(X_1, X_2) = 0.8$, which is more realistic value. Our aggregation process implicitly considers several distinctive features: (i) purchasing frequency or order numbers by customers, (ii) subsets of products which tend to be together and (iii) bestseller products. Such features may lead to better network representations, which eventually can be translated into better business insights.

Another important aspect when constructing the network refers to the presence of a large number of connections attached to each object [14]. However, many of these edges are meaningless as they represent spurious or weak associations (i.e., they do not necessarily imply a confirmed relationship between customers since they have low weights). A common practice to reduce the network is to establish a threshold δ such that the edge between two objects i and j is preserved if their similarity degree is greater or equal than such a threshold (i.e., edge weight $w_{ij} \geq \delta$). Clearly, there is no a universal recipe for choosing a threshold value [21] since an excessively low value of δ may lead to very dense representations which might comprise little information, while an excessively high value may cause the removal of interactions that can be relevant for the problem domain. In this work, we apply different threshold estimation strategies, which are given by Eqs. 4–7. The strategies *Alpha05*, *AveMax* and *AveMin* depend on the $\alpha \in [0,1]$ parameter, which can be estimated based on the frequency of connections obtained from the histogram analysis. We explore the properties of the resulting communities at different threshold values in Sect. 5. Notice that we adopt the notation *Min* for referring network construction avoiding the threshold reduction step.

$$Ave = \delta_{ave} = \frac{2}{N(N-1)} \sum_{i=1}^{N-1} \sum_{j=i+1}^{N} S(X_i, X_j) \tag{4}$$

$$Alpha05 = \delta_{\alpha=0.5} = \frac{\delta_{ave} + \delta_{max}}{2} \tag{5}$$

$$AveMax = \delta_{\alpha=f_{ave-max}} = \alpha\delta_{ave} + (1-\alpha)\delta_{max} \tag{6}$$

$$AveMin = \delta_{\alpha=f_{min-ave}} = \alpha\delta_{ave} + (1-\alpha)\delta_{min} \tag{7}$$

As a matter of closure, constructing the complex network for problems in which customers have associated with a bag of orders requires defining an MI similarity measure. The second challenge addressed in this research is related to the estimation of the δ value, but this issue is pertinent to any community detection problem. In the next section, our proposal is evaluated through real-world study cases concerning sales.

5   Performance Evaluation

We perform the experiments on 12 real-world transaction datasets provided by anonymous European stores. Table 1 outlines the number of transactions (T), products (P), customers (C), orders (O) and product categories (PC) for each dataset, which follows the structure shown in Table 2.

Aiming at quantifying how compact the communities are in terms of their purchasing behaviors, we propose a new *ItemQuality* measure in Eq. 8. In our experiments we use this measure because we need to evaluate for a specific topology how compact is a certain community structure based on the covering of the product characteristics over all customers inside each customer

Table 1. Characterization of the European store datasets.

Dataset	T	C	P	O	PC	Dataset	T	C	P	O	PC
D_1	3,546	339	1,853	372	1,034	D_7	29,201	1,286	4,811	855	1,915
D_2	25,934	1,171	4,387	2,483	1,816	D_8	30,217	1,279	4,847	1,805	1,935
D_3	26,158	1,190	4,475	2,465	1,854	D_9	29,339	1,303	4,886	900	1,952
D_4	25,858	1,218	4,512	2,487	1,845	D_{10}	28,212	1,219	4,919	2,336	1,938
D_5	27,050	1,238	4,546	2,466	1,850	D_{11}	33,407	1,353	5,061	2,800	1,974
D_6	29,285	1,266	4,729	2,666	1,913	D_{12}	33,012	1,333	5,040	2,737	1,967

Table 2. Dataset structure.

Customer	Order	Order date	Product	Amount	Revenue
1	1	2018/05/25	a	1	2.10 €
1	2	2018/05/26	b	2	4.50 €
2	1	2018/05/26	b	2	4.50 €

community. Notice that, a community detection result is denoted as $Com = \{com_1, \ldots, com_k\}$ and P_{com_i} denotes the collection of purchased products by customers in the community com_i. Also, $\tau(com_i, p_j)$ quantifies the number of customers in the community com_i, which purchase the product p_j.

$$ItemQuality = \frac{\sum_{k_1=1}^{|Com|} \frac{\sum_{k_2=1}^{|P_{com_{k_1}}|} \frac{\tau(com_{k_1}, p_{k_2})}{|com_{k_1}|}}{|P_{com_{k_1}}|}}{|Com|} \tag{8}$$

The customer segmentation is obtained based on the following community detection algorithms: the multi-level modularity optimization (LV) [1], walktrap (WT) [13], label propagation (LP) [15], infomap (IM) [17] and fast greedy modularity optimization (FGO) [4], which stand as popular ones. In general, our multiple instance similarity measure is independent of the clustering segmentation algorithm used. However, we choose community detection algorithms because they simultaneously exploit the network topology and the pair-wise similarities used in the classical clustering approaches. Besides, the network topology is able to encode in an elegant and systematic manner, interactions of the data items going from local to global structural information.

The experimental design for determining if there are significant differences between the compactness by using the measure *ItemQuality* consists of the following steps: (1) apply the Friedman test to identify if there are significant differences among approaches by using a significance value $\alpha = 0.05$. This test generates a ranking of the customer segmentation approaches, and (2) apply a Post-Hoc test to identify significant differences between each two approaches. For this, the Holm test with significance threshold $\alpha = 0.05$ was used.

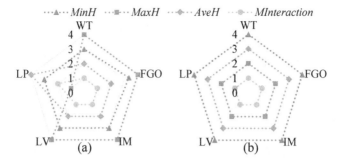

Fig. 2. Performance ranking of MI similarities for customer segmentation based on the *ItemQuality* measure: (a) Not weak interaction reductions and (b) Weak interaction reduction by using *Alpha*05 threshold approach.

Figures 2 and 3 show the results of the ranking of the network construction approaches, such that more internal result expresses a best ranking position.

The first experiment aims at exploring whether our multiple instance similarity leads to more compact customer segmentation or not. Firstly, for each dataset we compute the network construction stage by considering $MinH$, $MaxH$, $AveH$ and $MInteraction$ similarities avoiding both the threshold estimation and reduction steps. Secondly, we evaluate the CD algorithms' performance for each dataset and network construction. Figure 2(a) shows the performance ranking achieved by applying the CD algorithms from these four similarities based on the *ItemQuality* measure. Notice that, our multiple instance similarity is stable (i.e., the ranking position achieved by each algorithm over all dataset is stable) and produces the more accurate results independently of the CD algorithm used.

Aiming at exploring whether the weak interactions affect the community detection convergence over the four multiple instance similarities, we obtain customer segmentations by considering different network topologies. The second experiment explores the performance obtained by discovering communities from the network construction by using the different threshold estimations pointed out by Eqs. 4–7 for each multiple instance similarity. Figures 3(a)–3(d) show that *Alpha*05 threshold produces the more compact customer community results over all multiple instance similarities regardless of which community detection algorithm was selected. As expected, the lowest *ItemQuality* values are obtained when detecting communities in networks built by considering the thresholds δ_{min}, δ_{ave} and $\delta_{\alpha=f_{min-ave}}$, regardless of the fixed MI similarity. Thresholds δ_{min}, δ_{ave} and $\delta_{\alpha=f_{min-ave}}$ are less restrictive, therefore, they maintain weak interactions, which influences the community detection. In general, the variability was expected, since both measures $MinH$ and $MaxH$ are sensitive to outliers [7]. Besides, $MInteraction$ similarity better quantifies the similarity between customers; due to, this MI similarity only assigns the higher similarity value (i.e., 1) among customers, when they exactly have the same shopping patterns, as was illustrated in Sect. 4.

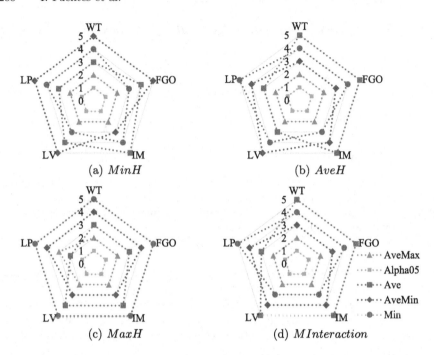

Fig. 3. Performance ranking of threshold reduction for customer segmentation based on the *ItemQuality* measure.

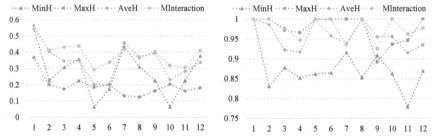

(a) Not threshold estimation and not weak interaction reductions.

(b) Threshold estimation and weak interaction reduction.

Fig. 4. Average of the *ItemQuality* measure by MI similarities for customer segmentation.

The third experiment aims at exploring whether our multiple instance similarity leads to more compact customer segmentation or not by considering the best network topology for each multiple instance similarity. Firstly, for each dataset we compute the network construction stage by considering *MinH*, *MaxH*, *AveH* and *MInteraction* similarities voiding the *Alpha05* threshold reduction. Secondly, we evaluate the community detection algorithms' performance for each dataset and network construction. Figure 2(b) shows the performance ranking achieved by applying the community detection algorithms from the above four network construction approaches based on the *ItemQuality* measure. Notice that, our multiple instance similarity is stable (i.e., the ranking position achieved by each algorithm over all dataset is stable) and produces more accurate results independently of the community detection algorithm used.

Figure 4 shows the results achieved by using the four MI similarities based on the average of the *ItemQuality* values over all community detection algorithms in both cases considering or avoiding the weak interactions (i.e., those that represent similarities less than a threshold). The x-axis denotes the dataset list, while the y-axis represents the average of the *ItemQuality* obtained by the previous community detection algorithms. The results confirm that *MInteraction* generally obtains the higher *ItemQuality* values. We conclude that *MInteraction* leads to the best compactness rates by considering or avoiding the threshold reduction overall tested community detection algorithms.

The above results show the feasibility of the proposed MI similarity to build high-quality customer communities by combining the main entities involving in the CPB problem. Thus, the detected communities by applying *MInteraction* similarity allow improving the efficiency of marketing actions by targeting the right customers with the right promotions/categories/products. It should be mentioned that the threshold estimation is an important step in the network construction stage, which contributes to faster convergence of the CD algorithms.

6   Conclusions

In this paper, we presented a new approach for performing CPB analysis by means of community detection. The proposed methodology comprises the computation of customer interactions using MI similarities and the threshold estimation for removing weak connections in the complex network construction. During the data gathering step, our proposal allows to determine the key transactions that will be included in the analysis, which should be in alignment with the envisaged business goals. The network analysis results have shown that the community performance reaches the best results using the proposed *MInteraction* similarity, which is capable of reflecting the structural interactions between the bags associated to customers. Moreover, we have illustrated the impact of the threshold estimation strategy towards reaching a specific level of compactness and performance. A closer inspection to the communities discovered by our algorithm allows determining key loyal customers to which the company can redirect its marketing efforts. Moreover, we can use the customer behavior summarized in each community to design more informed marketing campaigns.

References

1. Blondel, V.D., Guillaume, J.L., Lambiotte, R., Lefebvre, E.: Fast unfolding of communities in large networks. J. Stat. Mech. Theor. Exp. **2008**(10), P10008 (2008)
2. Chen, Y.L., Tang, K., Shen, R.J., Hu, Y.H.: Market basket analysis in a multiple store environment. Decis. Support Syst. **40**(2), 339–354 (2005)

3. Cheplygina, V., Tax, D.M., Loog, M.: Multiple instance learning with bag dissimilarities. Pattern Recogn. **48**(1), 264–275 (2015)
4. Clauset, A., Newman, M.E.J., Moore, C.: Finding community structure in very large networks. Phys. Rev. E **70**, 066111 (2004)
5. De Domenico, M.: Mathematical formulation of multilayer networks. Phys. Rev. X **3**(4), 041022 (2013)
6. Fuentes, I., Nápoles, G., Arco, L., Vanhoof, K.: Customer segmentation using multiple instance clustering and purchasing behaviors. In: Hernández Heredia, Y., Milián Núñez, V., Ruiz Shulcloper, J. (eds.) IWAIPR 2018. LNCS, vol. 11047, pp. 193–200. Springer, Cham (2018). https://doi.org/10.1007/978-3-030-01132-1_22
7. Herrera, F., et al.: Multiple instance learning. Multiple Instance Learning, pp. 17–33. Springer, Cham (2016). https://doi.org/10.1007/978-3-319-47759-6_2
8. Kaminskas, M., Bridge, D., Foping, F., Roche, D.: Product-seeded and basket-seeded recommendations for small-scale retailers. J. Data Semant. **6**(1), 3–14 (2017)
9. Kim, H.K., Kim, J.K., Chen, Q.Y.: A product network analysis for extending the market basket analysis. Exp. Syst. Appl. **39**(8), 7403–7410 (2012)
10. Linden, G., Smith, B., York, J.: Amazon. com recommendations: item-to-item collaborative filtering. IEEE Internet Comput. **7**(1), 76–80 (2003)
11. Miguéis, V.L., Camanho, A.S., Cunha, J.F.: Customer data mining for lifestyle segmentation. Exp. Syst. Appl. **39**(10), 9359–9366 (2012)
12. Monteserin, A., Armentano, M.G.: Influence-based approach to market basket analysis. Inf. Syst. **78**, 214–224 (2018)
13. Pons, P., Latapy, M.: Computing communities in large networks using random walks. J. Graph Algorithms Appl. **10**(2), 191–218 (2006)
14. Raeder, T., Chawla, N.V.: Modeling a store's product space as a social network. In: International Conference on Advances on Social Network Analysis and Mining, pp. 164–169. IEEE (2009)
15. Raghavan, U.N., Albert, R., Kumara, S.: Near linear time algorithm to detect community structures in large-scale networks. Phys. Rev. E **76**, 036106 (2007)
16. Reutterer, T., Hornik, K., March, N., Gruber, K.: A data mining framework for targeted category promotions. J. Bus. Econ. **87**(3), 337–358 (2016). https://doi.org/10.1007/s11573-016-0823-7
17. Rosvall, M., Axelsson, D., Bergstrom, C.T.: The map equation. Eur. Phys. J. Spec. Topics **178**(1), 13–23 (2009)
18. Shihab, S.H., Afroge, S., Mishu, S.Z.: RFM based market segmentation approach using advanced k-means and agglomerative clustering: a comparative study. In: 2019 International Conference on Electrical, Computer and Communication Engineering (ECCE), pp. 1–4. IEEE (2019)
19. Theobald, M., Siddharth, J., Paepcke, A.: Spotsigs: robust and efficient near duplicate detection in large web collections. In: 31st International Conference on Research and Development in Information Retrieval, pp. 563–570. ACM (2008)
20. Valero-Fernandez, R., Collins, D.J., Lam, K.P., Rigby, C., Bailey, J.: Towards accurate predictions of customer purchasing patterns. In: International Conference on Computer and Information Technology (CIT), August 2017, pp. 157–161. IEEE (2017)
21. Vörös, A., Snijders, T.A.: Cluster analysis of multiplex networks: defining composite network measures. Soc. Netw. **49**, 93–112 (2017)

Just Fun and Games? Utilitarian and Hedonic Chatbot Perceptions and Their Role for Continuance Intentions

Patrick Bedué[(⊠)]

Ulm University, Helmholtzstraße 16, 89081 Ulm, Germany
`patrick.bedue@uni-ulm.de`

Abstract. Conversational agents (CAs) offer huge potential for service companies by creating social closeness and enabling fast and scalable communication with customers. However, investigation of utilitarian and especially hedonic value as driving motivations for using CAs is still nascent. We found social presence to be an important predictor for hedonic and utilitarian value and subsequent continuance intention. Moreover, we reveal customers' continuance intention is determined primarily by hedonic value when expecting a CA, whereas focus shifts to utilitarian values if customers expect a human employee. With our results, CA services can be better tailored to customer needs and company service goals.

Keywords: Chatbot · Customer service · Social presence · Hedonic · Utilitarian

1 Introduction

In line with significant increases in electronic commerce (e-commerce) revenues and the growing number of internet users, more and more customers are seeking contact with companies via digital channels. It is forecasted that global revenues in e-commerce will grow decisively from $ 3.54 trillion in 2019 to $ 6.54 trillion by 2023 [1]. Furthermore, customer service expectations changed dramatically over the last years causing new challenges for companies. Consequently, new technological innovations in firm-customer interactions are required to ensure long-term success of companies and their business models. Therefore, companies are increasingly investing in new technology to automate their customer interactions thereby providing immediate service anytime and regardless of location. A prevalent method among companies especially in the service industry is the usage of conversational agents (CAs), whose distribution has gained importance recently [2, 3]. CAs – sometimes also referred to as chatbots, dialogue systems, or virtual assistants – enable users to interact with information technology similar to the communication with another human being, using natural language [4]. Among enhanced levels of convenience, reliability and efficiency, CAs are superior to classic self-service technologies (SST) as they can build a sense of human contact during the service encounter [5, 6]. Consequently, interest in these new technologies is tremendous, which can be seen for instance by a growing number of available CAs on the Facebook

© Springer Nature Switzerland AG 2020
W. Abramowicz and G. Klein (Eds.): BIS 2020, LNBIP 389, pp. 291–306, 2020.
https://doi.org/10.1007/978-3-030-53337-3_22

Messenger Platform from 11.000 in 2016 to over 300.000 in 2019 [7]. In addition, 7% of worldwide internet users mentioned that they already used CAs and 27% plan on using CAs [8]. Customers using CAs especially appreciate utilitarian values, such as effective processing of customer inquiries, independence of opening hours, avoidance of waiting time and quick answers to frequently asked questions [9]. However, more than 25% of users also mention fun as their driving hedonic motivation for using CAs [10]. Yet, research on human-computer interaction found that physical embodiments (avatars) or human-like appearances affect user perceptions and increases overall system acceptance [11]. Nevertheless, existing research still lacks an in-depth understanding of the relative importance of utilitarian and hedonic value for CA acceptance behavior. Indeed marketing related studies have explored impact of social presence for websites [12, 13], emails [6, 14] or recommendation agents [15, 16]. However, to the best of our knowledge, they do not examine the importance of social presence as an important predictor of both utilitarian and hedonic value and their significance for the development of continuance intentions in sufficient detail yet. We contribute to this gap, by creating a new model to explore why customers are willing not only to use, but rather continue to use CAs driven by their relationship building capabilities (social presence) and emphasize the practical relevance of this exciting topic. Moreover, we investigate the multidimensional nature of utilitarian and especially hedonic value and their relative importance in a joint model. This study has three major goals: *(1) Create an enhanced understanding of CA continuance intention considering both utilitarian and hedonic constructs and their relative importance. (2) Understand the relevance of social presence and its effect on utilitarian and hedonic value of the CA service. (3) Investigate differences in utilitarian and hedonic value if conversational partner expectations (CA vs. human) differ.* To investigate these research goals, we conduct a multi-group online experiment in which customers with different initial expectations (CA vs. human) interact with a CA. We analyze our results from the subsequent survey and outline the different aspects across both respective groups. We chose a service interaction from the airline industry, since many people have prior experience with air travel and some airlines already offer customers access to CAs [17, 18]. Our interesting results demonstrate the importance of creating social presence in CA conversations and more interestingly find customers expecting a CA develop continuance intention primarily from hedonic value, while customers expecting a human build continuance intention mostly from utilitarian value. With this study, we further aim to help to develop guidelines for the success of CA implementation in online customer service environments. Moreover, we provide managerial as well as research insights on this exciting topic. The remainder of the paper is organized as follows: We first give an overview of the related literature. Thereafter, we build our research model, followed by the results of our analysis. Subsequently, we discuss our results and present theoretical and practical implications. Finally, we conclude with a brief summary of our findings.

2   Theoretical Background

Originated from consumer behavior literature expectation-confirmation theory (ECT) can build an established basis to study consumer satisfaction, post-purchase behavior, or service usage and gain acceptance especially in the marketing domain [19]. Based on ECT Bhattacherjee [19] adapted and evaluated the theory in the information system (IS) context to create an expectation-confirmation model of IS continuance intention (IS ECM) [19]. With the frequent use of CAs especially in the service sector, investigation of causes for the development of continuance usage behavior becomes increasingly important. Consequently, derived from IS ECM, we use *continuance intention* as the main outcome variable of our model [19]. The variable reflects a user's *continuance intention* as the behavioral intend to continue using a CA [17, 19, 20]. According to IS ECM, customers' *continuance intention* is determined primarily by their *satisfaction* with prior IS usage and *perceived usefulness* [17]. Drawn from technology acceptance model (TAM), prior research found *perceived usefulness* to be an important driver for acceptance of IS technology and its positive influence has been shown across several studies such as websites [13], or recommendation agents [19, 21, 22]. According to research on SST *perceived usefulness* is an important utilitarian value and can be created by customer services which provide mostly functional benefits [23]. Following recent suggestions from marketing research, shopping behavior can create both utilitarian and hedonic value, which affects customers future purchase decisions through feedback loops into the decision process [24]. A commonly accepted hedonic value is *enjoyment* [25]. *Enjoyment* is the extent to which an activity of using a technology is perceived as enjoyable and which was found to be an important addition to the original TAM [25]. Moreover, its influence on customers technology adoption intentions has been shown in several empirical studies [15]. Both utilitarian and hedonic value were identified as important determinant for customers future behavior [24, 25]. Since CAs are successors of traditional SST, we believe that CAs are also dual technologies that create both hedonic- and utilitarian-oriented value [26]. As mentioned earlier the variable *satisfaction* has frequently been used in several ECM and TAM-based studies. Focusing on the context of SST adoption, Yan et al. [18] found different levels of customer *satisfaction*, when customers interact with SST compared to human agents. An important key advantage of CAs over traditional SST is their ability to make consumers feel that they interact with a social entity which is captured by our construct *social presence* [15]. The concept of *social presence* has long been employed to study the social aspects of technologies, such as websites [12, 27], recommendation agents [15], or CAs [3, 4]. According to social response theory, users respond to human characteristics in their interaction with a computer, even if they know that they are not interacting with a human being [5]. Among others, Hassanein and Head [27] empirically investigated the effects of manipulating website design elements to enrich perceptions of *social presence* and investigate its positive impact on both hedonic *(perceived enjoyment)* and utilitarian *(perceived usefulness)* values [12, 27]. Furthermore, Qiu and Benbasat [15] found evidence for the creation of stronger perceptions of *social presence* when users interacted with an embodied product recommendation agent [15]. Summing up, *satisfaction, social presence, perceived usefulness* and *enjoyment* were identified to be important variables for influencing customers' *continuance intention*.

3   Research Model

Based on the theoretical background, we create a new model to explore why customers are willing not only to use, but rather continue to use CAs driven by their relationship building capabilities (*social presence*) and further investigate the multidimensional nature of utilitarian (*perceived usefulness*) and especially hedonic (*perceived enjoyment*) value and their relative importance in a joint model. The proposed hypotheses are presented in Fig. 1 and explained below.

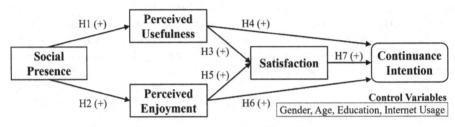

Fig. 1. Research model and hypotheses

While prior research on CAs shows that perceptions of *social presence* during user-system interaction can be increased [3, 5, 11], their effect on utilitarian and especially hedonic value has not been shown yet. Thus, we state the following hypotheses:

H1: Increased social presence will result in greater perceived usefulness.
H2: Increased social presence will result in greater perceived enjoyment.

Perceived usefulness is modeled as users' perception of the utilitarian benefits of the system [19, 22] which in turn affects their adoption decisions. We follow this argumentation by stating that CAs provide functional advantages, create higher levels of customer *satisfaction* and affect customers' *continuance intentions*. Thus, the following relationships are examined in our study:

H3: Increased perceived usefulness will result in greater satisfaction.
H4: Increased perceived usefulness will result in greater continuance intention.

Moreover, recent studies on CAs have shown that physical present systems can create an increased level of *enjoyment* and influence future use intentions [11, 28, 29]. Hedonic value and therefore higher usage intentions are created by services that look like being fun [23]. Thus, in addition to the utilitarian (*perceived usefulness*) value we extend the original model by capturing also the hedonic (*perceived enjoyment*) value. Therefore, we state the following hypotheses:

H5: Increased perceived enjoyment will result in greater satisfaction.
H6: Increased perceived enjoyment will result in greater continuance intention.

Besides its theorization and validation *satisfaction* was also identified as a key variable for customer relationship management [18, 19, 22]. Since we expect only satisfied customers to continue using CAs we assume that *satisfaction* with a CA will increase the likelihood for *continuance intention*. Therefore, we propose the following hypothesis:

H7: Satisfaction with the CA will result in greater continuance intention.

4   Research Methodology

4.1   Measurement Development

In order to measure the constructs **social presence (SP), perceived usefulness (PU), perceived enjoyment (EN), satisfaction (ST)** and **continuance intention (CI)** we used previously validated constructs from IS literature and adapted all constructs to the CA context (cf. Table 5 in the appendix). Our survey followed standard instrument construction procedures to foster reliability and validity of our measurements [30, 31]. *Satisfaction* items were rated on five-point semantic differential scale, consisting of adjective pairs [32], whereas all remaining items were rated to five-point Likert scales ranging from strongly disagree (1) to strongly agree (5). The items are provided in the appendix (cf. Table 5).

4.2   Method and Data Collection

The airline industry was one of the first seeking for operational efficiency by providing SST, significantly reducing the amount of required staff and helping customers to save time and money through faster and immediate interactions [17, 33]. Since CAs could complement these technologies and further improve automatic customer interactions, we focus on an airline customer service setting. To validate our research model, we chose an online survey capturing a dialog with a CA and a subsequent questionnaire with the constructs encompassed in the research model. Furthermore, we developed a CA able to conduct brief conversations during a flight booking process. The custom conversational interface is presented in Fig. 2 and was built by using JavaScript runtime environment Node.js and Alphabets Dialogflow for natural language processing and machine learning [34]. Dialogflow provides sufficient algorithms to identify intents in customers' questions and provide appropriate answers. Introduced in 2009 it is now widely used by companies like the KLM airline or Dominos and could be integrated in several platforms like Alexa, Slack or Facebook Messenger [34]. For our CA we chose a custom interface and implemented recent IS research suggestions in the area of human-computer interactions to give the illusion of a real booking process. Among others these suggestions include adding a response delay of 50 ms per character as recommended by Holtgraves and Han [35] since CAs may be perceived as unhuman-like when providing instant answers. During the delay, we used graphical indicators and a note informing the customer that their conversation partner is typing (cf. Fig. 2). In order to further increase the impression of a human employee we added variables (e.g. departure date) to enable the agent to adapt its responses to individual customer questions [4, 35]. To select the most sound dialog for the CA conversation we conducted a preliminary survey in which we asked participants to choose the most human-like dialog from several different service conversations. Moreover, we validated our final experimental design by conducting a pre-study with a group of undergraduate students and research assistant participants. Analyzing our results we found most participants to be either unable to differentiate between a CA and human or were at least unsure, which finally indicates that our measures to build the illusion of a real booking process were successful. Finally, we integrated our CA in our online-survey tool Lime Survey [36]. We chose an experimental design with two groups to measure the different outcomes when customers initially assumed a CA or a real human employee in their conversation with the agent. Recruitment of participants happened via clickworker.com, a marketplace for crowdsourcing [37]. We chose the clickworker marketplace since it is a fast and easy way to recruit participants. Moreover, compared to traditional methods, similar data quality and results can be expected, while at the same time threats to internal validity are reduced [38–40].

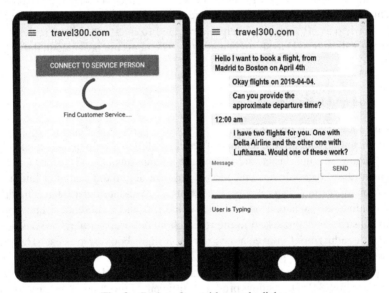

Fig. 2. CA interface with sample dialog

170 Participants with english skills were invited to the online experiment, since interactions with the CA required basic language knowledge. At the end of the survey, we were able to access a record of 170 questionnaires. 150 participants finished the conversation and completed the entire survey consciously. The experiment was divided into several tasks. First, participants had to confirm a privacy policy and got a short introduction to the service context. Afterwards, we asked questions about the demographic background which we captured by gender, age, education and hours of daily internet usage. Thereafter, participants were randomly assigned to one of two groups. While one group was told to await a conversation with a CA (n = 78), the other group

was promised a human service employee (n = 72). Apart from that, the survey and the technical capabilities of the CA did not differ in any way. The dialogue was designed in such a way that the CA followed a predefined but open conversation structure by giving the participants several degrees of freedom in their possible questions and answers. After finishing the conversation with the CA, participants answered questions about the proposed constructs. Table 1 provides demographic information about the participants.

Table 1. Demographic information about the respondents (n = 150)

Measure	Items
Gender	Male (n = 88; 58.67%) Female (n = 62; 41.33%)
Age	14–19 (n = 7; 4.67%) 40–49 (n = 33; 22.00%) 20–29 (n = 47; 31.33%) 50–59 (n = 11; 7.33%) 30–39 (n = 51; 34.00%) > 59 (n = 1; 0.67%)
Education	University (n = 92; 61.33%) Elementary School (n = 4; 2.67%) Second. Education (n = 54; 36.00%) No degree (n = 0; 0.00%)
Daily internet usage	1–2 h (n = 17; 11.33%) 6–9 h (n = 36; 24.00%) 3–5 h (n = 76; 50.67%) > 9 h (n = 21; 14.00%)

5 Data Analysis and Results

5.1 Measurement Model

To validate our research model, we applied a two-step approach by first examining the composition of the constructs (measurement model) and in a second step testing the structural relationships of the constructs among each other (structural model) [41, 42]. This ensures reliability and validity of the measures before examining the structural model parameters [41]. For our analysis, we chose Partial Least Squares (PLS) and the software package Smart PLS 3. PLS is recommended for studies that include formative constructs as it enables latent constructs to be modeled both as formative and reflective indicators [43] and offers great potential for early theory development and construct validation [41]. Moreover, a component-based approach for estimation is applied, placing minimal restrictions on sample size, measurement scales and residual distribution [44]. Based on minimum sample size recommendations each treatment group was supposed to be large enough to overcome the problem of biased results [44, 45]. To evaluate our measurement model we tested for convergent and discriminant validity as well as reliability and consistency of the measures in both experimental groups [45]. Reliability was ensured by investigating the composite reliability (CR) values, which are presented in Table 2. Our results show that CR values in both groups exceed the commonly acceptable threshold of 0.7 [46]. Convergent validity was ensured based on two requirements: First, indicator loadings need to exceed loadings of 0.5 [45], what applies to all indicators in our model. Second, the average variance extracted (AVE) should surpass a 0.5 threshold

[47]. As evident in Table 2, this condition is met. Furthermore the AVE of each construct should be higher than the variance due to the measurement error for that construct [45, 47]. Table 3 shows that all items have a loading higher than 0.5 on their respective construct.

Table 2. Descriptive statistics of variables

Construct	Items	Customers expect a CA			Customers expect a Human		
		CR	Mean	AVG	CR	Mean	AVE
CI	5	0.914	3.323	0.747	0.879	3.408	0.676
EN	5	0.934	3.462	0.790	0.917	3.403	0.751
PU	4	0.939	3.377	0.845	0.914	3.545	0.795
ST	4	0.944	3.565	0.856	0.950	3.644	0.870
SP	4	0.948	2.880	0.866	0.924	2.822	0.814

Table 3. Correlations among constructs and square root of AVE

Construct	Customers expect a CA					Customers expect a human				
	CI	EN	PU	ST	SP	CI	EN	PU	ST	SP
CI	**0.864**					**0.822**				
EN	0.686	**0.889**				0.572	**0.867**			
PU	0.711	0.778	**0.919**			0.712	0.612	**0.892**		
ST	0.748	0.736	0.820	**0.925**		0.636	0.785	0.525	**0.933**	
SP	0.460	0.670	0.604	0.589	**0.930**	0.419	0.518	0.501	0.469	**0.902**

To assess discriminant validity we checked for three requirements. First, indicator loadings should be higher than all of its cross-factor loadings [45]. Our results indicate that this condition is met. Second, the correlations among the constructs must be lower than 0.85 [48]. All values were observed below the recommended threshold. Third, the AVE of each latent construct should be higher than the construct's highest squared correlation with any other latent construct [46]. As shown in Table 3, these requirements also apply and therefore our measures demonstrate discriminant validity. Given the strong evidence for convergent and discriminant validity, the scales exhibit good internal consistency and reliability and the measurement model is deemed acceptable.

5.2 Structural Model

To test our hypotheses and examine the explanatory power of the structural model, we determined the structural paths and the R^2 score of our endogenous variable for both treatment groups. To analyze significance of structural paths we used bootstrap algorithm with 5,000 samples. The standardized path coefficients and path significance levels are presented in Fig. 3. Results from customers expecting the CA reveal that the model explains 63.1% of the variance of *continuance intention*. Furthermore, when applying the same bootstrap algorithm to customers expecting a human employee the model explains 61.1% of the variance of *continuance intention* for this group. Both R^2 values underline that the model explains the cohesion well and the model has a good fit for both groups [45]. Subsequently we will present and discuss our results in more detail by first investigating and comparing individual paths within the models and second by highlighting main differences between both respective groups.

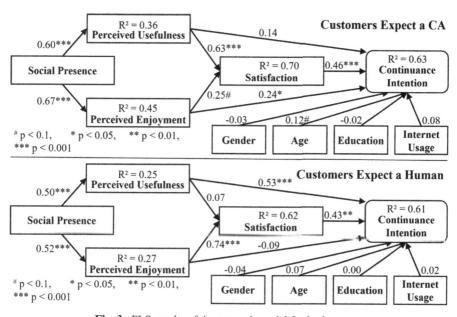

Fig. 3. PLS results of the research model for both groups

Results for Customers Who Expect a CA. Based on the data of respondents, *social presence* is positively related to *perceived usefulness* ($\beta = 0.60$, p $ < 0.001$) and *perceived enjoyment* ($\beta = 0.67$, p $ < 0.001$). Consequently, we confirm hypotheses 1 and 2. Furthermore, customers' *satisfaction* is positively influenced by both constructs – *perceived usefulness* ($\beta = 0.63$, p $ < 0.001$) and *perceived enjoyment* ($\beta = 0.25$, p $ < 0.1$) –, which is in line with our expectations and confirms hypotheses 3 and 5 (cf. Fig. 3). Regarding hypotheses 6 and 7 path analysis reveals a significant positive effect at the 0.001 significance level for *satisfaction* on *continuance intention* ($\beta = 0.46$, p $ < 0.001$) and an additional positive effect at the 0.05 significant level of *perceived enjoyment* on *continuance intention* ($\beta = 0.24$, p $ < 0.05$). Surprisingly, we could not confirm our hypothesis that *perceived usefulness* has a positive significant effect on *continuance intention*. Thus, based on data of users expecting a CA hypothesis 4 is declined, whereas hypotheses 6 and 7 are supported (cf. Fig. 3). Finally, *age* ($\beta = 0.12$, $p < 0.1$) exhibits a slightly significant influence on *continuance intention* while *gender, education* and *internet usage* were found to have no significant effect at all.

Results for Customers Who Expect a Human. The evaluation of our model with data from customers expecting a human employee reveals similar results regarding the link between *social presence, perceived usefulness* and *perceived enjoyment*. More specifically, we found a significant positive effect of *social presence* on *perceived usefulness* ($\beta = 0.50$, $p < 0.001$) and *perceived enjoyment* ($\beta = 0.52$, $p < 0.001$) which confirms hypotheses 1 and 2. Furthermore, as we look on the causal link with *satisfaction* more closely, we notice a strong positive significant effect of *perceived enjoyment* on *satisfaction* ($\beta = 0.74$, $p < 0.001$), however no significant positive effect of *perceived usefulness* on *satisfaction* was found. While these results indicate support for hypothesis 5, we could not find support for hypothesis 3 (cf. Fig. 3). Moreover, data of customers expecting a human also disclose different results for the effects of *perceived usefulness, satisfaction* and *perceived enjoyment* on CA *continuance intention*. While we observe a significant positive effect of *perceived usefulness* on *continuance intention* ($\beta = 0.53$, $p < 0.001$) and thus our results indicate support for hypothesis 4, we could not find a significant positive effect of *perceived enjoyment* on *continuance intention* (Hypothesis 6). However, we confirm our expectations that *satisfaction* has a positive effect on *continuance intention* ($\beta = 0.43$, $p < 0.01$). Finally, no significant influence of *gender, age, education* and *internet usage* on *continuance intention* was found at all (cf. Fig. 3).

Assessing Main Group Differences. While we first looked at the individual models by investigating path coefficients and their significance levels within the experimental groups, we afterwards performed a multi-group analysis (MGA) in SmartPLS 3 to gain deeper insights on main group differences [49]. Comparison across both groups reveals significant differences especially regarding the utilitarian (*perceived usefulness*) and hedonic (*perceived enjoyment*) constructs and their importance for *continuance intention* (cf. Table 4). The results for customers expecting a human employee show that utilitarian value is more important in respect to *continuance intention* than hedonic value. Surprisingly, we observe that the result is reversed for customers who expect a CA. Consequently, customers build their *continuance intention* primarily by hedonic value if they expect a CA. Thus, customers focus more on the fun factor of the CA compared to utilitarian benefits.

Overall, group comparison reveal significant path strengths differences for *perceived usefulness* ($p < 0.05$) and *perceived enjoyment* ($p < 0.05$) on *continuance intention*, as well as *perceived usefulness* ($p < 0.01$) and *perceived enjoyment* ($p < 0.01$) on *satisfaction* (cf. Table 4). Apart from these interesting results, we could not find any other significant differences between path coefficients strengths across both groups.

6 Discussion

With this study, we developed a research model that not only captures the utilitarian and hedonic value for *continuance intention*, but also includes CAs' relationship building

Table 4. SmartPLS MGA results

Causal path (H)	Supported (Expect a CA)	Supported (Expect a Human)	Significance level of group comparison
SP -> PU (H1)	Yes (p < 0.001)	Yes (p < 0.001)	–
SP -> PE (H2)	Yes (p < 0.001)	Yes (p < 0.001)	–
PU -> ST (H3)	Yes (p < 0.001)	No	p < 0.01
PU -> CI (H4)	No	Yes (p < 0.001)	p < 0.05
EN -> ST (H5)	Yes (p < 0.1)	Yes (p < 0.001)	p < 0.01
EN -> CI (H6)	Yes (p < 0.05)	No	p < 0.05
ST -> CI (H7)	Yes (p < 0.001)	Yes (p < 0.01)	–

capabilities (*social presence*) in a joint model. Moreover, we shed first lights on differing utilitarian and hedonic service perceptions and subsequent *continuance intentions* in CA service interactions. Our study contributes to theory in various ways:

First, we found that in the context of CAs, *perceived enjoyment*, *perceived usefulness* and *satisfaction* are crucial for *continuance intention*. More precisely, our data complements research by showing that aside from utilitarian value, hedonic value like *enjoyment* plays an important role explaining customers' behavioral intentions [15, 27]. However, to date the relative importance of these constructs has not been investigated in a multi-group experiment. Second, we found customers expecting a CA develop *continuance intention* primarily from hedonic value, while customers expecting a human build *continuance intention* mostly from utilitarian value. The reason may be cognitive dissonance theory originated from consumer behavior, which states that customers are entertained for their inherent *satisfaction*. However, creating dissonance by wrong partner perceptions (CA instead of the proposed human) shifts *enjoyment* away from an immediate reward to a functional (utilitarian) higher-level goal [50]. Furthermore, as unmediated effects of *perceived enjoyment* on adoption intentions are mostly manifested only among hedonic-oriented systems [15, 25] our results indicate decreasing perception of functional benefit when interacting with a CA. Based on these interesting findings we extend prior IS research by outlining similarities between consumer behavior theory and conclude that CAs are currently more seen just as a gimmick by customers. Third, we demonstrate the importance of creating *social presence* in CA conversations and show that our measures creating a human-like CA were successful. We are the first who seek to understand the importance of *social presence* for utilitarian and hedonic value in the context of CA conversations. Although existing research especially in the marketing domain examined the influence of *social presence* for websites [12, 13], emails [6, 14], or recommendation agents [15, 16], we are the first who seek to understand the importance of *social presence* for utilitarian and hedonic value in the context of CA conversations. We confirm the strong positive importance of *social presence* for utilitarian and hedonic value. Further, our results complement research of Lim et al. [12] and Rzepka and Berger [11] who found *social presence* result in higher utilitarian or hedonic value and increases future use intentions [11, 12].

Additional this study has several implications for practice. First, firms have to recognize that *social presence* is a particular important driver for utilitarian and hedonic value and consequently for developing *continuance intentions*. Therefore, companies are well advised to include CA features, such as avatars, or response delays to increase perceptions of *social presence* [5, 11]. Second, our results indicate that *continuance intention* is driven by both – hedonic and utilitarian value. Therefore, companies have to design their CA services to not only address task-related benefits, but also hedonic benefits like *enjoyment*. This especially holds true, as long as CAs are not yet mature technologies. Third, creating wrong perceptions of the conversational partner, increases cognitive dissonance and shifts the focus away from the hedonic motivation to a functional goal [50]. Consequently, it may be a good idea to create the perception of a human employee, if the service goal is primarily utilitarian and the technology considered mature. However, increasing customer *satisfaction* should be the overarching goal and companies are well advised to tell their customers the truth when using a CA in a conversation.

7 Limitations

We are confident that our results provide several interesting new insights with respect to customers' *continuance intention* towards using CAs. However, as with all research a number of limitations remain which could be interesting starting points for future research. First, the collected data solely resulted from a single specific customer service setting and contained a large amount of young, well-educated and technology affine participants, which may limit the generalizability of our findings. However, since our survey was carried out by interacting with a real CA implementation, we are confident that our rich data set provides a solid foundation for testing the research model. Second, we based our theory only on *social presence* and the utilitarian and hedonic dimension together with adapted constructs from the original IS ECM. We decided to do this in a first step, to prevent mixing effects that may arise from further potential drivers of *continuance intention*. Consequently, effects of trust, perceived humanness and IT self-efficacy may be interesting starting points for future research. Third, the short conversation with the CA happened in an experimental setting, although we have taken action by maintaining the illusion of a real booking process by building a CA appearing as human as possible. In the future, researchers may examine various participant samples, an already established CA with different maturity levels, or other (real-world) service settings to substantiate our findings.

8 Conclusion

The current study aims to improve our understanding on the willingness of customers not only to use, but rather continue to use CAs. Based on IS ECM, we investigated the multidimensional nature of relationship building capabilities together with utilitarian and especially hedonic value and their effect on *continuance intention* in a joint research model. We complement research by underline the importance of hedonic and utilitarian value for developing *continuance intentions* of CAs. Further important, this study demonstrated not only that creating perceptions of *social presence* is important in the context of CAs, but that hedonic and utilitarian value can be influenced by creating different initial perceptions of the conversational partner (CA vs. human). Overall, our work is the first to show the positive effect of *social presence* on hedonic and utilitarian value and confirms their relevance for subsequent *continuance intentions* when customers interact with CAs. Furthermore, we hope that our theoretical and practical contributions serve as a proper starting point for further research on this interesting topic.

Appendix – Questionnaire Items

Table 5. Constructs and questionnaire items

Continuance Intention (CI) (adapted from [19, 21, 51])	
CI1	I intend to continue using the online-chat rather than discontinue its use
CI2	My intentions are to continue using the online-chat than use any alternative means
CI3	If I could, I would like to discontinue my online-chat use (reverse coded)
CI4	Overall, I would intend to continue using online-chats
CI5	In the near future, I plan to keep using online-chats with travel agents
Perceived Enjoyment (EN) (adapted from [15, 24])	
EN1	Interacting with the travel agent is enjoyable
EN2	Interacting with the travel agent is exciting
EN3	Interacting with the travel agent is pleasant
EN4	Interacting with the travel agent is interesting
EN5	Interacting with the travel agent is fun
Perceived Usefulness (PU) (adapted from [13, 15, 19, 22, 33, 51])	
PU1	Using the online-chat improves my performance in interactions with travel agents
PU2	Using the online-chat increases my productivity in interactions with travel agents
PU3	Using the online-chat enhances my effectiveness in interactions with travel agents
PU4	Overall, online-chats are useful in interactions with travel agents
Satisfaction (ST) (adapted from [15, 19, 21, 51])	
ST1	displeased … pleased
ST2	dissatisfied … satisfied
ST3	frustrated … contented
ST4	terrible … delighted
Social Presence (SP) (adapted from [3, 13, 15, 27])	
SP1	I felt a sense of personalness in the conversation
SP2	I felt a sense of human warmth in the conversation
SP3	I felt a sense of sociability in the conversation
SP4	I felt a sense of human sensitivity in the conversation

References

1. eMarketer. https://www.statista.com/statistics/379046/worldwide-retail-e-commerce-sales/. Accessed 02 Dec 2019
2. Beverungen, D., Breidbach, C.F., Poeppelbuss, J., Tuunainen, V.K.: Smart service systems: an interdisciplinary perspective. Inf. Syst. J. **29**(6), 1201–1206 (2019)
3. Gnewuch, U., Morana, S., Adam, M., Maedche, A.: Faster is not always better. Understanding the effect of dynamic response delays in human-chatbot interaction. In: Twenty-Sixth European Conference on Information Systems (ECIS), Portsmouth (2018)
4. Schuetzler, R.M., Giboney, J.S., Grimes, G.M., Nunamaker, J.F.: The influence of conversational agents on socially desirable responding. In: Fifty-First Hawaii International Conference on System Sciences (HICSS), Hawaii (2018)
5. Schuetzler, R.M., Grimes, G.M., Giboney, J.S.: An investigation of conversational agent relevance, presence, and engagement. In: Twenty-Fourth Americas Conference on Information Systems (AMCIS), New Orleans (2018)
6. Gefen, D., Straub, D.W.: Gender differences in the perception and use of e-mail. An extension to the technology acceptance model. MIS Q. **21**(4), 389–400 (1997)
7. Facebook, Inc. https://de.statista.com/statistik/daten/studie/662144/umfrage/anzahl-der-verfuegbaren-chatbots-fuer-den-facebook-messenger/. Accessed 02 Dec 2019
8. Freedman International. https://www.statista.com/statistics/828101/world-chatbots-adoption-rate-intention/. Accessed 02 Dec 2019
9. Drift. https://www.statista.com/statistics/818859/chatbot-usage-benefits-customers-expect-to-enjoy-us/. Accessed 02 Dec 2019
10. LivePerson. https://de.statista.com/statistik/daten/studie/797057/umfrage/gruende-der-nutzung-von-chatbots-in-deutschland/. Accessed 02 Dec 2019
11. Rzepka, C., Berger, B.: User interaction with AI-enabled systems. A systematic review of IS research. In: Thirty-Ninth International Conference on Information Systems, San Francisco (2018)
12. Lim, E.T.K., Cyr, D., Tan, C.W.: Untangling utilitarian and hedonic consumption behaviors in online shopping. In: Sixteenth Pacific Asia Conference on Information Systems (PACIS), Ho Chi Minh City (2012)
13. Gefen, D., Straub, D.: Managing user trust in B2C e-Services. e-Serv. J. **2**(2), 7–24 (2003)
14. Karahanna, E., Straub, D.W., Chervany, N.L.: Information technology adoption across time: a cross-sectional comparison of pre-adoption and post-adoption beliefs. MIS Q. **23**(2), 183–213 (1999)
15. Qiu, L., Benbasat, I.: Evaluating anthropomorphic product recommendation agents. A social relationship perspective to designing information systems. J. Manag. Inf. Syst. **25**(4), 145–182 (2009)
16. Choi, J., Lee, H.J., Kim, Y.C.: The influence of social presence on customer intention to reuse online recommender systems: the roles of personalization and product type. Int. J. Electron. Commer. **16**(1), 129–154 (2011)
17. Chen, C.-F., Wang, J.-P.: Customer participation, value co-creation and customer loyalty - a case of airline online check-in system. Comput. Hum. Behav. **62**, 346–352 (2016)
18. Yan, A., Solomon, S., Mirchandani, D., Lacity, M., Porra, J.: The role of service agent, service quality, and user satisfaction in self-service technology. In: Thirty-Fourth International Conference on Information Systems (ICIS), Milan (2013)
19. Bhattacherjee, A.: Understanding information systems continuance. An expectation-confirmation model. MIS Q. **25**(3), 351–370 (2001)
20. Chen, S.-C., Chen, H.H., Chen, M.-F.: Determinants of satisfaction and continuance intention towards self-service technologies. Ind. Manag. Data Syst. **109**(9), 1248–1263 (2009)

21. Bhattacherjee, A.: An empirical analysis of the antecedents of electronic commerce service continuance. Decis. Supp. Syst. **32**(2), 201–214 (2001)
22. Davis, F.D.: Perceived usefulness, perceived ease of use, and user acceptance of information technology. MIS Q. **13**(3), 319–340 (1989)
23. Dabholkar, P.A.: Consumer evaluations of new technology-based self-service options: an investigation of alternative models of service quality. Int. J. Res. Mark. **13**(1), 29–51 (1996)
24. Babin, B.J., Darden, W.R., Griffin, M.: Work and/or fun: measuring hedonic and utilitarian shopping value. J. Consum. Res. **20**(4), 644–656 (1994)
25. van der Heijden, H.: User acceptance of hedonic information systems. MIS Q. **28**(4), 695–704 (2004)
26. Cetto, A., Klier, J., Klier, M.: Why should I do it myself? Hedonic and utilitarian motivations of customers' intention to use self-service technologies. In: Twenty-Third European Conference on Information Systems (ECIS), Münster (2015)
27. Hassanein, K., Head, M.: Manipulating perceived social presence through the web interface and its impact on attitude towards online shopping. Int. J. Hum Comput Stud. **65**(8), 689–708 (2007)
28. Mann, J.A., MacDonald, B.A., Kuo, I.-H., Li, X., Broadbent, E.: People respond better to robots than computer tablets delivering healthcare instructions. Comput. Hum. Behav. **43**, 112–117 (2015)
29. Lee, S., Choi, J.: Enhancing user experience with conversational agent for movie recommendation. Effects of self-disclosure and reciprocity. Int. J. Hum.-Comput. Stud. **103**, 95–105 (2017)
30. Atteslander, P.: Methoden der empirischen Sozialforschung. Erich Schmidt Verlag, Berlin (2010)
31. Moore, G.C., Benbasat, I.: Development of an instrument to measure the perceptions of adopting an information technology innovation. Inf. Syst. Res. **2**(3), 192–222 (1991)
32. Fishbein, M., Ajzen, I.: Belief, Attitude, Intention and Behavior, An Introduction to Theory and Research. Addison-Wesley, Reading (1975)
33. Lin, C.S., Wu, S., Tsai, R.J.: Integrating perceived playfulness into expectation-confirmation model for web portal context. Inf. Manag. **42**(5), 683–693 (2005)
34. Alphabet Inc. https://dialogflow.com/. Accessed 02 Dec 2019
35. Holtgraves, T., Han, T.-L.: A procedure for studying online conversational processing using a chat bot. Behav. Res. Methods **39**(1), 156–163 (2007). https://doi.org/10.3758/BF03192855
36. Schmitz, C. http://www.limesurvey.org. Accessed 02 Dec 2019
37. clickworker.com Inc. https://clickworker.com/. Accessed 02 Dec 2019
38. Buhrmester, M., Kwang, T., Gosling, S.D.: Amazon's mechanical turk. A new source of inexpensive, yet high-quality, data? Perspect. Psychol. Sci. **6**(1), 3–5 (2011)
39. Mason, W., Suri, S.: Conducting behavioral research on Amazon's mechanical turk. Behav. Res. Methods **44**(1), 1–23 (2012). https://doi.org/10.3758/s13428-011-0124-6
40. Paolacci, G., Chandler, J., Ipeirotis, P.G.: Running experiments on Amazon mechanical turk. Judgm. Decis. Mak. **5**(5), 411–419 (2010)
41. Anderson, J.C., Gerbing, D.W.: Structural equation modeling in practice. A review and recommended two-step approach. Psychol. Bull. **103**(3), 411 (1988)
42. Chiu, C.-M., Wang, E.T.G., Fang, Y.-H., Huang, H.-Y.: Understanding customers' repeat purchase intentions in B2C e-commerce. The roles of utilitarian value, hedonic value and perceived risk. Inf. Syst. J. **24**(1), 1–30 (2012)
43. Lowry, P.B., Gaskin, J.: Partial Least Squares (PLS) Structural Equation Modeling (SEM) for building and testing behavioral causal theory. When to choose it and how to use it. IEEE Trans. Prof. Commun. **57**(2), 123–146 (2014)
44. Chin, W.W.: Commentary. Issues and opinion on structural equation modeling. MIS Q. **22**(1), vii–xvi (1998)

45. Hair, J.F., Ringle, C.M., Sarstedt, M.: PLS-SEM. Indeed a silver bullet. J. Mark. Theor. Pract. **19**(2), 139–152 (2011)

46. Fornell, C., Larcker, D.F.: Evaluating structural equation models with unobservable variables and measurement error. J. Mark. Res. **18**(1), 39–50 (1981)

47. Nunnally, J.C.: Psychometric Theory. McGraw-Hill, New-York (1967)

48. Kline, R.B., Santor, D.A.: Principles & practice of structural equation modelling. Can. Psychol. **40**(4), 381 (1999)

49. Garson, G.D.: Partial Least Squares: Regression and Structural Equation Models. Statistical Associates Publishers, Asheboro (2016)

50. Botti, S., McGill, A.L.: The locus of choice: personal causality and satisfaction with hedonic and utilitarian decisions. J. Consum. Res. **37**(6), 1065–1078 (2011)

51. Bhattacherjee, A., Lin, C.-P.: A unified model of IT continuance. Three complementary perspectives and crossover effects. Eur. J. Inf. Syst. **24**(4), 364–373 (2015)

BIG-SWSDM: BIpartite Graph Based Social Web Service Discovery Model

Amal Hafsi[1,2]([⊠]), Youssef Gamha[1,2], Cheyma Ben Njima[1],
and Lotfi Ben Romdhane[1,2]

[1] MARS Research Laboratory LR17ES05, University of Sousse, Sousse, Tunisia
amal.hafsi@ensi-uma.tn, {youssef_gamha,bennjimacheima}@yahoo.fr,
lotfi.ben.romdhane@gmail.com
[2] Higher Institute of Computer Science and Telecom (ISITCom),
University of Sousse, Sousse, Tunisia

Abstract. With the increasing number of similar web services nowadays, the need to satisfy the complex user requirements and locate relevant services remain necessary. As a complex and challenging task, many approaches have been proposed. Nevertheless, they totally neglect the contribution of the social dimension. The mix of two domains social computing with service oriented computing opens the door to new discovery schemes. It gives birth to a new notion Social Web Services. In fact, integrating the social aspect into web services can benefit them to become active entities that can collaborate, compete or substitute each other. In this paper, we present the second step of our social web service discovery model that operates on a bipartite graph with user-user, user-service and service-service relationships and employs new metrics to evaluate the ability of a user to help the service requester to satisfy his needs.

Keywords: Social networks · Social web services · Social web services discovery

1 Introduction

With the rapid development of service oriented computing, an increasing amount of Web Services (WSs) have been published on the Web. In fact, it would be difficult to select relevant services to satisfy specific requirements. In the last few years, with the growing interest among social networks, many users use their social networks to find services or to offer services. In this way, WSs can be studied from a social computing perspective [5]. Social computing is a research area of computer science situated at the intersection of social behavior and computational systems. The most obvious examples of social computing would include social networks and blogs [17]. The result of integrating social network concept into WSs operation is Social Web Services (SWSs) [8]. A social web service is a service that lives in an interlinked network of services, where services can (i) collaborate to satisfy complex users' needs (ii) recommend the services to replace it, and (iii) compete with each other to be selected. In 2011, Maaradji et al. [1] classified

© Springer Nature Switzerland AG 2020
W. Abramowicz and G. Klein (Eds.): BIS 2020, LNBIP 389, pp. 307–318, 2020.
https://doi.org/10.1007/978-3-030-53337-3_23

the networks in which WSs can sign up to 3 groups: Competition, collaboration, and substitution. Recently, few studies have tried to discover WSs from a social perspective. Social network based web service discovery refers to use social network relationships of services or/and users in web service discovery. The majority of these research work, to address WSs discovery issue, develop either social network of users, or social network of WSs. The main purpose of this paper is integrating WSs and users in a common network, so we are interested on both of these categories. In this paper, we present the second process of our discovery model BIG-SWSDM. The first process, presented in our previous work [16], is based on extracting information from the service requester profile, while the second one, presented in this paper, aims to satisfy the query by exploiting the social relations of the service requester. Based on a bipartite graph, it takes advantage of the relationships between services and users to determine the services used previously, taking into account the date of invocation of these services, the feedback of users and the quality of service (QoS). We propose to compute a new metric helpfulness measure between users by aggregating two measures: similarity of interest and query domain usage rate. In order to compute these measures, we consider information extracted from the user profile which is modeled using the data-model SOAF (Service-Of-A-Friend) [13]. In fact, Individuals adopt recommendation that come from people with comparable profiles [2]. At first, we propose an arrangement of service' requester social relations according to helpfulness measure. Second, from each profile, starting with the profile which has the highest value of helpfulness measure, we extract SWSs previously invoked and evaluated as very satisfied in descending order of invocation date, and qualified SWSs that were collaborate with invoked services. This paper is organized as follows, Sect. 2 surveys related work, Sect. 3 describes the fundamentals of our model, the experimental evaluations will be presented in Sect. 4 and the final section concludes this paper and sheds light on future research directions.

2 Service Discovery Overview

2.1 Web Service Discovery

The goal of service discovery process is to find a web service that satisfies the consumer requirements. Three approaches are available for web service discovery. The first is syntactic such as discovery algorithms using UDDI or ebXML. Semantic approaches are based on the semantic annotation [3] or the WSs ontology such as OWL-S (Ontology Web Language for Services). The third approaches are based on integrating the paradigm of social networks on the discovery process.

2.2 Social Web Service Discovery

Social Networks

In just a few years, social network websites such as Facebook, Twitter and LinkedIn have become extremely popular. In fact, millions of Internet users have integrated these websites into their daily habits. The huge interest in social networks has opened up many new spaces of possible research in computing. A social network is a set of social entities connected by a set of social relationships, such as friendship, co-working, business or information exchange [4].

Social Networks of Web Services

The merging of service oriented computing and social computing produced SWSs which are different from regular WSs. SWSs can participate to three type of social networks: Collaboration, substitution, and competition. Through these networks, SWSs are mainly competing as each service wishes to be (i) a part of compositions (ii) a replacement in case of failure, and (iii) a competitor against others in case of selection. Thereby, the social networks of WSs differ from classical social networks which are based on collaboration between their members; there's no competition [5].

Social Web Services Discovery

Three categories of works discuss the WSs discovery in the context of social networks. A first category adopts social networks of persons. A second category develops social networks of WSs and the third one considers both WSs and persons in the same social network. In the first category, Kalai et al. [7] built a social network that integrates the users and their satisfactory WSs. The proposed discovery process extracts WSs from the FOAF profiles of the service requester and his similar friends. The similarity between users is based on number of common friend. Also, the invocation date and QoS properties are not taken into account to refine the discovery result. In [6], the authors developed a social network of providers and measured the relationships between the social networks nodes to calculate a trust degree of service provider. They introduced a new method to calculate a service trust rating using centrality measure of social networks to produce a discovery solution that consists of a set of WSs provided by trusted providers.

In the second category, we cite the works of Maamar et al. [8] and Chen et al. [9]. Maamar et al. defined two types of social networks in the particular context of WSs: collaboration when WSs participate in composition to satisfy complex users' needs and recommendation which is divided into partnership and robustness. In a partnership social network, a web service that previously participates in a composition can proposes to the user to add other WSs. In a robustness social network, a web service knows other WSs that can replace them in case of failure. In the second reference, Chen et al. proposed a framework to connect the isolated service according to social properties and built a global social service network to enhance the quality of service discovery.

In the third category that integrates social networks of WSs and social networks of users together, Fallatah et al. [10] proposed a new WSs discovery framework that describes diverse types of interaction between WSs and users inside the global social network. The QoS criteria are not taken into consideration. Duan et al. [11] proposed a collaborative service discovery approach based on four social links behind WSs and users: (i) the reputation degree of a user which is the number of the concepts in the user' personalization model divided by the quantity of the whole concepts in the ontology (ii) clustering link which is the set of services or users and their similarities (iii) preference link which is defined as the set of services that were consumed in the past by a user and (iv) trust link which describes the relations between the user and his trusted friends. But, the most of these approaches neglect the feedback of users, the user desired QoS properties and perform a simple user query. In fact, adding the set of non-functional properties such as QoS properties, to discover relevant services and increase the accuracy of the discovery process has still been poorly explored.

3 Our Propoasal: BIG-SWSDM

3.1 Architecture

In this section, we introduce our proposed model architecture that takes advantage of relations between user-service, service-service and user-user. Given a composite user query, the basic goal is to generate a list of relevant services that responds to user's requirements.

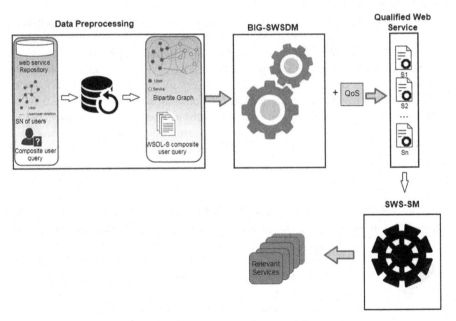

Fig. 1. Architecture of our model

Figure 1 illustrates our approach to discover and select SWSs which includes three modules: Data preprocessing, BIG-SWSDM (BIpartite Graph based social web service discovery model) and SWS-SM (Social Web Services Selection Model). In our model, all web services resulting from BIG-SWSDM are qualified services that match with composite user query and respond to user desired QoS. The first module consists of transforming the test data before feeding it to the second module. The outputs of this module are: Bipartite graph and a WSDL-S composite query. As mentioned in our previous work [16], our bipartite graph G is defined as a triplet (U, WS, R), where U is the nodes set of the users social network, N is the nodes set of atomic web services' social network and R represent the links between those nodes which show that a user has previously invoked a web service. In the social network of users, the nodes represent the users and the edges define the social relationships between them.

We used and extended SOAF user profile to describe the personalization of the user profile description. So, user profile (u ∈ U) is modeled by a triplet u = (I, F, {S}) where I is the set of user interests, F is the user' social relationships sets and {S} is the set of previously used web services. S is described by: S = (WSa, e, {WSc}) where WSa is the invoked service, e is the satisfaction degree after using service (very satisfied, satisfied or not satisfied) and {WSc} (WSc ∈ WS) is the set of services which collaborated with WSa. As regards the social network of WSs, in the literature, there are three types of social networks according to their functionalities: collaboration, competition and substitution [14]. These functionalities are either different so WSs work together on complex requests or similar, so SWs compete against each other or help each other when they fail. In our work, we are interested in collaboration social network, as mentioned in [16]. SWSs are described with WSDL-S [3] and QoS features such as reliability, cost, response time and availability. We used and extended WSDL-S to describe the personalization of the web service description.

An atomic web service (WSa ∈ WS) is described by: WSa = (I, O, E, P, d, QoS) where I is the service input parameters, O is the service output parameters, E is the finite set of effects, P is the finite set of Preconditions, d is the service domain, QoS is the finite set of QoS properties (QoS).

In BIG-SWSDM, user query is the second input, it is described with WSDL-S and desired QoS properties. User query (S0) is modeled as following: S0 = ({WSa}, σ, D, QoS) where {WSa} is the set of atomic services, σ is a control operator that defines the composition, D is the set of domain for atomic services and QoS is the set of the desired QoS properties.

3.2 BIG-SWSDM

As mentioned in Fig. 2, BIG-SWSDM performs two processes: Discovery process through service requester invocation history and discovery process through service requester' friend's invocation history. As explained in [16], the first process is done in two steps, a first step that extracts from the profile of the service requester all the:

1. SWSs that were consumed in the past and evaluated as very satisfied starting with the most recent services in terms of invocation date.
2. Qualified SWSs that were collaborated with invoked services

A web service can collaborate with several other services, in our approach we only consider qualified services. In fact, the QoS criteria are taken into account to discover qualified SWSs. Also, we take into consideration the date of invocation because a service which was recently used may be updated and available.

A second step evaluates the semantic similarity with user composite query in term of functionality by using the method of Liu et al. [12]. The discovery process stops if the query is satisfied. In this paper, we will focus on the second process of BIG-SWSDM: Discovery process through service requester' friend's invocation history.

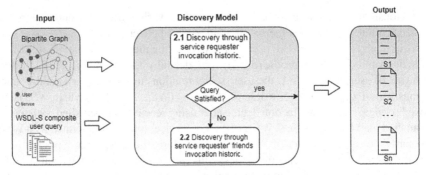

Fig. 2. BIG-SWSDM model

3.3 BIG-SWSDM: Discovery Process Through Service Requester' Friends Invocation History

Discovery process through service requester' friends' invocation history, as outlined by Fig. 3, is done in two steps and takes as input a composite user query and the social profile of the service requester and all his social relationships (direct friends in the social networks of users).

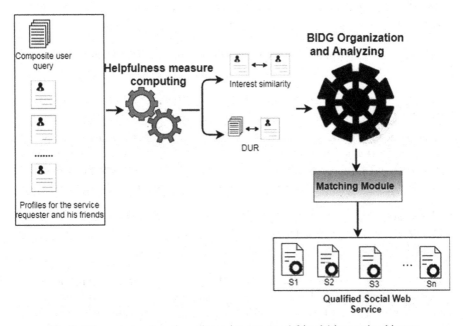

Fig. 3. Discovery process through service requester' friends' invocation history

A first step of this process is helpfulness measure computing. This later evaluates the ability of a user to help the user requester to satisfy his needs based on two measures: Interests similarity IS and domain usage rate DUR. Next, we extract the sub graph: Bipartite Discovery Graph BIDG of our bipartite graph which consists of service requester as root, his friends, the SWSs invoked by them and for each web service its egocentric collaboration network [16] (An egocentric collaboration network is defined as a sub-network consisting solely of a web service and all of its direct links in the collaborative social network of SWSs). In BIDG, with regard to the friendship relations between the users, only the relations between the service requester and his direct friends are considered. We apply depth-first traversal algorithm on the BIDG graph, the search progresses from the vertex, in our case it's the service requester, by starting with: The friend who has the highest value of helpfulness measure and the most recent SWSs in terms of invocation date which is evaluated very satisfied. This search algorithm is applied in depth until the request is satisfied. In the same way as the first step, it allows to extract SWSs that were consumed in the past and qualified SWSs that were collaborate with invoked services. The QoS properties are taken into account to refine the discovery result.

Scenario.
To more explain our proposal, we will present a scenario. Let us consider a simplified example of a BIDG depicted in Fig. 4. We assume that a user q wants to travel, books a room in a village hotel and looks for a restaurant. So, the query is as follows: S0 = ({S1, S2, S3}, sequence, {Travel Food} (response time, equal, 0.5, second)).

S1: Web service to book a flight ticket.
S2: Web service to book a room in a village hotel.
S3: Web service to look for a restaurant.

S1, S2, S3 ∈ WSn, sequence ⊂ σ, {Travel, Food} ⊂ D and as we presented above response time is a QoS properties. Our system starts by calculating the helpfulness measure HM for each friend of the BIDG vertex: user w, user i, user j.

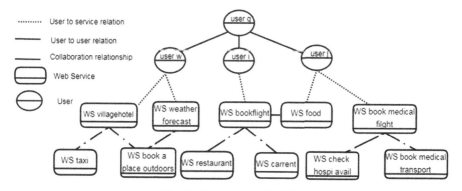

Fig. 4. An example of BIDG

In our scenario, for example, HM (user q, user w) = 0.68, HM (user q, user i) = 0.51 and HM (user q, user j) = 0.3. In fact, the system explores in order the invocation history of user w then user i then user j. Table 1 shows, for each user, the invoked service, his evaluation and the invoked date.

Table 1. Invoked service, evaluation and invocation date of each user

User	Invoked service	User's evaluation	Invocation date
User w	WS village hotel	Very satisfied	12-04-2019
User w	WS weather forecast	Not satisfied	03-02-2019
User i	WS book flight	Satisfied	07-10-2018

Table 2. Service response time

Service	Response time
WS food	0.6 s
WS restaurant	0.4 s

Table 2 give examples of response time values for some services. The system explores BIDG by starting with user w and extracts: WS taxi, WS book place outdoors and WS village hotel. The matching algorithm is executed to select services that match with composite user query and respond to user desired response time. So, our system continue to explore BIDG. WS restaurant, WS car rent, WS food, WS book flight are extracted. Therefore, the user query is satisfied. So, WS book flight WSvillagehotel and WS food are the qualified services resulting from BIG-SWSDM.

Helpfulness Measure Computing.
The helpfulness measure of a user U according to the service requester SR represents the ability of U to help SR to satisfy his needs based on two measures: Interests similarity and domain usage rate. The aim of Interests Similarity IS is to compare the interests of user requester to those of his directed friends in the social network. The set of interests is extracted from the SOAF profile of the target user. We adopted Wu and Palmers measure [15] to compute the semantic similarity between two concepts $c1$ and $c2$. DUR is the invocation rate of WSs in the query domain of a user. We admit that the more the user consumes WSs in the query domain, the more he will be able to help the service requester to satisfy his needs. The rate usage domain of a user U in the query domain DUR $(U, Dom(S_0))$ is shown in Eq. (1):

$$DUR(U, Dom(S0)) = \frac{|Dom(S0) \cap Dom(U)|}{|Dom(S0)|} \tag{1}$$

Where Dom (S0) is the domain set of query and Dom (U): is the domain set of invoked SWSs. The helpfulness measure of a user U according to the service requester SR HM (U, SR), in interval of [0, 1], is calculated by equation.

$$HM(U, SR) = \alpha \times IS(U, SR) + (1 - \alpha) \times DUR(U, Dom(S0)) \tag{2}$$

Where IS (U, SR) is the interest similarity between U and SR, and DUR (U, Dom(S0)) is the domain usage rate of U in the query domain Dom(S0) with α is in [0, 1].

Algorithm.

In this section, we present our algorithm named BIG-SWSDM-FR which implemented the second step of BIG-SWSDM. The details of the algorithm are as following:

Algorithm: BIG-SWSDM-FR

Input: q: user request
Output: SL: Qualified service list
1: List FL = sort list friend in descending order according to HM;
2: for each fr in FLdo
3: IWS = extract invoked WS with positive evaluation
4: IWSL = sort IWS list according to invocation date;
5: for each SI in IWSL do // query invoked services
6: calculate similarity Sim between q and SI
7: if Sim > Sim0 do //compare the matching degree
8: add SI to SL   //adding the service to the discover list
9: if q is satisfied do
10: exit;
12: end if
13: end if
10: CWS = extract qualified collaborate services
11: for each CS in CWS // query the collaborates services
12: calculate similarity Sim between q and CS
7: if Sim > Sim0 do
8: add CS to SL
9: if q is satisfied do
10: exit;
12: end if
13: end for
14: end for
15: end for
16: return SL;
7: End

---

4 Experimental Results

We used as test data: a social network of users SNAP STANFORD[1] real data sets "Epinions social network" which contains 75879 users and 508837 links, the service repository of test collection SemWebCentral[2]. There are 1083 semantic WSs and 42 requests advertised with OWL-S 1.1 and SAWSDL 1.1. There are seven service domains: Education, Medical, Food, Travel, Communication, Economy, and Weapon.

To evaluate the performances of BIG-SWSDM, we propose to compare the response rate provided by our model BIG-SWSDM and two other approaches: CSDRM [11] (Collaborative Web Service Discovery and Recommendation Mechanism) and SC-WSD [7] (Social Context based Web Service Discovery System). The response rate, as shown in Eq. (3), is in the interval of [0, 1] and used to compute the difference between the number of successful queries and the total number of launched queries.

$$\text{Response rate} = \frac{Successful\ queries}{\text{Total launched queries}} \tag{3}$$

Figure 5 illustrates the comparing result of the three different approaches.

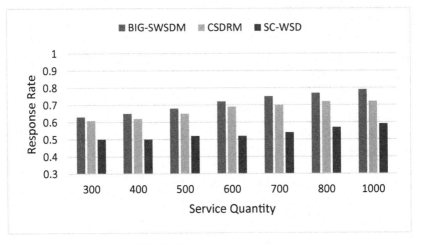

Fig. 5. Response rate variation

The best response rate of BIG-SWSDM is 0.79, while the best response rate of CSDRM and SC-WSD are respectively 0.72 and 0.59. SC-WSD and CSDRM approaches are less precise. In fact, SC-WSD neglect the social dimensions of services i.e. it doesn't consider relations between services. Moreover, the similarity degree between users based on number of common friends, it is limited to one measure. CSDRM that is the closest to our approach gives good response rate values since it considers relations between services. But, the similarity between users has no relation to the query launched. Indeed,

---

[1] SNAP STANFORD, www.snap.stanford.edu.

[2] http://semwebcentral.org/frs.

it is obvious to find a user who has a high similarity degree with the service requester but he/she has not consumed any service in the query domain what decreases the success rate degree. This demonstrates that taking into account the domain query in the similarity measure between users can improve the discovery results.

Figure 6 describes the variation of the completion time for the discovery process of two version of our model BIG-SWSDM. The first one adopts the HM measure. The second one ignore the step of social relationship ordering, in fact it considers that all friends have the same importance i.e. all the social relations of the service requester have the same ability to help him to satisfy his needs. In fact, the system explores invocation history of service requesters' friends by starting with any friend.

Fig. 6. Completion time variation

The response time is getting higher as the number of WSs grow. The results show also that BIG-SWSDM costs less time. This is due to the following reason: exploring the invocation history of a user who has the highest value of HM can help the system to satisfy the query rapidly and thus save time. Indeed, users who have a high value of HM, consumed necessarily WSs in the same domain of the query and have interests semantically similar to those of the service requester.

5   Conclusion

This paper details the second process of our BIG-SWSDM. This process consists, while the query is unsatisfied, in extracting the most recent SWSs in terms of invocation date which is evaluated as very satisfying and invoked by service requester friends, and qualified SWSs that were collaborate with invoked services. We propose a new metric to evaluate the ability of a user to help in satisfying the launched query which is based on analyzing his social profile. Our model takes into consideration the feedback of users, the invocation date of a service and QoS criteria for refinement of the discovery result. We conducted a series of experimentation which have demonstrated that connecting SWSs in a social network can improve the efficiency of service discovery.

References

1. Maaradji, A., Hacid, H., Daigremont, J., et al.: Towards a social network based approach for services composition. In: 2010 IEEE International Conference on Communications, pp. 1–5. IEEE (2010)
2. Ziegler, C.-N., Lausen, G.: Analyzing correlation between trust and user similarity in online communities. In: Jensen, C., Poslad, S., Dimitrakos, T. (eds.) iTrust 2004. LNCS, vol. 2995, pp. 251–265. Springer, Heidelberg (2004). https://doi.org/10.1007/978-3-540-24747-0_19
3. Akkiraju, R., Farrell, J., Miller, J.A., et al.: Web service semantics – WSDL-S (2005)
4. Shekarpour, S., Katebi, S.D.: Modeling and evaluation of trust with an extension in semantic web. J. Web. Seman. **8**(1), 26–36 (2010)
5. Maamar, Z., Wives, L.K., Badr, Y., et al.: Even web services can socialize: a new service-oriented social networking model. In: 2009 International Conference on Intelligent Networking and Collaborative Systems, pp. 24–30. IEEE (2009)
6. Bansal, S.K., Bansal, A., Blake, M.B.: Trust-based dynamic web service composition using social network analysis. In: 2010 IEEE International Workshop on: Business Applications of Social Network Analysis (BASNA), pp. 1–8. IEEE (2010)
7. Kalai, A., Zayani, C.A., Amous, I.: User's social profile-based web services discovery. In: 2015 IEEE 8th International Conference on Service-Oriented Computing and Applications (SOCA), pp. 2–9. IEEE (2015)
8. Maamar, Z., Wives, L.K., Badr, Y., et al.: LinkedWS: a novel web services discovery model based on the metaphor of social networks. Simul. Model Pract. Theory **19**(1), 121–132 (2011)
9. Chen, W., Paik, I., Hung, P.C.K.: Constructing a global social service network for better quality of web service discovery. IEEE. Trans. Services Comput. **8**(2), 284–298 (2013)
10. Fallatah, H., Bentahar, J., Asl, E.K.: Social network-based framework for web services discovery. In: 2014 International Conference on Future Internet of Things and Cloud, pp. 159–166. IEEE (2014)
11. Duan, L., Tian, H.: Collaborative web service discovery and recommendation based on social link. Future Internet **9**(4), 63 (2017)
12. Liu, M., Shen, W., Hao, Q., Yan, J.: An weighted ontology-based semantic similarity algorithm for web service. Exp. Syst. Appl. **36**, 12480–12490 (2009)
13. Treiber, M., Truong, H.-L., Dustdar, S.: SOAF – design and implementation of a service-enriched social network. In: Gaedke, M., Grossniklaus, M., Díaz, O. (eds.) ICWE 2009. LNCS, vol. 5648, pp. 379–393. Springer, Heidelberg (2009). https://doi.org/10.1007/978-3-642-02818-2_31
14. Maamar, Z., Hacid, H., Huhns, M.N.: Why web services need social networks. IEEE. Int. Comput. **15**(2), 90–94 (2011)
15. Wu, Z., Palmer, M.: Verbs semantics and lexical selection. In: Proceedings of the 32nd Annual Meeting on Association for Computational Linguistics, pp. 133–138. Association for Computational Linguistics (1994)
16. Hafsi, A., Gamha, Y., Romdhane, L. B: SWSDM: Social Web Service Discovery Model. Int. Conf. Appl. Comput. (2019). https://doi.org/10.33965/ac2019_201912L013
17. Parameswaran, M., Whinston, A.B.: Social computing: an overview. Commun. Associat. Inform. Syst. **19**(1), 37 (2007)

Novel Version of PageRank, CheiRank and 2DRank for Wikipedia in Multilingual Network Using Social Impact

Célestin Coquidé[1]([⊠]) [ID] and Włodzimierz Lewoniewski[2] [ID]

[1] Institut UTINAM, Observatoire des Sciences de l'Univers THETA, CNRS,
Université de Bourgogne Franche-Comté, Besançon, France
`celestin.coquide@utinam.cnrs.fr`
[2] Department of Information Systems, Poznan University of Economics,
Poznan, Poland
`Wlodzimierz.Lewoniewski@ue.poznan.pl`

Abstract. Nowadays, information describing navigation behaviour of internet users are used in several fields, e-commerce, economy, sociology and data science. Such information can be extracted from different knowledge bases, including business-oriented ones. In this paper, we propose a new model for the PageRank, CheiRank and 2DRank algorithm based on the use of clickstream and pageviews data in the google matrix construction. We used data from Wikipedia and analysed links between over 20 million articles from 11 language editions. We extracted over 1.4 billion source-destination pairs of articles from SQL dumps and more than 700 million pairs from XML dumps. Additionally, we unified the pairs based on the analysis of redirect pages and removed all duplicates. Moreover, we also created a bigger network of Wikipedia articles based on all considered language versions and obtained multilingual measures. Based on real data, we discussed the difference between standard PageRank, Cheirank, 2DRank and measures obtained based on our approach in separate languages and multilingual network of Wikipedia.

Keywords: PageRank · Wikipedia · CheiRank · Clickstream · Pageviews · Google matrix · Centrality measures

1 Introduction

For the last 10 years, Wikipedia has been one of the most popular source of knowledge. In different countries, this online encyclopedia is in the top 10 most visited websites [1]. Wikipedia even influences the language used by scientists in their publications [33]. Nowadays, this free encyclopedia contains over 51 million articles in more than 300 languages [36]. Content on a separate subject can be created and edited independently in each language version of Wikipedia [26]. Often a Wikipedia article contains links to other pages, which can be used to find more information related to the subject. Based on links between articles,

© Springer Nature Switzerland AG 2020
W. Abramowicz and G. Klein (Eds.): BIS 2020, LNBIP 389, pp. 319–334, 2020.
https://doi.org/10.1007/978-3-030-53337-3_24

it is possible to identify important places, persons, products in specific area or language community. To find such important Wikipedia articles, different methods can be used. One of the well-known algorithms for this purpose is PageRank [30].

The google matrix and the PageRank algorithm are at the foundation of the very famous search browser Google [2]. They describe a random internet user, surfing on the World Wide Web (WWW). Indeed WWW can be seen as a directed network where nodes are Web pages and links are hyperlinks allowing an internet user to navigate. By knowing a network's topology we can construct its adjacency matrix A as well as its stochastic matrix S. Its mathematical aspect is well described in [24]. Centrality measures in complex network theory are important, especially as nowadays networks are very large and complex (high occurrence of nodes and links). In case of directed networks such as WWW, eigenvector-based centrality measure is helpful and efficient as demonstrated by Google's web search engine. However, this simple random walk model may be limited for describing a human user's behaviour. PageRank algorithm, applied to the Wikipedia network, leads to robust ranking lists of articles and articles of countries tend to be well ranked. Using a simple random walk, such algorithm didn't allow us to observe a trend of interest from readers.

In this study, we extend the usual PageRank algorithm with clickstream's and pageview's data from Wikipedia October 2019's dumps. In this paper, we first mention works related to Wikipedia and PageRank algorithm, then we present you the "WikiClick" (*wc*) and "WikiClick Plus View" (*wcpv*) models and finally we discuss how this social information affects Wikipedia articles ranking.

2   Related Works

Pageviews statistics in Wikipedia [11] can show which article is popular over a specific period. Therefore, we can observe a topic trend over time. Using such data, it is possible to forecast stock market moves [29], movies' success [25, 28], demand for services in the tourism sector [21], cryptocurrencies price and market performance [7,22], epidemics in specified territory [20] and can be used in electoral prediction [37]. Moreover, it can be used to assess the quality of Wikipedia articles alongside other measures (such as text length, number of references, images, sections and others) [26,27].

Google matrix and PageRank algorithm have been applied to Wikipedia for ranking articles related to historical figures [14] and Universities [23]. We can also find studies on the World influence for Universities [4], cancers [31] and infectious diseases [32] using the Wikipedia network. The Wikipedia articles ranking evolution over time using PageRank and CheiRank for countries, historical figures, physicist and chess players have been investigated in [15]. Spectral analysis of the Google matrix permits us to retrieve communities of articles [17]. Moreover the google matrix analysis have been used in business oriented networks such as the World trade network [5] and cryptocurrencies network [3].

The use of the internet users' behaviour information such as pageviews and clickstream data has been studied for the biomedical ontology repository BioPortal [18]. Pageviews data is used by pantheon group to rank famous individuals from Wikipedia [38]. Clickstream data has been used to study the navigational phase space of Wikipedia [19].

3   Proposed Methods

3.1   Wikinetwork Models

In such network, nodes are Wikipedia articles and the directed link $A \rightarrow B$ exists if article B is reachable from article A by using an intra-wiki hyperlink. Usually, we consider the unity rule for such links, it means that we count once such hyperlink in article A. It follows that the corresponding adjacency matrix will be asymmetric and its elements are equal to 1 or 0. We call this model the standard wikinetwork model (***nowc***).

Information from wikidata such as clickstream for intra-wiki hyperlinks provides an interesting social bias of the Wikipedia network. We construct two different models of Wikipedia network: wikiclick (***wc***) and wikiclick plus view (***wcpv***). We compare results obtained using these methods with ***nowc***.

3.2   Google Matrix and Ranking Algorithms

Google Matrix. The google matrix G is a Perron-Frobenius operator based on the stochastic matrix of a network and a teleportation term. We detail here the construction of G for each model.

The general formula for the google matrix is

$$G = \alpha S + (1 - \alpha)\mathbf{v}\mathbf{e}^{\mathbf{T}} \tag{1}$$

where α is the so-called damping factor, S the stochastic matrix, $\mathbf{e}^{\mathbf{T}}$ a row vector with N ones and $\mathbf{v}$ is a preferential vector such as $\sum_j v_j = 1$. $\alpha = 0.85$ is a standard value for different real complex networks, such as WWW.

The teleportation term $\mathbf{v}\mathbf{e}^{\mathbf{T}}$ may simply be a matrix of elements equal to $1/N$.

The google matrix for the standard model is computed with this trivial teleportation term.

S is computed from the adjacency matrix describing the network topology

$$A_{ij} = \begin{cases} 1 & \text{if edge } j \rightarrow i \text{ exists} \\ 0 & \text{else} \end{cases} \tag{2}$$

In case of ***wc*** version of Wikipedia network, we simply use W the matrix of clicks where W_{ij} element is the number of clicks received, article i from article j. The Wikidata for clickstream only counts clicks ≥ 10, $W_{ij} = 0$, therefore elements are replaced by a standard A_{ij} element representing the possibility of click because the link exist in the network, this final weighted adjacency matrix is noted A_{wc}. From A_{wc} we define the stochastic matrix S_{wc} representing the probability to reach node i from j by:

$$S_{wc_{ij}} = \begin{cases} \frac{A_{wc_{ij}}}{\sum_{i'} A_{wc_{i'j}}} & \text{if } \sum_{i'} A_{wc_{i'j}} \neq 0 \\ 1/N & \text{if } \sum_{i'} A_{wc_{i'j}} = 0 \end{cases} \tag{3}$$

We also use pageviews information as a teleportation matrix instead of $\mathbf{ve^T}$. In that way, the preferential vector component $v_j = \#$views for article j. $\tilde{\mathbf{v}}$ is the normalized vector computed from $\mathbf{v}$ such as $\sum_j \tilde{v}_j = 1$.

Finally, we obtain:

$$G_{ij} = \begin{cases} \alpha S_{wc_{ij}} + (1-\alpha)/N & \text{for } \boldsymbol{wc} \\ \alpha S_{wc_{ij}} + (1-\alpha)\tilde{v}_i & \text{for } \boldsymbol{wcpv} \end{cases} \tag{4}$$

PageRank and CheiRank Algorithm. The leading right eigenvector of G with corresponding eigenvalue $\lambda = 1$ corresponds to the steady state of a random walker moving through the network for an infinite time. We have the relation

$$\mathbf{GP} = \mathbf{P} \tag{5}$$

where $\mathbf{P}$ is called PageRank vector, its ith component represents the probability that a random surfer reaches node i after an infinite journey. By sorting components in decreasing order, we obtain the nodes ranking. PageRank measures ingoing links efficiency, seen as importance ranking. Let $K_1, K_2, ..., K_N$ be the PageRanks of the nodes such as $P_{K_1} > P_{K_2} > P_{K_3}$ and so on.

To measure the efficiency of outgoing links, we simply reverse the direction of all the network's links. This leads to a new adjacency matrix $A_{wc}^* = A_{wc}^T$ and then we compute the corresponding google matrix noted G^*.

CheiRank vector ($\mathbf{P}^*$) is defined as the eigenvector of G^* such as $G^*\mathbf{P}^* = \mathbf{P}^*$ and note $K_1^*, K_2^*, ..., K_N^*$ the obtained ranking of nodes.

2DRank Algorithm. PageRank and CheiRank algorithms are two sides of the same coin, the first one describes relevance of an article within Wikipedia's network and the last one represents its communicability. As described in [16,39], we can place nodes of a network in the two-dimensional (K, K^*) plane. The 2DRank algorithm uses both rankings and is defined as following:

- Firstly, for each node let, $K_{max}(i) = Max(K_i, K_i^*)$.
- Secondly, we sort nodes in ascending order according to their K_{max}.
- Finally, we sort couple of nodes with the same K_{max} regarding the increasing K^* ordering.

4   Datasets and Extraction Methods

To extract data about links between Wikipedia articles (wikilinks or intra-wiki), we use Wikipedia dumps from October of 2019 [34]. We focused on two separate approaches to identify these links:

- **XML** - directly from the Wikicode [10] (XML dumps)
- **SQL** - rendered versions of the articles (SQL dumps).

In the case of Wikicode, for each article, we searched internal links (wikilinks) [8] placed in doubled square brackets in code for each considered language version (below example for English Wikipedia):

- "enwiki-20191001-pages-articles.xml.bz2" - recombined articles, templates, media/file descriptions, and primary meta-pages.
- "enwiki-20191001-redirect.sql.gz" - redirect list.

Among the extracted links were also the ones that led to other types of Wikipedia pages (other namespaces [9]). Therefore, we only kept those belonging to article namespace (ns 0). We also removed links leading to nonexistent articles (so called "red links" [12]). Additionally, we took into account other names of the same articles based on redirects [13].

Rendered version of the articles usually have more links to other pages than we can find in their source (in Wikicode). It comes from additional elements placed in the article. For example, we can find the same template with certain set of links in articles related to similar topic (such as French cities, cryptocurrencies, processors, Nobel Prize laureates and others). In order to analyze links in rendered version of the Wikipedia articles, we took into account other dumps files for each language version (below is an example for the English Wikipedia):

- "enwiki-20191001-pagelinks.sql.gz" - wiki page-to-page link records.
- "enwiki-20191001-page.sql.gz" - base per-page data (id, title, old restrictions, etc).
- "enwiki-20191001-redirect.sql.gz" - redirect list.

The Table 1 shows statistics of the extraction source destination pairs of links for each considered language version of Wikipedia. We extracted over 1 billion pairs from the SQL dump and over 500 million pairs from XML dumps. For every pair, we only took Wikipedia articles from ns 0. After excluding duplicate pairs, redlinks (nonexistent pages) and unification of the articles names (based on redirects), the total pair number is reduced.

For both methods we additionally extracted pageviews statistics [11] and clickstream data [35] (click counts of source-destination pairs of articles) from September 2019.

5   Application of Methods and Discussion

In this section, we detail the application of our method using different dumps of Wikipedia from October 2019. We have built the network from English edition

Table 1. All and unified source-destination pairs of links from Wikipedia in different languages. Source: own calculations based on Wikimedia dumps in October 2019

Lang.	XML		SQL	
	All	Unified	All	Unified
de	86 242 247	63 618 326	111 288 696	108 762 081
en	228 373 266	165 832 345	500 144 739	479 163 241
es	54 878 393	39 369 961	53 948 827	50 625 623
fa	15 298 097	7 427 045	74 249 078	71 867 560
fr	80 719 270	61 576 083	156 691 399	153 108 004
it	52 731 642	40 857 564	117 826 190	115 641 441
ja	63 112 674	50 122 887	92 626 350	90 901 975
pl	36 838 878	27 240 200	76 958 914	76 318 086
pt	31 311 443	22 167 152	61 269 416	58 843 986
ru	52 646 408	37 922 206	99 995 034	95 706 281
zh	30 253 747	18 718 463	86 343 098	83 272 015

of Wikipedia from XML and SQL dumps as well as a multilingual version of the network. In the multilingual network, we took all language editions with available clickstream and pageviews data.

5.1   English Edition

XML Dumps. Table 2 shows the top 10 articles using **wcpv** method and PageRank algorithm. For each of these articles, we also show its rank among K_{wc}, K_{nowc}, clickstream (K_{cR}) and pageviews ranking lists (K_{vR}).

We found articles related to sovereign states for **wcpv** method, which is usually the case for **nowc** PageRank. By comparing the top 10 from K_{wcpv} with K_{nowc} we see that this set of articles is a reordering of K_{nowc} except for "Wikipedia", "List of Queen of the South episodes" and "Queen of the South (TV series)". These three articles are badly ranked in K_{nowc} as well as in K_{wc}. The two first elements are well ranked because of their Pageviews. The third one is very interesting because "Queen of the South (TV series)" is badly ranked in all other ranking list. According to our method, we found a top 10 PageRank containing two articles related to the same TV series. This last result is not common for Wikipedia PageRank.

The top 10 articles from CheiRank according to **wcpv** method is detailed in Table 3. Usually, using the CheiRank method applied to Wikipedia network, articles related to a list of articles have a better rank than others. We do not see that in **wcpv**. Indeed, we only have 4 lists ("Deaths in 2019, Lists of deaths by year", "2019 in film" and "List of Bollywood films of 2019") whereas **nowc** shows 100% of lists in its top 10. In the other articles of top 10 **wcpv** CheiRank, some are related to social interest such as "Joker (2019 film)" and "2019 FIBA

Table 2. First 10 articles obtained by PageRank with **wcpv** model from English Wikipedia. Source: own calculations based on Wikimedia dumps in October 2019.

Name	K_{wcpv}	K_{wc}	K_{nowc}	K_{cR}	K_{vR}
United States	1	1	1	15	24
Wikipedia	2	11665	3542	25013	1
List of Queen of the South episodes	3	5170889	5128933	4455336	2
United Kingdom	4	9	5	81	63
New York City	5	23	10	150	139
World War II	6	12	3	181	78
Germany	7	7	7	727	118
India	8	10	9	138	68
France	9	5	2	1432	197
Queen of the South (TV series)	10	166342	744237	28297	5871

Basketball World Cup". **wcpv** CheiRank is similar to **wc**. We think that with the use of clickstream and pageviews, CheiRank algorithm gives us a list of entry points of Wikipedia.

Table 3. First 10 articles obtained by CheiRank with **wcpv** model from English Wikipedia. Source: own calculations based on Wikimedia dumps in October 2019.

Name	K^*_{wcpv}	K^*_{wc}	K^*_{nowc}	K_{cR}	K_{vR}
Deaths in 2019	1	2	909	406	4
Lists of deaths by year	2	1	10	55939	7874
It Chapter Two	3	10	334575	2	12
2019 in film	4	9	382	365	34
List of Bollywood films of 2019	5	12	19249	640	54
Wikipedia	6	421	11031	25013	1
It (2017 film)	7	19	106231	18	37
Joker (2019 film)	8	20	95145	33	10
Mindhunter (TV series)	9	17	310462	147	35
2019 FIBA Basketball World Cup	10	11	20498	44	13

In order to quantify changes in ranking of Wikipedia articles, coming from the used model, we computed two overlap measures η_N and η_O. The first one measures the presence of same articles in two ranking lists and η_O measures the exact rank similarity. Quantitatively, by regarding Fig. 1 differences between rankings are related to rank switching rather than exact similarity. As we can see η_N is higher than η_O (inset plots). Moreover, at short range $j \in [1, 20]$, we have the highest overlap measures, with $\eta_N = 0.75(K_{nowc}$ vs. $K_{wc}), 0.7(K_{nowc}$ vs. K_{wcpv} for $j = 20$. In case of CheiRank overlap, we respectively have $\eta_N = 0.35$ and $\eta_N = 0.15$ comparing K^*_{wc} and K^*_{wcpv} with K^*_{nowc}. Social information seems to change more drastically CheiRank than PageRank. When comparing **wcpv** and **wc** methods, we have an overlapping very close regarding both PageRank and CheiRank. The highest value for $j = 100$ is for K_{nowc} vs. K_{wc} with $\eta_N = 0.75$. **wcpv** method gives us more differences in the top 100 with respectively 53% and 6% of similarity for PageRank and CheiRank algorithm. Exact overlap η_O is very low with 5% and 2% regarding K_{wcpv} and K_{wc} with K_{nowc} for $j = 100$. Exact similarity regarding CheiRank is 0.0 for **wc**, **wcpv** and **nowc** methods. The left panel shows us how similar K_{wc} (Resp. K^*_{wc}) and K_{wcpv} (Resp. K^*_{wcpv}) are, with clickstreams and pageviews statistical rankings. Highest measures are for CheiRanks, **wcpv** method has the highest overlap with vR (0.5 and 0.24) for both PageRank and CheiRank.

The overlap measures show us that when we take into account social impact in PageRank and CheiRank algorithm, we have a drastic change in the final ranking. This change is mainly due to pageviews information. We are interested in **wcpv** method, because it brings new elements to both PageRank and CheiRank.

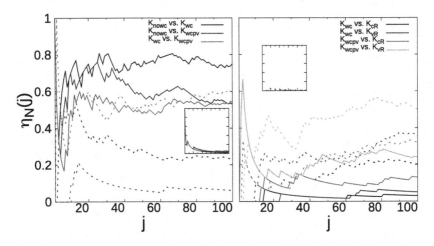

Fig. 1. Overlap η_N versus rank j for doublet of ranking lists computed from **wc**, **wcpv** and **nowc** models (left panels), from **wc**, **wcpv**, cR and vR (right panels) considering English edition. Inset plots correspond to exact overlap η_O. Solid and dotted lines are for PageRanks and CheiRanks.

Figure 2 shows the distribution of articles in the (K, K^*) plane. In case of **nowc**, articles with good PageRank have bad CheiRank and reversely. The **nowc** method doesn't represent the social interest and is robust with time. In case of both **wc** and **wcpv** methods, the articles in (K, K^*) plane related to top cR and vR have a lower K and K^* value. As we can see with bottom panel, **wcpv** method brings the top 100 of vR and cR at the left bottom corner of (K, K^*)

plane. Also the articles are much more along the diagonal. We think that using a nontrivial matrix teleportation, **wcpv** method tends to give a PageRank and a CheiRank for an article that are much more similar compared to previous methods.

2Drank algorithm, based on both K and K^*, gives a higher rank to an article that is central in term of incoming and outgoing links. In case of **wcpv** method applied on English edition of Wikipedia, the top 10 contains only one sovereign country, which is "United States". The first element is "Wikipedia", which is expected, but also missing from **nowc** 2Drank's top 10 (478th). The top 10 of **wcpv** 2Drank is far from the **nowc** and **wc** ones but closer to top vR. Articles of social interests present in this top 10 are "September 11 attacks", "Donald Trump", "Greta Thumberg" and "It Chapter Two".

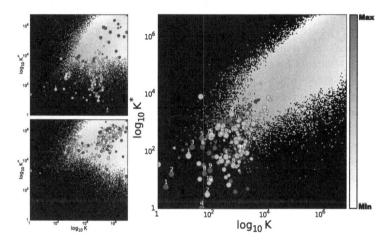

Fig. 2. Density distribution of Wikipedia articles $W(K, K^*) = dn/dKdK^*$ in case of **nowc** (bottom left), **wc** (top left) and **wcpv** (right) models. We have divided K, K^* plane into 200×200 decimal logarithmic cells. For each cell of area $dKdK^*$ we compute the corresponding articles density. We colored boxes using a decimal logscale. Blue and orange circles respectively represent the top 100 articles from vR and cR. Top 10 of both vR and cR are labeled. Circle radius is decreasing with the rank. (Color figure online)

SQL Dumps. In case of the network based on SQL data, we see a difference when we keep information related to the "Main Page". For English edition SQL, this article isn't a dangling node anymore. Wikipedia regularly suggests to users a set of articles in this main page. By removing "Main Page" related information, some interesting high PageRanked articles are missing: "Brexit", "Impeachment inquiry against Donald Trump", "2019 Southeast Asian haze" for **wcpv** PageR-

Table 4. First 10 articles obtained by PageRank with **wcpv** model from the Multilingual Wikipedia network obtained with XML and SQL dump. Source: own calculations in October 2019

Name		K_{wcpv}	K_{wc}	K_{nowc}	K_{cR}	K_{vR}
XML	continent	1	1	1	4	25
	United Kingdom	2	6	5	27	49
	endemic to	3	3	4	81	43
	France	4	2	2	161	23
	Wikipedia	5	5606	2680	12647	1
	English	6	13	3	865	203
	World War II	7	11	6	73	39
	People's Republic of China	8	9	10	77	50
	list of Queen of the South episodes	9	9296110	9299151	8992585	2
	headquarters location	10	5	8	112	70
SQL	List of Queen of the South episodes	1	152700	9402560	1	1
	International Standard Book Number	2	2	1	12422	6820
	United States of America	3	4	5	3	596
	Queen of the South	4	128752	782421	33017	32562
	Geographic coordinate system	5	1	4	27224	4161
	Wikidata	6	3	2	166656	578987
	Virtual International Authority File	7	5	6	215168	77748
	English	8	6	3	571	2227
	Library of Congress Control Number	9	8	7	171325	51054
	Japan	10	26	19	175	1992

ank and "English Wikipedia" and "QR code" for **wcpv** CheiRank. Obviously Main Page is the top 1 for **wcpv** PageRank and CheiRank.

In case of applying our method to SQL dumps, keeping "Main Page" leads to new information. Larger tables of top 100 articles are available at [6].

5.2 Multilingual Network

A simple way to build a multilingual Wikipedia network is to aggregate networks corresponding to each considered edition (11 languages). **wcpv** method weights the link $A \rightarrow B$ with clicks summed over all considered editions.

Here, we show results and discuss the case of both XML and SQL based multilingual Wikipedia network considering 11 languages: Chinese, English, French, German, Italian, Japanese, Spanish, Persian, Polish, Portuguese, Russian.

Regarding Table 4, we see that the **wcpv** PageRank's top 10 presents more differences for XML than for SQL version of the network. In context of SQL, the top 10 **wcpv** PageRank is very different from both vR and cR. Moreover, top 10 related to wcpv shows different countries, "United Kingdom" and "France" for XML and "United States of America" and "Japan" for SQL. As for the case of only English, K^*_{wcpv} is very far from K^*_{nowc} as we can see with Table 5, whether

Table 5. First 10 articles obtained by CheiRank with **wcpv** model from the Multilingual Wikipedia network obtained with XML and SQL dump. Source: own calculations in October 2019

Name		K^*_{wcpv}	K^*_{wc}	K^*_{nowc}	K_{cR}	K_{vR}
XML	Deaths in 2019	1	1	571	1029	5
	Lists of deaths by year	2	2	12	88018	20254
	It: Chapter Two	3	6	318108	6	15
	It	4	12	87569	18	30
	Once Upon a Time in Hollywood	5	9	78393	15	37
	2019 FIBA Basketball World Cup	6	3	22945	47	11
	Joker	7	16	247252	35	12
	Greta Thunberg	8	23	94431	869	7
	September 11 attacks	9	10	10665	21	13
	2019 in film	10	15	249	750	83
SQL	Deaths in 2019	1	1	388	284	914
	Queen of the South	2	7495	219841	33017	32562
	List of Queen of the South episodes	3	433980	5166620	1	1
	List of Bollywood films of 2019	4	17	12084	1640	4
	Llists of deaths by year	5	2	1231	64356	1070417
	2019	6	4	2195	821	4652
	It: Chapter Two	7	5	388604	7	360
	Once Upon a Time in Hollywood	8	7	78976	17	66
	2019 in film	9	13	2578	262	592877
	September 11 attacks	10	9	12762	23	320

it's XML or SQL. We also have fewer articles related to lists in case of the multilingual Wikipedia network. There are 2 articles related to social trend in common: "It: Chapter Two" and "Once Upon a Time in Hollywood".

The top 10 Wikipedia articles **wcpv** using 2Drank for XML multilingual network is much closer to vR compared to other rankings. SQL dumps leads to a top 10 **wcpv** 2DRank far from vR. Top 10 **wcpv** 2DRank elements with rank in $[2,4]$ are very unexpected, their corresponding rank in other lists are at least equal to 5287 (cR) and at last 605770 (vR). These articles are of social interests "Queen of the South", "Alice Braga", "La Reina del Sur". Note that Alice Braga is a main character of this TV series. In case of SQL **wcpv** top 10 2DRank, we found 4 sovereign countries "United States", "Japan", "Italy" and "Russia" which are not present in **nowc** 2DRank top 10.

Regarding Fig. 3, as for the English version of Wikipedia network, we see that CheiRanks' overlapping with other CheiRanks and with both cR and vR are lower than for PageRanks. In case of multilingual version, the exact overlap η_O is higher but still low compared to η_N values. Regarding both SQL and XML dumps, **wc** is the most similar to **nowc** with respectively $\eta_N = 0.8$ and 0.7. While PageRanks are more similar and CheiRanks related overlaps are almost

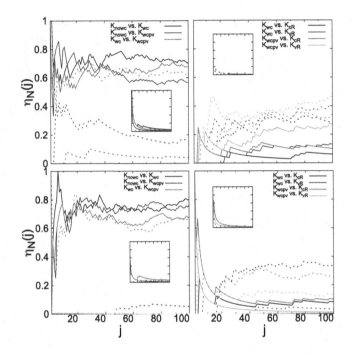

Fig. 3. Overlap η_N versus rank j for doublet of ranking lists computed from **wc**, **wcpv** and **nowc** models (left panels), from **wc**, **wcpv**, cR and vR (right panels). Inset plots correspond to exact overlap η_O. Top row is for multilingual Wikipedia network built with XML dumps and bottom row is for network built with SQL dumps. (Color figure online)

0 for SQL, XML version leads to more different tops 100 for PageRank. cR and vR have the same similarities with K_{wc} (resp. K_{wc}^*) and K_{wcpv} (resp. K_{wcpv}^*) for both XML and SQL. In case of XML K_wcpv and **vR** are much closer than K_wc and **vR**.

We present in Fig. 4 the articles' distribution in (K, K^*) plane for **wcpv** in case of both XML and SQL multilingual Wikipedia network. The articles are still organized along the diagonal line and the top 100 from cR and vR are gathered in the bottom left corner of the plots. Unlike SQL dumps, in case of XML dumps, the green circles corresponding to top 100 vR have better PageRank than articles from top 100 cR. The top 100 vR articles are more scattered in case of XML than SQL dumps.

In case of a multilingual Wikipedia network, **wcpv** method brings articles of social interest to the top of PageRank and CheiRank. Moreover, by changing the dump from SQL to XML, articles from both top vR and top cR have better PageRanks.

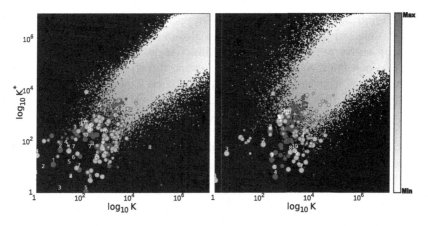

Fig. 4. Density distribution of Wikipedia articles in case of **wcpv** XML dumps (left) and SQL dumps (right). The same color code as in Fig. 3. (Color figure online)

6 Conclusion and Future Works

Wikipedia gives us plenty of free information related to a large spectrum of knowledge. Articles in this free encyclopedia are edited, checked and corrected by various users (even anonymous). A standard use of PageRank and other related ranking algorithms give a time robust ranking of articles whereas clickstream and pageviews based ranking reflects statistical social trends. In this study, we present an altered version of PageRank, CheiRank and 2DRank by using both clickstream and pageviews data together with connections between Wikipedia articles. "WikiClick Plus View" (**wcpv**) model of the Wikipedia network gives different rankings of Wikipedia articles. With **wcpv** we measured the centrality of articles in Wikipedia network regarding the actual social interest. We showed that the two type of Wikipedia dumps (XML and SQL) may give different results in final rankings. **wcpv** model gives the top PageRank articles related to actual social interest and the top CheiRank articles that are not related to lists (as they usually are) but rather related to entry point of interest. Instead of roughly aggregating individual language edition based Wikipedia networks to build a multilingual network, the use of **wcpv** method permits us to define a more realistic linkage between articles, by using clickstream data as weight and pageviews as teleportation matrix. Merging both the ephemeral aspect of social trends with time robustness of links based on historical truth, we think that this method can be used for further interesting results.

In future works, we plan to provide a deeper analysis on advantages that can give proposed novel versions of PageRank, CheiRank and 2DRank using social impact in different fields. In this work, we only used data from October of 2019. In our next works, we plan to investigate differences between results of the measures from various time periods of the content of the Wikipedia articles, pageviews and clickstream data. Based on such time dependent data, we would be able to find out the correlation between social impact and evolution of the articles on selected language versions and also on multilingual level. Such data can also be useful to analyse how the indirect social flow between articles lead to the creation of new links. Another interesting direction of the research would be to find out how the proposed measures influences quality of the content in Wikipedia. Using these measures as additional predicting variables could also improve existing prediction models of stock market moves and performances (including price of cryptocurrencies), success of the products or demand for services, as well as electoral predictions and forecasting epidemics in the specified territory.

Wikipedia is one of the representatives of wiki services. Therefore methods proposed in the paper can be also valuable for any knowledge base created using MediaWiki open source software, including corporate ones. These knowledge bases can contain information about customers, products and other business oriented content. Therefore our method can provide new information to companies allowing them to understand social trend's evolution and help to improve products recommendations for customers, as well as improve existing prediction models related to stock market moves, demand for services, elections results and others.

Acknowledgments. We thank D. Shepelyansky and J. Lages for useful discussions. This work was supported by the French "Investissements d'Avenir" program, project ISITE-BFC (contract ANR-15-IDEX-0003) and by the Bourgogne Franche-Comté Region 2017-2020 APEX project (conventions 2017Y-06426, 2017Y-06413, 2017Y-07534; see http://perso.utinam.cnrs.fr/~lages/apex/).

References

1. Alexa: Wikipedia.org traffic, demographics and competitors. https://www.alexa.com/siteinfo/wikipedia.org
2. Brin, S., Page, L.: The anatomy of a large-scale hypertextual Web search engine. Comput. Netw. ISDN Syst. **30**(1), 107–117 (1998). https://doi.org/10.1016/S0169-7552(98)00110-X
3. Coquidé, C., Lages, J., Shepelyansky, D.L.: Contagion in bitcoin networks. In: Abramowicz, W., Corchuelo, R. (eds.) Business Information Systems Workshops, pp. 208–219. Springer, Cham (2019). https://doi.org/10.1007/978-3-030-36691-9_18
4. Coquidé, C., Lages, J., Shepelyansky, D.L.: World influence and interactions of universities from Wikipedia networks. Eur. Phys. J. B **92**(1), 1–20 (2019). https://doi.org/10.1140/epjb/e2018-90532-7
5. Coquidé, C., Lages, J., Shepelyansky, D.L.: Crisis contagion in the world trade network (2020). https://arxiv.org/abs/2002.07100
6. Coquidé, C., Lewoniewski, W.: Supplementary materials. http://data.lewoniewski.info/pagerank2020/
7. ElBahrawy, A., Alessandretti, L., Baronchelli, A.: Wikipedia and digital currencies: Interplay between collective attention and market performance. arXiv preprint arXiv:1902.04517 (2019)

8. English Wikipedia: Help: Link. https://en.wikipedia.org/wiki/Help:Link
9. English Wikipedia: Help: Namespace. https://en.wikipedia.org/wiki/Wikipedia:Namespace
10. English Wikipedia: Help: Wikitext. https://en.wikipedia.org/wiki/Help:Wikitext
11. English Wikipedia: Wikipedia: Pageview statistics. https://en.wikipedia.org/wiki/Wikipedia:Pageview_statistics
12. English Wikipedia: Wikipedia: Red link. https://en.wikipedia.org/wiki/Wikipedia:Red_link
13. English Wikipedia: Wikipedia: Redirect. https://en.wikipedia.org/wiki/Help:Redirect
14. Eom, Y.H., Aragón, P., Laniado, D., Kaltenbrunner, A., Vigna, S., Shepelyansky, D.L.: Interactions of cultures and top people of wikipedia from ranking of 24 language editions. PLOS One **10**(3), e0114825 (2015). https://doi.org/10.1371/journal.pone.0114825
15. Eom, Y.-H., Frahm, K.M., Benczúr, A., Shepelyansky, D.L.: Time evolution of Wikipedia network ranking. Eur. Phys. J. B **86**(12), 1–9 (2013). https://doi.org/10.1140/epjb/e2013-40432-5
16. Ermann, L., Chepelianskii, A.D., Shepelyansky, D.L.: Toward two-dimensional search engines. J. Phys. A: Math. Theor. **45**(27), 275101 (2012). https://doi.org/10.1088/1751-8113/45/27/275101
17. Ermann, L., Frahm, K.M., Shepelyansky, D.L.: Spectral properties of Google matrix of Wikipedia and other networks. Eur. Phys. J. B **86**(5), 193 (2013). https://doi.org/10.1140/epjb/e2013-31090-8
18. Espín-Noboa, L., Lemmerich, F., Walk, S., Strohmaier, M., Musen, M.: HopRank: how semantic structure influences teleportation in pagerank (a case study on bioportal). In: The World Wide Web Conference, WWW 2019, pp. 2708–2714. ACM, New York (2019). https://doi.org/10.1145/3308558.3313487. Event-place: San Francisco, CA, USA
19. Gildersleve, P., Yasseri, T.: Inspiration, captivation, and misdirection: emergent properties in networks of online navigation. In: Cornelius, S., Coronges, K., Gonçalves, B., Sinatra, R., Vespignani, A. (eds.) Complex Networks IX, pp. 271–282. Springer, Heidelberg (2018). https://doi.org/10.1007/978-3-319-73198-8_23
20. Hickmann, K.S., Fairchild, G., Priedhorsky, R., Generous, N., Hyman, J.M., Deshpande, A., Del Valle, S.Y.: Forecasting the 2013–2014 influenza season using Wikipedia. PLoS Comput. Biol. **11**(5), e1004239 (2015)
21. Khadivi, P., Ramakrishnan, N.: Wikipedia in the tourism industry: forecasting demand and modeling usage behavior. In: Twenty-Eighth IAAI Conference (2016)
22. Kristoufek, L.: Bitcoin meets google trends and wikipedia: quantifying the relationship between phenomena of the internet era. Sci. Rep. **3**, 3415 (2013)
23. Lages, J., Patt, A., Shepelyansky, D.L.: Wikipedia ranking of world universities. Eur. Phys. J. B **89**(3), 1–12 (2016). https://doi.org/10.1140/epjb/e2016-60922-0
24. Langville, A., Meyer, C.: Google's PageRank and beyond: the science of search engine ranking. Princeton University Press, Princeton (2006)
25. Latif, M.H., Afzal, H.: Prediction of movies popularity using machine learning techniques. Int. J. Comput. Sci. Netw. Secur. (IJCSNS) **16**(8), 127 (2016)
26. Lewoniewski, W., Węcel, K., Abramowicz, W.: Multilingual ranking of Wikipedia articles with quality and popularity assessment in different topics. Computers **8**(3), 60 (2019). https://doi.org/10.3390/computers8030060
27. Lewoniewski, W.: Measures for quality assessment of articles and infoboxes in multilingual Wikipedia. In: International Conference on Business Information Systems, pp. 619–633. Springer, Heidelberg (2019)

28. Mestyán, M., Yasseri, T., Kertész, J.: Early prediction of movie box office success based on Wikipedia activity big data. PloS One **8**(8), e71226 (2013)

29. Moat, H.S., Curme, C., Avakian, A., Kenett, D.Y., Stanley, H.E., Preis, T.: Quantifying Wikipedia usage patterns before stock market moves. Sci. Rep. **3**, 1801 (2013)

30. Page, L., Brin, S., Motwani, R., Winograd, T.: The pagerank citation ranking: bringing order to the web. Technical report, Stanford InfoLab (1999)

31. Rollin, G., Lages, J., Shepelyansky, D.L.: Wikipedia network analysis of cancer interactions and world influence. PLOS One **14**(9), e0222508 (2019). https://doi. org/10.1371/journal.pone.0222508

32. Rollin, G., Lages, J., Shepelyansky, D.L.: World influence of infectious diseases from Wikipedia network analysis. IEEE Access **7**, 26073–26087 (2019). https:// doi.org/10.1109/ACCESS.2019.2899339

33. Thompson, N., Hanley, D.: Science is shaped by Wikipedia: evidence from a randomized control trial. Soc. Sci. Res. Netw. (2018)

34. Wikimedia Downloads: English Wikipedia latest database backup dumps. https:// dumps.wikimedia.org/enwiki/latest/

35. Wikimedia Meta-Wiki: Research: Wikipedia clickstream. https://meta.wikimedia. org/wiki/Research:Wikipedia_clickstream

36. Wikipedia Meta-Wiki: List of Wikipedias. https://meta.wikimedia.org/wiki/List_ of_Wikipedias

37. Yasseri, T., Bright, J.: Wikipedia traffic data and electoral prediction: towards theoretically informed models. EPJ Data Sci. **5**(1), 1–15 (2016). https://doi.org/ 10.1140/epjds/s13688-016-0083-3

38. Yu, A.Z., Ronen, S., Hu, K., Lu, T., Hidalgo, C.A.: Pantheon 1.0, a manually verified dataset of globally famous biographies. Sci. Data **3**(1), 1–16 (2016). https:// doi.org/10.1038/sdata.2015.75

39. Zhirov, A.O., Zhirov, O.V., Shepelyansky, D.L.: Two-dimensional ranking of Wikipedia articles. Eur. Phys. J. B **77**(4), 523–531 (2010). https://doi.org/10. 1140/epjb/e2010-10500-7

Smart Infrastructures

Internet-of-Things Marketplaces: State-of-the-Art and the Role of Distributed Ledger Technology

Daniel Noll[(✉)] and Rainer Alt

Leipzig University, Grimmaische Straße 12, 04109 Leipzig, Germany
`noll@wifa.uni-leipzig.de`

Abstract. The advent of the Internet-of-Things (IoT) generates increasing data with the majority being gathered for a single purpose and staying unused after serving this purpose. With IoT platforms, cross-domain use cases, combining data from different sources, become possible. Accordingly, the need for marketplaces to trade data arises. This paper examines existing IoT platforms to frame the current opportunities for an IoT marketplace. In a second step, it analyzes the potentials of Distributed Ledger Technology (DLT) regarding transaction costs and efficiencies. In doing so, a classification regarding the functional distribution of IoT marketplaces is developed.

Keywords: Internet-of-Things (IoT) · Platform · Marketplace · Distributed Ledger Technology (DLT) · Blockchain

1 Introduction

1.1 Motivation

The Internet-of-Things (IoT) is defined as the interlinking of devices connected via a network to collect and exchange data. Currently, a variety of such devices (e.g., sensors) measure various attributes and produce an amount of data every day [1]. The quick deployment of IoT devices will lead to more than 20 billion connected devices gathering data by 2020 [2]. All this data is mostly stored in so-called 'data silos'. A data silo is characterized by a closed environment with little to no sharing with outside environments, and thus, data does not contribute to any additional revenue streams for its creator. The data usually leave the silo once only to be transferred to an IoT platform for analysis and visualization for a single purpose [3]. The idea of a cross-domain data scenario is the combination of data from different single-purpose data silos to achieve value by serving an additional purpose in an additional context. The consideration of external, non-domain data promises an optimization of forecasts and an improvement of decisions [4].

For this paper, nine semi-structured interviews with CEOs, managing directors, one head of sales, one head of product management, and one consultant of small and midsize

W. Abramowicz and G. Klein (Eds.): BIS 2020, LNBIP 389, pp. 337–350, 2020.
https://doi.org/10.1007/978-3-030-53337-3_25

companies (SMC) from different branches in Germany were conducted. The interview procedure follows an unstructured interview guideline. It provides an introduction to the cross-domain data approach and its goals, according to Bär et al. [4]. It is followed by the question of whether cross-domain data projects are currently being conducted in a particular company or have already been carried out in the past? Since this question was answered in the negative in all interviews, it was followed by a survey on the reasons for these obstacles.

Although all SMCs believe cross-domain data scenarios have promising potential for their own company, none of the companies already have or had in the past a cross-domain data project. The obstacles mentioned by the interviewees are synthesized and may be organized in the following clusters: technological, economic, organizational, and legal. This paper adopts a technological and partly an economic perspective. Thus, the main questions for SMCs in these clusters are: Where is data from external domains offered? At what price is this data offered? Where are the essential analytic and presentation services offered and at which price?

One possible solution for answering these challenges is an IoT platform that offers data for sale via an integrated data marketplace [5]. Such an IoT data marketplace follows the concept of electronic marketplaces by Schmid and Lindemann [6] where data vendors from different sources and branches offer their data. Buyers of data pay the vendor and acquire the data set or data stream [7]. However, data is not the only conceivable commodity on IoT marketplaces. Services for the analysis and visualization of data could also potentially be made available via such a marketplace by various participants. Especially for complex cross-domain use cases that require multiple data from different vendors and branches as well as different analytic functionalities, an IoT marketplace would become relevant [4]. The hypothesis is that this approach – opening those data silos and trading the data – enables the creation of new business models.

The fact that the Distributed Ledger Technology (DLT) is suitable for electronic markets as an underlying technology, is acknowledged in the literature (e.g., [8, 9]). Especially the combination of DLT and IoT marketplaces holds great potential for innovation [7, 10, 11]. DLT enables micropayments and serves as a 'tool' for handling transactions on an IoT marketplace. In addition, DLT facilitates the use of smart contracts, enabling the autonomous checking of predefined events and the automatic execution of transactions. Thus, a smart contract can autonomously control, monitor, and document actions depending on digitally verifiable events. Based on the DLT, an information and value transfer (e.g., between data vendor and data purchaser) could be designed efficiently without the need for an intermediary [12]. A DLT-based platform not only promises to minimize the risk of vendor lock-ins and monopolization but also to reduce transaction costs for its users [13].

1.2 Methodology

The overall aim of this paper is to advance the understanding of the technical tools with which cross-domain data projects can be carried out in companies and which economic potentials can be realized. Figure 1 summarizes the research methodology in this paper. In the first step, the key IoT platform functionalities are identified by a scientific literature search, according to vom Brocke et al. [14]. These key functionalities are used in a second step to conduct an online market examination, according to Hileman and Rauchs [15], to provide an impression of traditional as well as DLT-based IoT platform solutions currently available on the market. Building on an established electronic market model by [6] and the dimensions of digital platforms by Blaschke et al. [16], a classification of the functional distribution of IoT marketplaces is developed. Subsequently, it is used to analyze the opportunities for synergies of IoT marketplaces and DLT as a potential enabling technology for some of the key functionalities. Thereby the paper highlights the suitability of the DLT to create value for such an IoT marketplace.

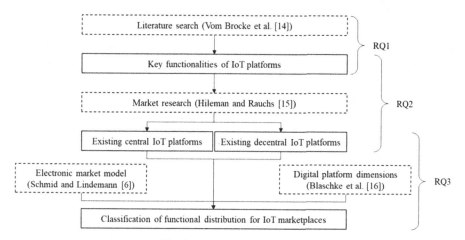

Fig. 1. Research methodology.

Three research questions (RQ) shall be formulated: (RQ1) What are the key functionalities of IoT platforms? (RQ2) Which IoT platform solutions, using which of these functionalities, currently exist for companies to carry out cross-domain IoT data projects? (RQ3) How can the functional distribution of IoT marketplaces be conceptually delineated for analyzing the potential of DLT?

The remainder of this paper is structured as follows. In Sect. 2, a brief overview of the terms IoT platform, IoT marketplace, and DLT, is presented. A literature review is conducted in Sect. 3 to identify the key functionalities of IoT platforms. The state-of-the-art based on the market examination in traditional and DLT-based IoT platforms is outlined in Sect. 4. The developed classification for the functional distribution of IoT marketplaces is presented in Sect. 5. Finally, a brief conclusion, limitations, and further research lines are provided in the last section.

2 Foundations

2.1 Internet-of-Things, IoT Platform, and IoT Marketplace

The Internet-of-Things (IoT) enables 'things' in terms of devices like sensors and everyday items, typically not considered as computers to become active participants in business, information, and social processes. This facilitates devices to generate data for communicating or interacting among themselves and with its environment independently [4].

However, the capacity of IoT devices is typically limited. Therefore, the storage and processing of the captured data are usually not taking place within the device. In a generic IoT infrastructure, the device itself is only a minor part of the IoT landscape [17]. The significant part in which data converges and the generation of added value occurs is the so-called IoT platform [18].

Gawer [19] describes platforms as foundational products, services, or technologies serving as a basis for developing supplementary products, services, or technologies. In the case of an IoT platform, it handles different hardware and software communication and authentication protocols for IoT devices and users. The primary purpose of IoT platforms may be summarized as gathering, analyzing, and visualizing data [4]. Further, an electronic platform may be seen as an electronic market, focusing on the coordination of multiple participants that aim to interact with each other [20]. Schmid and Lindemann [6] developed a reference model and elaborated two key functionalities for electronic markets: (1) Nodes are linking multiple participants, e.g., producers or vendors that offer services or products to customers. (2) An infrastructure, used for offering services and products [6, 21]. This is in accordance with the four layers of digital platforms: the digital infrastructure, the technical platform core, the ecosystem (containing the participants), and the service dimension [16]. These layers are transferable to IoT marketplaces since platform businesses by nature are intermediaries on an electronic marketplace [20]. Traditionally, there is one owner as the central provider of a platform or rather a marketplace. The platform owner is responsible for providing access to the platform (e.g., via web), governing the platform (e.g., user registration), and offering coordination tools on the platform (e.g., directories) [22]. Besides providing these basic platform services, the owner decides whether other services (e.g., analytic functionalities) are offered exclusively by the owner or also by third providers [6, 23]. In addition, data may be traded on such an electronic market. Due to the fact, every electronic marketplace is a platform, but not every platform is an electronic marketplace [20], the following refers to an IoT platform with an integrated marketplace as an IoT marketplace.

2.2 Distributed Ledger Technology

DLT describes a distributed and digital ledger that is built on a peer-to-peer (P2P)-network of independent nodes. Upon this ledger, transaction information of digital assets or digital values is immutably grouped into transaction sets. These sets are cryptographically linked to the previous transaction sets by a consensus mechanism. Leading to a chronological ordering of all transactions. As the transactions are immutable, they cannot be altered or deleted afterward – only new transactions may be added. Finally, the ledger is shared and synchronized among all nodes of the P2P-network, which increases the fault tolerance of the entire ledger [11, 24, 25].

The specifics of storing data and processing transactions in a DLT generates challenges in the IoT context [1], such as confidentiality, autonomous behavior and fault tolerance [7, 10, 26]. DLT enables the automation of complete or partial processes and services and thus improves the coordination among electronic markets. Moreover, the decentralized nature of DLT empowers the individual users by allowing them more control of data. The consolidation of data located within a network of different actors is achieved by DLT while offering properties to avoid vendor lock-ins at the same time

[8, 25]. In conclusion, the technology may advance the generic advantages of electronic markets [21] towards autonomously managed multi-sided electronic marketplaces [9].

3 Functionalities of IoT Platforms

3.1 Literature Review

Using the systematic literature search according to [14], the current state of research on reference architectures and key functionalities of IoT marketplaces and platforms, as well as possible DLT-based solutions, will be elaborated. First, the scope of the review is defined, and the conceptualization of the topic is elaborated. Based on this, the following search term pairs in English and German are derived for obtaining comprehensive and relevant search results: The first pair is 'IoT platform' and 'IoT marketplace' and the second pair is 'functionality' and 'feature'. For identifying relevant literature, common scientific databases (AIS eLibary, EBSCOhost, IEEE Xplore, and Springer Link) are searched for journals and conference proceedings with a peer-review procedure. The mentioned keywords are used in various combinations for the full-text search, and only German and English literature is considered. To adequately reflect the continuous developments in the field of IoT and DLT, the literature review is extended by a forward and backward search, according to [14]. This search procedure is based on further relevant publications of the authors as well as on the sources used in the articles found. Publications without a peer-review procedure are also considered here to elaborate on the latest state. After the literature search for each database and after removing duplicates, a set of 41 selected papers is analyzed. In total, 11 relevant papers are identified, which are used for synthesizing key functionalities of IoT platforms in the following.

3.2 Synthesized Functionalities

Based on the conducted literature review, various proposals for overviews of functionalities of IoT platforms (e.g., [1, 3]) could be found. Although the platforms are partly equipped with similar functionalities, there are different technological approaches and terminologies [27]. However, the descriptions of these functionalities in the literature partly focus on single functionalities in detail. As shown in Table 1, key functionalities of IoT platforms have been synthesized from the literature to compensate these shortcomings. They will be used in the following examination of this paper.

Apart from the basic platform services mentioned in 2.1 (access, governance, and coordination, which are provided as cross-layer functionalities), these key functionalities are:

1. Connectivity: Includes identification management and dedicated device management, supporting users to deploy and configure heterogeneous IoT devices for collecting data [3, 28].
2. Data Storage: Data may be stored at a cloud server provided by the platform provider, on local databases/servers by the user or via an internal interface on a dedicated DLT infrastructure [7, 27].

3. Marketplace: A platform-integrated marketplace for data, analytics, or presentation services of third-party service providers [6, 20].
4. Analytics: Tools including edge analytics and machine learning skills, like cognitive, anomaly, and predictive analytics capabilities to extract insights from IoT data [28, 29]. Edge analytic describes a first simple data analysis taking place at the point of gathering (e.g., sensor device). After that, only a pre-selection of the data is transferred to the platform for further analysis [17]. Cognitive analytics include capabilities like natural language processing, text mining, or video and image recognition [29]. Anomaly or stream analytics describes a low latency analysis in real-time for anomaly detection [28]. Predictive analytics generate forecasts using historical data. In particular, this is also available in real-time [30].
5. Presentation: Empowers to create and deploy an IoT application or smart service rapidly. Another form are visualizations, e.g., dashboards, diagrams, or graphs [4, 27].

Table 1. Overview of IoT platform key functionalities

Functionality	Functionality characteristics			
Presentation	Visualization	Smart service		IoT application
Analytics	Edge	Cognitive	Anomaly	Predictive
Marketplace	Data	Analytic functionality as a service		Presentation functionality as a service
Data storage	Local database	Cloud server		Distributed ledger
Connectivity	Device management	Deployment configuration		Identification management

4   State-of-the-Art in IoT Platforms

The previous literature search identified papers that focused on individual platforms (e.g., IBM Watson IoT), on certain IoT platform aspects (e.g., edge analytics [17]) or on individual use cases, but not on the overall structure of IoT platform concepts. To obtain this information, an online market research to identify current IoT platform solutions is conducted according to [15]. To identify smaller IoT platform or IoT marketplace projects, a search for whitepapers, technical documentations, and postings in non-scientific media like branch-specific websites (e.g., cryptoslate.com) or blogs (e.g., medium.com) was conducted. The key functionalities and its characteristics elaborated in Table 1 serve as the underlying research design for the market research. From an architectural perspective, there are two types of IoT platform solutions – centralized and decentralized – which are examined in the market research separately.

4.1   Traditional IoT Platforms as Centralized Solutions

A variety of different IoT platforms have been created for a wide range of use cases by open source communities (e.g., FIWARE, OpenMTC) as well as commercial companies (e.g., IBM, Siemens). This analysis is limited to commercial providers for serving better comparability. Table 2 summarizes the key functionalities for selected commercial IoT platforms available on the market. Since the field of IoT is quite complex and fast-moving, it cannot be excluded that individual IoT platforms either are missing or are outdated yet. The set of examined IoT platforms in this paper has to be considered as a snapshot of current market insights.

Table 2. Comparison of traditional IoT platforms currently available on the market

Platform name	Connectivity	Serial interface to DLT	Cloud-server	Local database	Marketplace	Edge analytics	Cognitive analytics	Anomaly analytics	Predictive analytics	Presentation functionalities
Atos Codex IoT	●	○	●	●	○	○	○	○	●	○
AWS IoT Core	●	○	●	○	○	●	●	●	●	●
Bosch IoT Suite	●	○	●	○	○	○	○	●	○	●
Cumulocity IoT	●	○	●	●	○	●	○	●	●	●
GE Predix	●	○	●	○	○	●	○	●	●	●
Google IoT Core	●	○	●	○	○	●	●	●	●	●
HPE Universal IoT	●	○	●	●	○	●	○	●	●	●
IBM Watson IoT	●	Hyperledger	●	●	○	●	●	●	●	●
Microsoft Azure IoT	●	Azure Blockchain	●	●	○	●	○	●	●	●
Oracle IoT Cloud	●	○	●	○	○	○	○	●	●	●
PTC ThingWorx	●	○	●	●	○	○	○	●	●	●
Relayr IoT Middleware	●	○	●	●	○	●	○	●	●	●
SalesForce IoT	●	○	●	○	○	○	●	●	●	●
SAP Leonardo IoT	●	Hyperledger Fabric, MultiChain, Quorum	●	●	○	●	○	●	●	●
Siemens MindSphere	●	○	●	●	○	●[A]	●*	●*	●*	●

● - yes | ○ - no | [A] - Analytics services | * only in combination with AWS or Azure services

All examined platforms provide connectivity functionality, and all but one platform offer presentation functionalities. Both seem to be standard functionalities of traditional IoT platforms. This is possibly since these functionalities affecting the end-user directly and, therefore, are suitable sales arguments. The analytic functionalities are partly given. Only Siemens MindSphere offers the possibility to use third-party analytic functionalities. However, as MindSphere lacks in-house analytic functionalities, it offers AWS and Microsoft Azure services. None of the further platforms examined feature an integrated marketplace. Moreover, all platforms relied on data storage in the cloud, and three providers even had a serial interface to a DLT framework – namely IBM Watson with Hyperledger, Microsoft Azure with the Ethereum-based Azure Blockchain and SAP Leonardo with Hyperledger Fabric, MultiChain or Quorum.

Overall, the key functionalities suggested in this paper may be regarded as provisionally validated in order to accurately map and compare available IoT platform solutions in their functionalities.

4.2 DLT-Based IoT Platforms as Decentralized Solutions

The majority of the reviewed traditional platforms rely on conventional database architectures and are centralized operated by one provider. This evokes the question of whether a market of IoT platforms using DLT and pursuing functional distribution is already existing. Two main fields of application may be distinguished for DLT: First, it is used for processing purchase transactions on the IoT marketplaces and the autonomous operation of this marketplace. Second, it may be used for data verification and storage. Accordingly, the key functionalities in Table 1 may also be applied to DLT-based platforms, with a small change by introducing the distinction between on-chain and off-chain data storage locations. There are two ways to connect or integrate IoT platforms with DLT. (1) As mentioned above, a distributed ledger is integrated into a traditional IoT platform via a serial interface, e.g., IBM Watson with Hyperledger. (2) Another solution is a standalone DLT framework providing a completely DLT-based IoT platform, e.g., IOTA.

Although a small number of IoT platforms already implement DLT as the underlying architecture, this industry is continuously progressing. These platforms partly address specific niches, e.g., some operate open platforms for storing, sharing, and trading sensor data [10]. Table 3 gives an impression of a sample of commercial DLT-based IoT platforms currently available. Since it is a fast-moving and fast-developing industry, this may only be seen as a current snapshot of market insights without claiming completeness.

It may be viewed that DLT-based IoT platforms offer more often the possibility of marketplaces, while analytic and presentation functionalities are less often represented compared to traditional platforms. Mainly the marketplaces are used for trading data. Streamr additionally offers presentation functionalities as a service in the form of a visualization dashboard on its marketplace. Analytic services are not offered as a service on a marketplace within the examined platforms. All IoT marketplaces used DLT for processing purchase transactions. IOTA and Datum are the only IoT platforms that offer the possibility of on-chain storage of the sensor data. The remainder of the platforms examined uses DLT to store markers or hashes for verifying the sensor data. An initial validation of the elaborated functionalities in Table 1 may be assumed by this market examination, as all the solutions examined are appropriately described.

Table 3. Comparison of DLT-based IoT platforms currently available on the market

Platform name	Connectivity	DLT framework	On-chain	Off-chain: Local database	Off-chain: Cloud-server	Marketplace	Edge analytics	Machine Learning: Predictive analytics	Machine Learning: Anomaly analytics	Machine Learning: Cognitive analytics	Presentation functionalities
Atonomi	●	Ethereum	○	●	●	●D	●	●	●	○	○
Cyber Physical Chain	○	Ethereum	○	●	●	●D	○	●	●	●	○
Databroker	○	Ethereum	○	●	●	●D	○	○	○	○	○
Datum	●	Ethereum	●	●	●	●D	○	○	○	○	○
Flowchain	●	Ethereum	○	●	●	○	○	○	○	○	○
IOTA	●	IOTA	●	●	●	●D	○	○	○	○	●
Machine eXchange Coin	●	Ethereum	○	●	●	●D	○	○	○	○	○
Orbis Mesh	●	NEO	○	●	●	○	○	○	○	○	●
SDChain	○	Ethereum	○	●	●	●D	○	○	○	○	●
Streamr	●	Ethereum	○	●	●	●D, P	○	●	●	●	●
Weeve	●	Ethereum, Hyperledger, IOTA	○	●	●	●D	○	●	●	○	●
XBR Network	○	Ethereum	○	●	●	●D	○	○	○	○	○

● - yes | ○ - no | D - Data | P - Presentation services

5 Classifying Functional Distribution

Based on the electronic market reference model [6], the architecture dimensions of digital platforms [16], and the setting of IoT cross-domain scenarios [4], the typical pipeline of a multi-sided IoT marketplace for a cross-domain sensor data scenario is shown in Fig. 2.

It starts with generating and gathering sensor data. By using the connectivity functionality, a sensor owner records its sensor and transfers it on the IoT platform. First, a pre-processing or filtering of the sensor data is executed. After that step, the prepared data is forwarded to a specific storage location, and the sensor owner, as an end-user, can do both, use its data for own purposes and offer it on the platform-integrated marketplace. Service providers can also offer analytic and presentation functionalities as a service on the marketplace. As an end-user, a customer or smart service may obtain both data and functionality services on the marketplace and, therefore, within the platform. The advantage for both service providers and customers is the fact that the functionality service is offered at the point where the actual need of the customer arises. Each participant interacting on the platform uses the basic platform services.

Functional distribution is possible in this outlined scenario in analogy to other fields such as business process outsourcing [31], enterprise resource planning systems [32], or service lifecycle management [33]. Functional distribution in the context of IoT marketplaces denotes single functionalities that may be distributed horizontally or vertically.

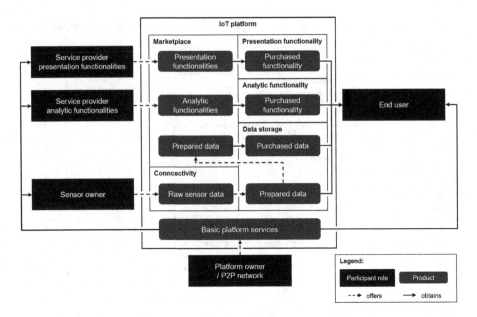

Fig. 2. The setting of an IoT marketplace with roles and products

Thus, they are either provided by a central hierarchical party or by one or more external parties. Basic platform services may also be considered in the functional distribution.

Building on the results of this paper so far enables to create a morphological box for classifying the functional distribution of different IoT platform architectures (see Table 4). Depending on the degree of functional distribution, two elementary business models and hence, two basic architecture designs exist for an IoT marketplace. As mentioned above, a platform may be designed either centralized or decentralized. Based on the results of this paper, three additional subcategories can be identified for IoT platforms: Traditional IoT platforms, IoT platforms with a serial interface to a distributed ledger, and DLT-based IoT platforms. In a centralized architecture, the provider assumes all processes – namely basic platform services, operation of the marketplace, and offering analytics and presentation services. In a decentralized architecture, the provider obtains at least some services by external providers or is entirely obsolete due to external providers or the usage of DLT. Basic platform services and the operation of the marketplace may be distributed via the P2P-network by using DLT. Transactions are processed by using smart contracts, which automate payment processing, data verification, and releasing access to data and services. A trusted third-party is no longer necessary in a fully decentralized IoT platform. Except for providing basic platform services, each participant can take every role – even more than one at the same time.

The developed classification in Table 4 can be used to react to different requirements for different application cases and to carry out a corresponding functional distribution. Thus, it supports a first high-level recommendation about functional distribution and platform architecture design in the information system engineering, in IT departments and for business executives. The following example illustrates the classification process

Table 4. Classifying functional distribution of IoT platforms

	IoT platform		
	Centralized		Decentralized
	Traditional	Serial interface to DLT	DLT-based
Presentation functionalities	platform provider		various service providers
Analytic functionalities	platform provider		various service providers
Marketplace	platform provider		distributed into P2P-network
Data storage location	possessed	platform provider	distributed into P2P-network
Device access	possessed		various third parties
Basic platform services	platform provider		distributed into P2P-network

for a cross-domain data scenario using a decentralized IoT marketplace. A farmer owns and operates sensors to measure the soil moisture of his fields (*device access: possessed*). He captures the data in the platform and enriches it with weather data purchased at the IoT marketplace (*device access: various third parties*). In the next step, he uses services from various providers for analysis and visualization purposes, obtained from the marketplace (*analytic and presentation functionalities: various service providers*). At the same time, he stores his own gathered sensor data locally and offers it at the marketplace for sale, e.g., for research institutes (*device access: various third parties and Data storage location: distributed in P2P-network*). Thus, the farmer is at the same time a sensor owner, a data vendor, and a customer purchasing data and services. Besides, he does not have to fear any vendor lock-ins because he is able to change his current vendors and providers platform-internally at any time. By decentralizing and shifting the provision of basic platform services into the P2P-network, the intermediary 'platform provider' becomes obsolete (*basic platform services: distributed into P2P-network*). All participants using the marketplace benefit from this as it prevents the risk of vendor lock-ins, and a reduction in transaction costs may be anticipated [9]. For cross-domain use cases, it can be postulated a DLT-based IoT marketplace is highly recommendable.

6 Conclusion

In order to answer the first research question (RQ1), a systematic literature review was conducted to derive key functionalities of IoT platforms, which served as the basis for the market research to answer RQ2. This market research yields the state-of-the-art regarding currently existing IoT platforms, both traditional and DLT-based. It provides an impression of available solutions and may serve businesses in conducting cross-domain IoT data projects. For answering RQ3, a classification was developed to point out functional distribution within IoT marketplaces. The classification shows that functional distribution from centralized to decentralized is possible across all the elaborated key functionalities of an IoT marketplace. Selective use of DLT may also be reasonably evaluated using the classification developed. The role of DLT may be seen as an enabler of IoT marketplaces for further developments. By supporting micropayments, smart contracts, and the establishment of trust across a network of untrusting participants, DLT allows the operation of an entirely autonomously managed electronic marketplace. It thus enables the development of new business models like the sale of data previously disappearing into data silos, as well as the possibilities for third-party service providers of analytics and presentation functionalities to offer their services directly within the platform – the point where the customer's need for such services occurs.

This paper is not without limitations, and there are three to mention. (1) First, DLT and IoT are still growing exploration fields and are still in progress. Consequently, it has to be emphasized that integrating IoT and DLT introduces new complexity, vulnerabilities, and hazards into platform architectures. (2) Some use cases handle private or sensitive data from closed ecosystems (e.g., sensitive production and machine data of factory environments), which are intended for internal use only. If no intention to share data with third-parties exists, a traditional IoT platform is recommendable, while storing data distributed in a P2P-network is not target-oriented here. It may even be legally questionable under certain circumstances because the data owner might demand full control over the data and its processing at any time. (3) The developed classification needs to be proved by qualitative expert interviews in a further research step.

Three main findings of the paper can be summarized. (1) The paper shows that although many of the functionalities of IoT platforms are still centrally designed today, these functionalities may be designed in a decentralized way as well. Depending on the use case, new business models and approaches may emerge through this functional distribution. (2) Thereby DLT is suitable as an infrastructure technology, especially for marketplace functionalities. The examination shows that IoT platforms also use DLT for data storage and immutable data verification. For developers, this is a key aspect to consider when designing IoT platforms. (3) All in all, IoT marketplaces enable new approaches such as cross-domain sensor data scenarios by making available previously imprisoned data. This emerging business model promises to be profitable and offers completely new opportunities for companies and other organizations in IoT.

The contributions, based on these findings, to the scientific community and practitioners, form a triad. It consists of (1) an overview of state-of-the-art traditional and DLT-based IoT platform solutions, (2) key functionalities of IoT platforms, which have been elaborated in the form of a morphological box, and (3) the developed classification for describing functional distribution within IoT marketplaces.

These findings may prove helpful to address the technical obstacles elaborated by expert interviews. For practitioners, IoT marketplaces are presented as a part of the solution, and a classification for rating their functional distribution is provided. For the realization of DLT-based IoT marketplaces for data, analytic and presentation functionalities, few research and findings exist so far. Thus, it will be a required field of research and practical application in the years to come. For the scientific community, this paper forms a starting point for further research directions.

Acknowledgment. *The authors received financial support of this research by the Development Bank of Saxony (SAB) and the European Social Fund (ESF) within the project S2DES.*

References

1. Gubbi, J., Buyya, R., Marusic, S., Palaniswami, M.: Internet of Things (IoT): a vision, architectural elements, and future directions. Future Gener. Comput. Syst. **29**(7), 1645–1660 (2013)
2. Gartner: Gartner Says 8.4 Billion Connected "Things" Will Be in Use in 2017, Up 31 Percent From 2016
3. Miorandi, D., Sicari, S., de Pellegrini, F., Chlamtac, I.: Internet of Things: vision, applications and research challenges. Ad Hoc Netw. **10**, 1497–1516 (2012). https://doi.org/10.1016/j.adhoc.2012.02.016
4. Bär, S., Reinhold, O., Alt, R.: The role of cross-domain use cases in IoT – a case analysis. In: Proceedings of the 52nd Hawaii International Conference on System Sciences, pp. 390–399 (2019)
5. Andersen, J.V., Khan, D.S.: Value flows in IoT ecosystems: towards an IoT data business model. In: Proceedings of the 27th European Conference on Information Systems (2019)
6. Schmid, B.F., Lindemann, M.A.: Elements of a reference model for electronic markets. In: Proceedings of the 31st Hawaii International Conference on System Sciences, vol. 4, pp. 193–201 (1998). https://doi.org/10.1109/hicss.1998.655275
7. Panarello, A., Tapas, N., Merlino, G., Longo, F., Puliafito, A.: Blockchain and IoT integration: a systematic survey. Sensors **18**, 2575 (2018). https://doi.org/10.3390/s18082575
8. Subramanian, H.: Decentralized blockchain-based electronic marketplaces. Commun. ACM **61**, 78–84 (2017). https://doi.org/10.1145/3158333
9. Alt, R.: Electronic markets and current general research. Electron. Markets **28**(2), 123–128 (2018). https://doi.org/10.1007/s12525-018-0299-0
10. Elsden, C., Manohar, A., Briggs, J., Harding, M., Speed, C., Vines, J.: Making sense of blockchain applications: a typology for HCI. In: Proceedings of the 2018 Conference on Human Factors in Computing Systems, pp. 1–14 (2018). https://doi.org/10.1145/3173574.3174032
11. Florea, B.C.: Blockchain and Internet of Things data provider for smart applications. In: 7th Mediterranean Conference on Embedded Computing, pp. 1–4. IEEE (2018)
12. Beck, R., Stenum Czepluch, J., Lollike, N., Malone, S.: Blockchain – the gateway to trust-free cryptographic transactions. In: Proceedings of the 24th European Conference on Information Systems (2016)
13. Alt, R.: Electronic markets on transaction costs. Electron. Markets **27**(2), 1–5 (2017). https://doi.org/10.1007/s12525-017-0273-2
14. Vom Brocke, J., Simons, A., Riemer, K., Niehaves, B., Plattfaut, R., Cleven, A.: Standing on the shoulders of giants: challenges and recommendations of literature search in information systems research. CAIS **37**(1), 9 (2015). https://doi.org/10.17705/1CAIS.03709
15. Hileman, G., Rauchs, M.: Global Blockchain Benchmarking Study. Cambridge University Press, Cambridge (2017)
16. Blaschke, M., Haki, K., Stephan, A., Robert, W.: Taxonomy of digital platforms: a platform architecture perspective. In: Proceedings of the 14th International Conference on Wirtschaftsinformatik, pp. 572–586 (2019)
17. Singh, S.: Optimize cloud computations using edge computing. In: Proceedings of the 2017 International Conference on Big Data, IoT and Data Science, pp. 49–53 (2017)
18. Tan, L., Wang, N.: Future internet: the Internet of Things. In: 3rd International Conference on Advanced Computer Theory and Engineering, vol. 5, pp. 376–380 (2010)
19. Gawer, A.: Bridging differing perspectives on technological platforms: toward an integrative framework. Res. Policy **43**, 1239–1249 (2014)

20. Alt, R., Zimmermann, H.-D.: Electronic markets on platform competition. Electron. Markets **29**(2), 143–149 (2019). https://doi.org/10.1007/s12525-019-00353-y

21. Bakos, Y.: The emerging role of electronic marketplaces on the internet. Commun. ACM **41**, 35–42 (1998). https://doi.org/10.1145/280324.280330

22. Daiberl, C., Oks, S., Roth, A., Möslein, K., Alter, S.: Design principles for establishing a multi-sided open innovation platform: lessons learned from an action research study in the medical technology industry. Electron. Markets **29**(4), 711–728 (2019). https://doi.org/10.1007/s12525-018-0325-2

23. Parker, G., Van Alstyne, M., Choudary, S.P.: Platform Revolution: How Networked Markets are Transforming the Economy – and How to Make Them Work for You. W.W. Norton & Company, New York (2016)

24. Beck, R., Müller-Bloch, C.: Blockchain as radical innovation: a framework for engaging with distributed ledgers. In: Proceedings of the 50th Hawaii International Conference on System Sciences, pp. 5390–5399 (2017)

25. Glaser, F.: Pervasive decentralisation of digital infrastructures: a framework for blockchain enabled system and use case analysis. In: Proceedings of the 50th Hawaii International Conference on System Science, pp. 1543–1552 (2017)

26. Reyna, A., Martín, C., Chen, J., Soler, E., Díaz, M.: On blockchain and its integration with IoT. Challenges and opportunities. Future Gener. Comput. Syst. **88**, 173–190 (2018). https://doi.org/10.1016/j.future.2018.05.046

27. Guth, J., Breitenbucher, U., Falkenthal, M., Leymann, F., Reinfurt, L.: Comparison of IoT platform architectures: a field study based on a reference architecture. In: 2016 Cloudification of the Internet of Things, pp. 1–6. IEEE (2016), https://doi.org/10.1109/ciot.2016.7872918

28. Khan, R., Khan, S.U., Zaheer, R., Khan, S.: Future internet: the Internet of Things architecture, possible applications and key challenges. In: Proceedings of the 10th International Conference on Frontiers of Information Technology, pp. 257–260 (2012). https://doi.org/10.1109/fit.2012.53

29. Alpaydin, E.: Machine Learning. The New AI. The MIT Press, Cambridge (2016)

30. Shmueli, G., Koppius, O.: Predictive analytics in information systems research. MIS Q. **35**, 553–572 (2011)

31. Mani, D., Barua, A., Whinston, A.B.: An empirical analysis of the contractual and information structures of business process outsourcing relationships. Inf. Syst. Res. **23**, 618–634 (2012). https://doi.org/10.1287/isre.1110.0374

32. Gattiker, T.F., Goodhue, D.L.: What happens after ERP implementation: understanding the impact of interdependence and differentiation on plant-level outcomes. MIS Q. **29**, 559–585 (2005)

33. Fischbach, M., Puschmann, T., Alt, R.: Service lifecycle management. Bus. Inf. Syst. Eng. **5**, 45–49 (2013). https://doi.org/10.1007/s12599-012-0241-5

Complementor Satisfaction with Boundary Resources in IIoT Ecosystems

Dimitri Petrik$^{(\boxtimes)}$ and Georg Herzwurm

Graduate School of Excellence Advanced Manufacturing Engineering (GSaME),
University of Stuttgart, Nobelstraße 12, 70569 Stuttgart, Germany
{dimitri.petrik,georg.herzwurm}@gsame.uni-stuttgart.de

Abstract. Fostering partnerships and generativity, the Industrial Internet of Things (IIoT) platforms change the way of value creation, enabling platform-based ecosystems. Platform Boundary Resources (BR) provide a recognized concept to foster third-party innovation, enabling various hardware- and software-developing companies to use the functionalities of the platform. Despite the high importance of BR to open the platform and foster the innovation, their design and quality aspects, as well as their influence on the satisfaction of complementors, remain under-researched. To understand how complementors value different BR in IIoT ecosystems, we conducted a complementor satisfaction survey, addressing developers in an IIoT ecosystem, who utilize various BR. The study is based on the case of the IIoT platform MindSphere, developed by Siemens. Our findings include the calculation of the weighted complementor satisfaction with BR. Adding the complementor satisfaction perspective to BR research, our study shows how to apply a structured quality improvement to the BR concept and supports platform providers, highlighting which BR should be in focus during the ecosystem development through the quality improvement of prioritized BR.

Keywords: Industrial IoT · IIoT platforms · IIoT ecosystems · Boundary Resources · Quality Management · Complementor satisfaction · Satisfaction survey

1 Introduction

Digital platforms in the context of the Industrial Internet of Things (IIoT) and the fourth industrial revolution are used to support software and non-software companies to create added value through customer-specific digital solutions. The integration of information and communication technologies in machines, plants, and manufacturing processes requires interoperable, and flexible digital infrastructure, usually offered by cloud-based digital platforms [1, 2]. Such platforms build the core technology to launch new business models, and exploit additional revenue streams through digital services [2, 3], modifying the functionality of the previously offered hardware products, and increasing their value through specific end-to-end (E2E) solutions [4, 5]. IIoT platforms provide interoperability, and foster the generativity, recombining previously unrelated components, overall

© Springer Nature Switzerland AG 2020
W. Abramowicz and G. Klein (Eds.): BIS 2020, LNBIP 389, pp. 351–366, 2020.
https://doi.org/10.1007/978-3-030-53337-3_26

resulting in the value creation [4, 6]. The complexity constraints of the E2E solutions in IIoT, require platform providers to collaborate with third-party companies that offer niche solutions or have advantages as local specialists. From the perspective of a platform provider, third-party companies are complementors, and essential partners, because they increase the value of the used platform. Such an understanding of the partners also facilitates the development of IIoT ecosystems.

IIoT ecosystems usually include numerous stakeholder types, such as mechanical engineering companies, toolmakers, automation companies, software developers, and others [7]. This variety increases the complexity for platform providers to control such ecosystems, and create competitive advantages for the participating companies, known as "ecosystem health" [8–10]. For that reason, platform providers are required to design technical and organizational routines attractively to enhance the launch and the growth of IIoT ecosystems [11, 12]. Boundary Resources (BR) offer a suitable concept for this purpose (see Sect. 2) [13–15]. Due to his keystone advantage, a platform provider is in a power position to use BR as determinants to enable third parties, and end customers to use the functionalities of the platform, simultaneously controlling the output, provided by the complementary third parties [14, 16–18]. The quality of BR can either hinder or support the use of the platform, ultimately affecting the complementary innovation [19]. Thus, we argue that the quality of BR is important, for the satisfaction of complementors, who already use the platform, and potential complementors, who consider the quality of BR in their decision which platform to use. High fragmentation of the IIoT platform market, and the variety of market-ready IIoT platforms [20], indicate the establishment of platform-based ecosystems remaining a challenge for platform providers. Prior research on BR and platform evolvement mechanisms mainly focuses on business-to-consumer (B2C) domains [14, 18, 22]. Only one preliminary paper identified an even larger number of technical, and non-technical BRs used in IIoT, compared to business-to-consumer B2C markets [21]. However, the question remains mostly unanswered, which BRs from the platform provider perspective are essential for the satisfaction of the complementors and which are primarily to be improved.

Motivated by this lack of scientific literature, our research goal is to find out *how BR are valued, and prioritized by the complementors in IIoT ecosystems based upon their perceived importance, and satisfaction.* To achieve this goal, we applied the customer satisfaction survey and surveyed 19 technical experts in complementary companies about their experience using BR around the MindSphere IIoT platform, provided by Siemens. We have selected MindSphere from the many IIoT platforms available on the market [20] due to its openly communicated provision of BR, its growing complementor numbers [12], and its maturity, as it has been developed since November 2015. This paper is a continuation of two papers on the use of BR in IIoT, using the example of Mind-Sphere IIoT ecosystem as well. Firstly, prior work identified a list of 15 existing BR, which were initiated by Siemens to support the MindSphere ecosystem [21]. Secondly, prior work identified a strong correlation between the provision and maintenance of BR, and the established business partnerships in the MindSphere ecosystem, uncovering the ecosystem development dynamics [12]. In the current study, we asked complementors to rate each one of the previously identified 15 BR, aiming to capture their satisfaction during the use. Following the idea that BR quality is linked with the complementor satisfaction, and results in an increased attractiveness of the IIoT ecosystem, we combined the complementor engagement perspective [23] with the BR concept, simultaneously addressing the lack of empirical data for platform design research [15]. As a part of ongoing research, this study provides first results on how BR are perceived and valued for their usefulness and design quality by the complementors. Using rankings and building an action diagram, we show important steps for the planning of a structured quality improvement process for BR to increase the satisfaction of complementors (i.e., as users of BR). The results support IIoT platform providers, creating transparency in the variety of existing BR, and proposing how to foster the attractive design of platform-based IIoT ecosystems.

2 Related Concepts and Foundation

Our study grounds in differing research streams, including platforms, software ecosystems, boundary resources, and stakeholder engagement, and embeds them into the complex enterprise setting of IIoT. The existing IIoT reference architectures classify IIoT platforms as cloud-based integration middleware layer to connect digitized assets (e.g., machine tools, industrial robots, or single components), and process their received data, monitoring the physical processes, and parameters in platform-based applications [4, 24]. The platform concept used in this paper mainly emphasizes their technical and organizational roles. Thus, platforms form an extensible codebase of a software-based system that provides core functionality and interfaces, shared by interoperable apps, and heterogeneous hardware assets [25, 26]. The definition of software ecosystems adds the organizational aspect, whereby companies working as a unit, and interacting in a shared market for software, and services, and being interconnected by a unifying technology (e.g., IIoT platform) [27, 28], to fuel the collaboration and create E2E services. Collaborative value creation through E2E solutions depends on the availability of the open interfaces, and the possibilities for the third party to utilize the platform [29]. The interfaces, their legal conditions of availability, and comprehension aids are included in the concept of BR, which aims to open up the platform technology [19]. Technical (e.g., APIs), and social (e.g., documentation) BR are generic artifacts and may be used by various ecosystem partners for individual purposes, simultaneously allowing the platform provider to stay in a formal control of the core of the ecosystem and to adjust the platform interfaces. The design of BR helps to create a delicate balance between the generativity, and the prevention of fraud, thus launching the governance mechanisms through the BR [13, 14, 16]. Ungoverned platform-based ecosystems have a negative impact on the attractiveness, preventing the complementors from joining the ecosystem [23]. Furthermore, BR are used to foster knowledge transfer about the platform to complementors, and end customers [17, 18]. If well designed, they may also create value-driven lock-in effects for complementors, resulting in a positively perceived governance mechanism [22]. Previous work conceptualizes BR into three distinctive clusters and assigns them to an onion model [21, 30]. Application BR (**ABR**) include program resources, which enable software extensions, and applications, and the hardware assets to interact with the platform. Development BR (**DBR**) include program resources, which support the third-party developers in programming, and testing the platform-based applications or integrating or integrating the hardware assets with the IIoT platform. Social BR (**SBR**) include measures, and tools to transfer information about the platform to third parties and enable them to utilize the platform functions, and the offered technical BR. Using these three clusters to discover BR used in IIoT we determined the following 15 BR, implemented by Siemens to open up MindSphere [21]:

Table 1. Frequently used BR in IIoT ecosystems.

BR classes	Boundary resources	Ecosystem-related aspects
ABR	APIs	Application programming interfaces (APIs) are HTTP-Methods, usually utilizing the Representational State Transfer (REST) style. They enable third-party applications and other enterprise software to interact with the platform core
	Connectivity Libraries	Connectivity libraries enable the connection between the platform and various hardware devices (such as PLCs). In the case of MindSphere, third-party devices and Siemens devices are both supported, lowering the connectivity efforts with the platform
	Support of the Machine Protocols	Support of proprietary and open protocols is necessary to connect various industrial assets. Due to the variety of existing protocols in industrial assets, a platform should provide support for multiple protocols to allow the connectivity with reasonable efforts, and complexity. Exemplary open protocols are OPC Unified Architecture (OPC UA), MTConnect, ModbusTCP, ModbusRTU, MQTT. S7 is an exemplary proprietary protocol, supported by MindSphere
	Support of Different Cloud Infrastructures	Although abstracted away from the third-party developer, support of multiple cloud infrastructures is important to fulfill legal requirements (e.g., Alibaba in China) or overcome the geographical constraints. Besides, supporting the same cloud infrastructure by the platform, as the one already used by the end customer, facilitates its adoption. IIoT platforms are usually cloud-based. Hence differing infrastructures may cause functional constraints for the operated software (e.g., specific functionalities of the platform, resulting in differing API versions)
	DevOps Metrics	Metrics allow third-parties to monitor the behavior of the application during its usage (such as failure rates of the traffic caused by the application) and improve the operation of applications
DBR	SDKs	Software Development Kits (SDKs) support the development of third-party applications and reduce the efforts to create platform-based applications. In the case of MindSphere, Siemens officially offers SDKs for programming languages such as Java, Node.JS, and Python

(*continued*)

Table 1. (*continued*)

BR classes	Boundary resources	Ecosystem-related aspects
	Cloud Foundry	CloudFoundry integrates the container idea into the MindSphere, and helps third-parties to standardize the deployment process routines, and submit the applications to the cloud in a simple way
	App Store	The platform-related app store supports the distribution of applications to make them available for the entire ecosystem. However, most of the IIoT use cases are very specific, so the app store is only relevant for rather generic applications
	MDD	Model-driven development (MDD) enables specialist departments to create applications without writing code. The integration of the low-code platform Mendix enabled MDD for MindSphere
SBR	Developer Portal + Forum	The developer portal fosters knowledge about the technical components and building blocks (i.e., ABR + DBR), which are supported by the platform. The aligning developer forum is an additional source of knowledge, simultaneously fostering the communication in and with the ecosystem
	Partner Programs	Partner programs offer various benefits to the participants. Siemens operates two partner programs for MindSphere. The three-tier partner program provides access to development resources, and grants free trainings and discounts. The user organization "MindSphere World" allows third-parties to discuss the specific requirements, and to influence which functionalities should be included in the next releases, thus transforming the user organization in a standardized channel for requirements discussions and submissions
	Onsite Demonstrators	Siemens operates onsite demonstrators around the world, such as excellence factories, and numerous application centers, to introduce how they use MindSphere in its own factories so that the visitors may discover potential platform-based use cases
	Events	Different types of events fuel the ecosystem development related to an IIoT platform. Platform providers usually exhibit at industrial fairs to attract new customers. They also conduct hackathons, bridging the distance between the software companies and the industrial companies. Platform providers also host developer meet-ups to intensify the information flows to the ecosystem participants
	Trainings + Workshops	Another way to transfer knowledge on how to utilize the platform and to evaluate the platform for intended industrial use cases is the provision of trainings or workshops
	Start-Up Support	Such programs offer help for start-ups to master the platform, and develop an application. A platform provider may also forward the participating start-ups to the industrial end customers, who may adopt the idea or the platform-based software of the start-ups

After a careful examination of the existing literature, we notice a lack of research on the quality aspects of BR and their impact on stakeholder satisfaction. The only known study about the correlation between the enjoyment with the quality of BR and the loyalty of developers [19] examined a generic application development framework (Qt). Our study is inspired by this work and aims to link the "developer journey" with complementor satisfaction in the context of IIoT ecosystem development. Accordingly, this paper is guided by the user satisfaction, to evaluate the complementor satisfaction, as one of the principles of the Total Quality Management (TQM) approach. TQM is applicable for software, requiring the platform provider to perform beyond the quality standards [31, 32]. Agreeing with this approach, we argue that a structured quality management approach, which takes into account the measurement of complementor satisfaction, could improve the quality of BR, and increase the perceived satisfaction, fostering the attractiveness of the ecosystem in the end.

3   Methodology

This paper follows an inductive research approach, based upon the collection of empirical data using an evaluation sheet, which combined quantitative rating scales, and qualitative input forms. As mentioned in Sect. 1, our goal is to explore which BR are valued, linking their usefulness, and quality to the experienced importance, and satisfaction from the complementor perspective. For this purpose, we collected primary data through surveying technical experts, such as software developers or software engineers, in complementor companies (from the platform provider perspective). Therefore, the collected primary data contains multiple experiences, gathered during the realization of platform-based projects using BR. They include digitization of hardware assets and the development of platform-based applications either for own use cases or on behalf of the customers. To ensure the validity of the findings, we applied a case study research methodology to our empirical study and chose to survey the partner ecosystem of the open IIoT platform MindSphere. Interpretive case study research is an accepted research method to understand complex circumstances in their specific context due to its deep understanding [33]. That fits our goal to explore BR used in IIoT ecosystems. In addition, the case study setting allowed us to increase the comparability of the ratings because we surveyed the complementors, who utilized the same MindSphere-related BR according to their specific IIoT project.

In previous studies, we identified, MindSphere as an open IIoT platform, as indicated by the offered BR, to cope with the diverse complementors, and enable them to use MindSphere (see Table 1). Moreover, we discovered the efforts Siemens undertook to establish MindSphere in various industries, and how these efforts paid off, highlighting the ongoing ecosystem growth to this date [12]. Therefore, we have chosen the MindSphere IIoT ecosystem as a case. The data collection and analysis were conducted between the 31st of May, and 04th of December in 2019. The overall sample includes n = 19 surveyed companies, including four hardware-providing complementors (e.g., component manufacturers, and automation suppliers), one retrofitting company and 15 software, and analytics companies. The evaluation sheets were completed either independently or over the phone, clarifying possible questions on the survey immediately.

The synthesis of the last results showed that the evaluation scores of the utilized BR are not changing, indicating a theoretical saturation as no new insights are generated. According to the previously conducted analysis of the development of the MindSphere ecosystem, based upon the business relationships between Siemens and its partners, and end customers, the current size of the MindSphere ecosystem included at least 300 partners [12]. Consequently, our study covered around 7% of the whole IIoT ecosystem. The full list of surveyed complementary companies is available online: https://bit.ly/2DXue0V.

During the preparation of the evaluation sheet, we combined two evaluation methods. The proposed mechanism of the customer satisfaction survey (CSS) [34] was used for the ratings of each BR. CSS is a valid technique to retrieve the customer's voice, according to the ISO 16355 standard for product development [35], and has already been used in the evaluation of software in the past [36]. In total, the evaluation sheet consists of 17 questions, starting with the opening question to capture the context of the platform-based project, and to understand why and when the IIoT platform was used, as the number of BR, and their quality changed over time. The remaining 15 questions refer to the specific BR, and the last question aims to discover BR, which have not been identified in advance. Table 2 presents an exemplary excerpt of the survey (full survey sheet is available online at: https://bit.ly/2DXue0V):

Table 2. Excerpt of the evaluation sheet

	Did you use MindSphere APIs?		yes			no		
1	How important is this resource for your MindSpere project?	1	2	3	4	5	NU
	How is your experience using it?	1	2	3	4	5	NU
	Can you remember a positive and/or negative event/incident related to the use of the APIs?	Please fill out:					
	Which criterion (characteristic) contributed to your perception of the critical event:	Please fill out:					
	Optional reasons / comments:	Please fill out:					

The survey uses five-point nominal Likert scales [32, 34] to measure the dimension of perceived importance, and the dimension of experienced satisfaction for each used BR in earlier or currently ongoing MindSphere projects. Altogether, the survey is consistent with the SERVIMPERF (Service Importance Performance) approach:

- Importance: "1" should be checked if the BR is not important at all for the implementation of the project
- Importance: "5" should be checked if the BR is very important for the implementation of the project
- Satisfaction: "1" should be checked if the interviewee was very dissatisfied with the BR
- Satisfaction: "5" should be checked if the interviewee was very satisfied with the BR
- "0" should be checked on both scales if the BR was not used

From the complementor perspective, BR are perceived as satisfaction factors. Taking into account the differently perceived importance, some BR affect the overall satisfaction with the IIoT ecosystem more and some less. Capturing the stated importance in the survey allowed us to include weights for each of the 15 BR, which are used as "satisfaction sub-criteria" [32]. For a better visualization of both dimensions, we transformed the values assigned to the evaluation items to a scale from one to 100 (e.g. 0 = 0; 1 = 20; 2 = 40; 3 = 60; 4 = 80; 5 = 100). The qualitative information fields are used to include the critical incident technique in the survey, which exceeds the scope of this paper. In the next section, we describe how we used the individual ratings to analyze the complementor satisfaction with the used BR.

4   Results

Table 3 represents the data set of the study (see also: https://bit.ly/2DXue0V):

By calculating performance indicators, we create rankings based upon both measured dimensions for each BR: Let (S_{mean}) be the arithmetic mean satisfaction, and (W_{rel}) be the relative importance for each attribute (i $\triangleq$ for each BR). The required values for these calculations are available online: https://bit.ly/2DXue0V. The surveyed ratings help us to rank the BR according to the stated importance, and experienced satisfaction, as shown in Fig. 1. The figure also includes heat maps as graphic indicators for the urgency of the improvement and development actions. The red and orange colors highlight the most important BR or the ones that caused lower satisfaction, thus making them critical and requiring the platform provider to focus on them.

Next, we combine the relative importance (percentage of value achieved for each BR related to a total of 100%), and the arithmetic means of satisfaction values to obtain the significance for BR improvement, taking into account importance and satisfaction. It is not advisable to multiply the importance and satisfaction (as supposed by the customer satisfaction index [32, 34], as BR with already achieved high quality and satisfaction get more points. This leads to a wrong estimation of which BR should be prioritized and improved with higher priority. Therefore, to get correct values to improve each BR the importance values are divided by the satisfaction values.

$$SWCS_i = \frac{W_i^{rel}}{S_i^{mean}}$$

The resulting quotients for each BR are then compared with its mean value, which equals 0,90619864. If the resulting SWCS value is higher than its mean value, such BR should be considered with higher priority for improvement, and monitoring activities. If the SWCS value is lower than the mean value, these BR should be improved with a lower priority, as they either cause high satisfaction or have lower importance (indicated in red). Figure 2 shows the SWCS values, and the resulting ranking for BR to develop a prioritized action plan for the operations of a platform provider.

For a better understanding of which BR require urgent improvement actions, we depict the input values for SWCS in the action diagram. The diagram is designed as a two-dimensional map and includes the two measured dimensions of importance and satisfaction. The diagram is divided into four quadrants to categorize the BR-related actions

Table 3. Obtained ratings for importance and satisfaction with BR

Weight / Importance																			
Complementor ID	1	2	3	4	5	6	7	8	9	10	11	12	13	14	15	16	17	18	19
APIs	80	100	90	100	100	100	100	90	40	100	20	100	100	100	100	60	100	100	100
Connectivity Libraries	100	100	0	0	0	0	0	100	0	0	0	0	100	100	100	0	0	100	90
Support of Open Protocols	100	0	0	0	0	80	0	100	0	0	0	0	100	20	0	0	0	0	90
Infrastructure Support	100	0	0	100	60	60	100	80	20	0	0	0	100	100	60	100	40	0	30
DevOps Metrics	0	0	0	0	0	0	0	0	0	0	0	0	0	40	0	80	0	0	0
SDK	60	0	0	0	0	80	100	0	0	0	60	100	0	60	100	0	0	0	0
Cloud Foundry	100	0	0	100	60	80	60	80	80	0	0	100	100	80	80	100	100	60	80
App Store	80	0	90	0	0	0	0	50	0	80	0	0	0	100	100	40	0	0	30
Model-Driven Development	0	0	0	100	0	0	0	30	20	0	0	0	0	20	0	0	0	0	0
Documentation	80	100	90	60	100	0	0	100	80	60	0	100	80	0	60	100	100	100	100
Partner Programs	90	40	40	80	0	60	100	0	0	0	80	100	60	40	0	60	40	0	60
Onsite Demonstrators	0	0	0	0	0	0	0	60	0	0	0	0	0	80	0	0	0	0	0
Events	60	40	0	0	0	60	80	0	0	40	0	0	100	60	0	90	0	0	70
Workshops	40	0	0	60	0	0	100	90	0	0	100	0	0	20	0	0	0	0	0
Start-Up Support	0	0	0	0	0	0	0	0	0	100	0	0	0	100	0	0	0	100	0
Rating	890	380	310	600	320	520	640	780	240	380	260	500	740	920	600	630	380	460	650
Satisfaction																			
Complementor ID	1	2	3	4	5	6	7	8	9	10	11	12	13	14	15	16	17	18	19
APIs	80	100	70	80	80	40	80	70	40	80	60	60	80	60	100	60	80	40	70
Connectivity Libraries	80	100	0	0	0	0	0	80	0	0	0	0	100	80	80	0	0	60	90
Support of Open Protocols	60	0	0	0	0	100	0	60	0	0	0	0	100	0	0	0	0	0	90
Infrastructure Support	100	0	0	100	40	40	80	70	40	0	0	0	100	40	80	80	60	0	70
DevOps Metrics	0	0	0	0	0	0	0	0	0	0	0	0	0	0	0	0	0	0	0
SDK	60	0	0	0	0	80	100	80	0	0	60	100	0	0	100	0	0	0	0
Cloud Foundry	80	0	0	100	60	80	60	60	80	0	0	100	80	60	80	60	60	80	90
App Store	60	0	0	0	0	0	0	0	0	80	0	0	60	0	100	40	0	0	30
Model-Driven Development	0	0	0	100	0	0	0	0	40	0	0	0	0	0	0	0	0	0	0
Documentation	80	100	80	100	20	0	0	0	60	80	0	20	100	40	40	80	80	40	60
Partner Programs	100	60	0	100	0	40	100	0	0	0	80	80	80	60	0	60	60	0	0
Onsite Demonstrators	0	0	0	0	0	0	0	0	0	0	0	0	0	80	100	0	0	0	0
Events	100	0	0	0	0	80	100	0	0	80	0	0	80	0	0	90	0	0	90
Workshops	0	0	0	80	0	0	80	60	0	0	80	0	0	20	0	0	0	0	0
Start-Up Support	0	0	0	0	0	0	0	0	0	100	0	0	0	0	0	0	0	60	0
Rating	800	360	150	660	200	460	600	480	260	420	280	360	780	440	680	470	340	280	590

and is also suitable for monitoring of changing levels of importance and satisfaction. The diagram is methodically based on the action diagram, proposed by Motorola [32], and the customer satisfaction portfolio, applied by Herzwurm in the field of enterprise software [36]. Figure 3 shows the application of the action diagram:

It is recommended to set the upper limit of the scale of the dimension importance in such a way that the particularly important BRs are highlighted. The boundary between the quadrants in terms of satisfaction is deliberately set higher than the mean value of 50%, since we assume that measures are already required at a satisfaction level of 70% and below. The quadrants specify actions of varying urgency for the platform provider:

Importance			Satisfaction		
List of BR	Ranking	Relative Importance	List of BR	Ranking	Mean Satisfaction
APIs	1	16,47058824	Onsite Demonstrators	1	90
Developer Portal + Forum + Blog	2	12,84313725	Events	2	88,57142857
Cloudfoundry	3	12,35294118	Connectivity Libraries	3	83,75
Cloud Infrastructure Support	4	9,31372549	SDKs	4	82,85714286
Partner Programs	5	8,333333333	Open Machine Protocols	5	82
Connectivity Libraries	6	7,745098039	Start-Up Support	6	80
Events	7	5,882352941	Cloudfoundry	7	75,33333333
App Store	8	5,588235294	Partner Programs	8	74,54545455
SDKs	9	5,490196078	APIs	9	70
Open Machine Protocols	10	4,803921569	Model-driven Development (Mendix)	9	70
Workshops + Trainings	11	4,019607843	Cloud Infrastructure Support	11	69,23076923
Start-Up Support	12	2,941176471	Developer Portal + Forum + Blog	12	65,33333333
Model-driven Development (Mendix)	13	1,666666667	Workshops + Trainings	13	64
Onsite Demonstrators	14	1,37254902	App Store	14	61,66666667
DevOps Metrics	15	1,176470588	DevOps Metrics	15	0

Fig. 1. BR rankings based on the mean importance and the mean satisfaction (Color figure online)

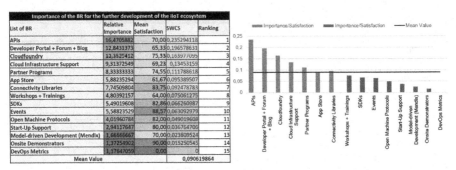

Fig. 2. Action plan for BR improvement based on weighted complementor satisfaction (Color figure online)

- **Quadrant A:** BR in this quadrant should be improved with a high priority, as they are of relatively high importance for the customer, but at the same time cause low customer satisfaction.
- **Quadrant B:** BR in this quadrant cause high satisfaction among complementors, and have a relatively high importance. An action plan should consider their monitoring to maintain a high level of quality.
- **Quadrant C:** BR in this quadrant achieve high satisfaction but are of relatively little importance for the complementors. An action plan should consider the improvement of the BR in this quadrant only if the BR in the other three quadrants are no longer in need of improvement.
- **Quadrant D:** BR in this quadrant indicate quite low satisfaction, while at the same time, they have quite low importance. These BR should be considered for improvement only if the BR from the upper left quadrant "A" have already been improved to an adequate satisfaction level.

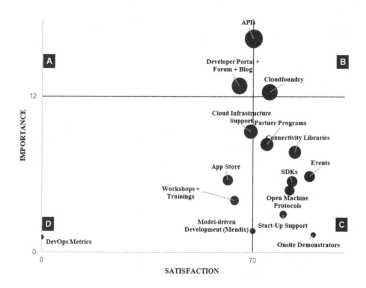

Fig. 3. Action diagram for the BR improvement for MindSphere IIoT platform

5 Discussion of Key Insights, Limitations and Outlook

The survey identifies the APIs and the aligning online documentation (e.g., Developer Portal + Forum + Blog) as potential "satisfaction drivers" being the most important BR for the complementors. However, these two BR do not exceed the satisfaction level of 70%, requiring the most attention from Siemens. The retrieved classification of Cloud-Foundry in quadrant "B" provides another interesting insight. This is one of the most valued BR among the surveyed developers, and the integration of CloudFoundry can be considered as a competitive advantage of the MindSphere platform. Furthermore, it is interesting to note that the app store and the support of different cloud infrastructures represent less important BR than expected. However, these two BR achieve only relatively low satisfaction levels, requiring Siemens to improve them (see Fig. 2). The same applies to the support of MDD, which currently has little relevance in the surveyed target group, despite a costly acquisition of Mendix by Siemens and the integration of Mendix in MindSphere. The supported partner programs, and the numerous events, conducted by Siemens, ensure a high level of partial complementor satisfaction with these two BR. Furthermore, the zero-points satisfaction rating for DevOps Metrics is very surprising and indicates that no complementors have used this resource so far. Depending on the reasons for non-use of this BR, Siemens operations may use this insight to increase the level of awareness about the existence and the usefulness of this specific BR. Overall, the results may help Siemens as a platform provider to include the complementor perspective in future ecosystem development. In terms of urgency, indicated actions are to focus on the quality of APIs, and the online documentation, while monitoring, and keeping up the high level of quality for the integration of CloudFoundry. Considering the perceived weights for each BR, new platform providing competitors may provide similar BR to catch up to the more mature platforms. In addition, more mature platforms (e.g., particularly Siemens) may use the BR-related insights to strengthen their position and develop systematic action plans for the improvement of existing BR.

Capturing the complementor perspective, our study adds to the body of knowledge on BR, providing first **empirical data on the complementor satisfaction with the IIoT ecosystem**, considering the quality and importance of the provided BR. In general, our results confirm that complementors can perceive and evaluate the quality of BR. Consequently, the **perceived quality of BR, and the resulting satisfaction may play a role during the platform selection**. Considering the level of competition in the highly fragmented market for IIoT platforms, the study proposes to consider BR as a driving concept for ecosystem development in the enterprise domain of IIoT. Following the theory on ecosystems that the ecosystem development process is controllable by the keystone company [9], our results highlight the areas of action for decision-makers in IIoT platform companies to foster the ecosystem development by improving perceived satisfaction with BR.

Secondly, our study implements the idea of structured quality improvement processes, **bridging the components of the TQM approach with the BR concept**. Due to the feasibility of the study, we propose to pay more attention to the quality measuring of BR, applying empirical software engineering methods in future research on digital platforms, and BR. Our study shows how to apply satisfaction surveys to BR and benefit from monitoring the changing perception of satisfaction and importance for the total number of BR offered. Improving the quality of BR during their evolution is an additional perspective for platform providers to consider, in addition to the previously researched power exercising, and formal governance [14, 16].

Thirdly, we achieve a theoretical contribution to the current state of research on the BR concept. The results show **how interdependent the distinctive classes of BR are**. This is indicated by the three most important BR (see Fig. 2), from all three clusters, defined and structured in an onion model by Dal Bianco et al. [30]. If the onion model is extended by the perceived complementor satisfaction, and importance, the strict classification between ABR, DBR, SBR is challenged, and the classes are mixed up. However, the onion model proposed by Dal Bianco et al. is still very important to get a holistic understanding of the concept of BR. Consequently, the perceived quality of different BR (affecting the platform selection) goes beyond the consideration and evaluation of single BR (e.g. APIs). Current research streams such as API quality, for example, focus only on single BR types, missing "the big picture". Overall, the MindSphere case study transfers Saadatmand's proposition of considering the interplay between the architecture, and governance to create complementor engagement (e.g., consistent with a holistic view on interdependent BR) in the emerging, and dynamic domain of IIoT [23]. Against this background, a holistic consideration of relevant technical, and non-technical BR should be included in the strategic development of the platform architecture, and the aligning ecosystem, outlining areas of action. IIoT platform companies should consider the interdependence of various BR types, and the relationship between the quality of BR, the resulting complementor satisfaction, and the efforts required to reach a desired level of quality, compared to competing IIoT platforms. Customer satisfaction surveys originate from the research on service quality, and the use of these methods in our study shows their suitability for BR improvement in the IIoT context. For instance, platform companies can use the proposed matrix to define the least quality goals and monitor the BR improvement to reach a defined level of quality (which may differ from the assumed 70%). In fact, our results do not provide complete solutions for platform providers on how to proceed, but rather highlight areas of improvement, and their priority in the IIoT context.

Threats to Validity. The relation between the subjective perceived ratings and the theoretical contributions may threaten the construct validity because explicitly stated rankings might differ from the implicitly derived ones [37]. This has to be considered in future research, using correlation and regression analysis. However, it should be noted that the allocation of the BR in the action diagram and the action plans may change if we use the arithmetic mean of importance [32] as an input value (see also: https://bit.ly/2DXue0V). Moreover, the consideration of critical events as a valuable addition to understand the rankings, and test their consistency is missing in this paper. Threats to external validity result from the single case study design, containing the generalization of the results for technologically differing IIoT platforms. IIoT platforms such as tapio focus a specific industry and consider complementors only in terms of asset connection, and integration of the platform into specific production processes, without providing software development resources. In that case, most identified BR, and their satisfaction ratings do not apply. Due to the domain-specific focus of the study, the generalization of the results for other enterprise software platforms might be limited as well. Furthermore, we do not explore the differences in ratings between the heterogeneous stakeholder types in the MindSphere ecosystem [12]. In addition, the long-term validity of the MindSphere-related results is also questionable, as the ratings only build a snapshot of the current state of the platform. The IIoT platform market is highly dynamic, and either Siemens or its numerous competitors may introduce new BR, which are not included in the study but cause particularly high levels of satisfaction, building a source for competitive advantage.

Research Outlook. As a part of an ongoing research approach, we were not able to carry out the complete analysis of the survey data. We plan to calculate additional key figures, such as the complementor satisfaction index with BR. Using the customer satisfaction index (CSI), which proved to be an effective instrument in corporate environments to monitor the improvement of attributes continuously or to compare own performance with the competition [36, 38], we aim to develop additional diagrams (e.g., customer satisfaction portfolio, as proposed by Herzwurm [36]) to monitor the quality of BR. Furthermore, we plan to integrate the aligning critical incidents (CI), which were also part of the study. CI may help to develop a better understanding for the creation of satisfaction through BR, and provide explanations for the evaluation results. Critical incidents will also help to identify relevant requirements [34] for each of the 15 BR, based upon positive or negative experiences during their utilization. The combination of ratings with CI will also help to apply the Kano model, and discover how to achieve attractive quality through BR design, and significantly increase the complementor satisfaction. These results will be used to develop a holistic and structured quality improvement approach for BR in IIoT. Altogether, this paper may also serve as an inspiration to apply further quality management methods (such as ISO 16355) to BR, and the linked ecosystem development. In addition, we aim to analyze whether there are significant differences in the measured satisfaction between the different complementor types in IIoT ecosystems (e.g., mechanical engineering, and software companies), since this can be achieved with the obtained data.

References

1. Hodapp, D., Gobrecht, L.: Towards a coherent perspective: a review on the inter-play of the internet of things and ecosystems. In: Proceedings of the 27th European Conference on Information Systems, Stockholm-Uppsala, Sweden (2019)
2. Endres, H., Indulska, M., Ghosh, A., Baiyere, A., Broser, S.: Industrial internet of things (IIoT) business model classification. In: Proceedings of the 40th International Conference on Information Systems, Munich, Germany (2019)
3. Herterich, M.M., Brenner, W., Uebernickel, F.: Stepwise evolution of capabilities for harnessing digital data streams in data-driven industrial services. MIS Q. Executive **15**(4), 299–320 (2016)
4. Porter, M.E., Heppelmann, J.E.: How smart connected products are transforming competition. Harvard Bus. Rev. **92**(11), 64–88 (2014)
5. Schermuly, L., Schreieck, M., Wiesche, M., Krcmar, H.: Developing an industrial IoT platform – trade-off between horizontal and vertical approaches. In: 14th International Conference on Wirtschaftsinformatik Proceedings, Siegen, Germany, pp. 32–46 (2019)
6. Nambisan, S., Wright, M., Feldman, M.: The digital transformation of innovation and entrepreneurship: progress, challenges and key themes. Res. Policy **48**(8), 103773 (2019)
7. Petrik, D., Herzwurm, G.: Stakeholderanalyse in plattformbasierten Ökosystemen für industrielle IoT-Plattformen. In: Becker, S., Bogicevic, I., Herzwurm, G., Wagner, S. (eds.) Software Engineering and Software Management 2019, pp. 113–114. Bonn, Gesellschaft für Informatik e.V. (2019)
8. Adner, R.: Ecosystem as structure: an actionable construct for strategy. J. Manag. **43**(1), 39–58 (2017)
9. Jacobides, M.G., Cennamo, C., Gawer, A.: Towards a theory of ecosystems. Strateg. Manag. J. **39**(8), 2255–2276 (2018)
10. Iansiti, M., Levien, R.: The keystone advantage: What the new dynamics of business ecosystems mean for strategy, innovation, and sustainability. Harvard Bus. Sch. Press, Boston Mass. (2004)
11. Schreieck, M., Wiesche, M., Krcmar, H.: Design and governance of platform ecosystems - key concepts and issues for future research. In: Proceedings of the 24th European Conference on Information Systems, Istanbul, Turkey (2016)
12. Petrik, D., Herzwurm, G.: Towards an understanding of iIoT ecosystem evolution - MindSphere case study. In: Hyrynsalmi, S., Suoranta, M., Nguyen-Duc, A., Tyrväinen, P., Abrahamsson, P. (eds.) ICSOB 2019. LNBIP, vol. 370, pp. 46–54. Springer, Cham (2019). https://doi.org/10.1007/978-3-030-33742-1_5
13. Ghazawneh, A., Henfridsson, O.: Governing third - party development through platform boundary resources. In: Proceedings of the 31st International Conference on Information Systems, St. Louis, USA (2010)
14. Ghazawneh, A., Henfridsson, O.: Balancing platform control and external contribution in third-party development: the boundary resources model. Inf. Syst. J. **23**(2), 173–192 (2013)

15. De Reuver, M., Sorensen, C., Basole, R.C.: The digital platform: a research agenda. J. Inf. Technol. **33**(2), 124–135 (2018)
16. Eaton, B.D., Elaluf-Calderwood, S., Sorensen, C., Yoo, Y.: Distributed tuning of boundary resources: the case of Apple's iOS service system. MIS Q. **39**(1), 217–243 (2015)
17. Hein, A., Weking, J., Schreieck, M., Wiesche, M., Böhm, M., Krcmar, H.: Value co-creation practices in business-to-business platform ecosystems. Electron. Markets **29**, 503–518 (2019). https://doi.org/10.1007/s12525-019-00337-y
18. Skog, D.A., Wimelius, H., Sandberg, J.: Digital service platform evolution: how Spotify leveraged boundary resources to become a global leader in music streaming. In: Proceedings of the 51st Hawaii International Conference on System Sciences, pp. 4564–4573. Hawaii (2018)
19. Myllärniemi, V., Kujala, S., Raatikainen, M., Sevon, P.: Development as a journey: factors supporting the adoption and use of software frameworks. J. Softw. Eng. Res. Dev. **6**(1), 6 (2018)
20. Krause, T., Strauß, O., Scheffler, G., Kett, H., Lehmann, K., Renner, T.: IT-Plattformen für das Internet der Dinge (IoT). Basis intelligenter Produkte und Services. Fraunhofer Verlag, Stuttgart (2017)
21. Petrik, D., Herzwurm, G.: IIoT ecosystem development through boundary resources: a Siemens MindSphere case study. In: Proceedings of the 2nd ACM SIGSOFT International Workshop on Software Intensive Business: Start-ups, Platforms and Ecosystems (IWSiB 2019). ACM, New York (2019)
22. Ofe, H.A., Sandberg, J.: Platform establishment: navigating competing concerns in emerging ecosystems. In: Proceedings of the 52nd Hawaii International Conference on System Sciences (HICSS), Waikoloa Village, USA, pp. 1425–1434 (2019)
23. Saadatmand, F., Lindgren, R., Schultze, U.: Configurations of platform organizations: implications for complementor engagement. Res. Policy **48**(8), 103770 (2019)
24. Guth, J., et al.: A detailed analysis of IoT platform architectures: concepts, similarities, and differences. In: Di Martino, B., Li, K.-C., Yang, L.T., Esposito, A. (eds.) Internet of Everything, pp. 81–101. Springer, Singapore (2018)
25. Tiwana, A.: Platform Ecosystems: Aligning Architecture, Governance, and Strategy. Morgan Kaufmann, Amsterdam (2014)
26. Lakhani, K.R., Iansiti, M., Herman, K.: GE and the industrial internet. Harvard Business School Case, 614-032 (2014)
27. Jansen, S., Brinkkemper, S., Finkelstein, A.: Business network management as a survival strategy: a tale of two software ecosystems. In: Proceedings of the First International Workshop on Software Ecosystems, Falls Church, USA, pp. 34–48 (2009)
28. Mazhelis, O., Luoma, E., Warma, H.: Defining an Internet-of-Things Ecosystem. In: Andreev, S., Balandin, S., Koucheryavy, Y. (eds.) NEW2AN/ruSMART -2012. LNCS, vol. 7469, pp. 1–14. Springer, Heidelberg (2012). https://doi.org/10.1007/978-3-642-32686-8_1
29. West, J.: How open is open enough? Melding proprietary and open source platform strategies. Res. Policy **32**(7), 1259–1285 (2003)
30. Dal Bianco, V., Myllärniemi, V., Komssi, M., Raatikainen, M.: The role of platform boundary resources in software ecosystems: a case study. In: IEEE/IFIP Conference on Software Architecture. IEEE (2014)
31. Li, E.Y., Chen, H.-G., Cheung, W.: Total quality management in software development process. J. Qual. Assur. Inst. **14**(1), 4–6 (2000)
32. Grigoroiudis, E., Siskos, Y.: Customer Satisfaction Evaluation. Methods for Measuring and Implementing Service Quality. Springer, New York (2010)
33. Walsham, G.: Interpretive case studies in is research: nature and method. Eur. J. Inf. Syst. **4**(2), 74–81 (1995)

34. Hayes, B.E.: Measuring Customer Satisfaction and Loyalty, 3rd edn. ASQ Quality Press, Wisconsin (2008)
35. ISO 16355-1: Application of statistical and related methods to new technology and product development process – Part 1: general principles and perspectives of Quality Function Deployment (QFD). ISO 16355-1:2015, Genf (2015)
36. Herzwurm, G.: Kundenorientierte Softwareproduktentwicklung. Vieweg + Teubner Verlag, Wiesbaden (2000)
37. Matzler, K., Sauerwein, E.: The factor structure of customer satisfaction: an empirical test of the importance grid and the penalty-reward-contrast analysis. Int. J. Serv. Ind. Manag. **13**(4), 314–332 (2002)
38. Eklof, J.A., Westlund, A.: Customer satisfaction index and its role in quality management. Total Qual. Manag. **9**(4–5), 80–85 (1998)

Avoiding Vendor-Lockin in Cloud Monitoring Using Generic Agent Templates

Mathias Mormul[(⊠)], Pascal Hirmer, Christoph Stach, and Bernhard Mitschang

Institute for Parallel and Distributed Systems, University of Stuttgart,
70563 Stuttgart, Germany
{mathias.mormul,pascal.hirmer,christoph.stach,
bernhard.mitschang}@ipvs.uni-stuttgart.de

Abstract. Cloud computing passed the hype cycle long ago and firmly established itself as a future technology since then. However, to utilize the cloud optimally, and therefore, as cost-efficiently as possible, a continuous monitoring is key to prevent an over- or under-commissioning of resources. However, selecting a suitable monitoring solution is a challenging task. Monitoring agents that collect monitoring data are spread across the monitored IT environment. Therefore, the possibility of vendor lock-ins leads to a lack of flexibility when the cloud environment or the business needs change. To handle these challenges, we introduce *generic agent templates* that are applicable to many monitoring systems and support a replacement of monitoring systems. Solution-specific technical details of monitoring agents are abstracted from and system administrators only need to model generic agents, which can be transformed into solution-specific monitoring agents. The transformation logic required for this process is provided by domain experts to not further burden system administrators. Furthermore, we introduce an agent lifecycle to support the system administrator with the management and deployment of generic agents.

Keywords: Vendor lock-in · Cloud monitoring · Monitoring agents · Genericity

1 Introduction

Cloud computing passed the hype cycle long ago and firmly established itself as a future technology since then. Studies still predict a further increase of revenue of the worldwide public cloud services market by more than 15% in 2019, whereby Infrastructure-as-a-Service (IaaS) is the fastest growing segment [1]. The main advantages of cloud computing are the self-service based commissioning and decommissioning of resources (e.g., virtual machines) as needed, a flexible pay-per-use model, and seemingly infinite scalability to enable a perfectly fitted IT

This work is partially funded by the BMWi project IC4F (01MA17008G).

W. Abramowicz and G. Klein (Eds.): BIS 2020, LNBIP 389, pp. 367–378, 2020.
https://doi.org/10.1007/978-3-030-53337-3_27

infrastructure for each company [18]. However, to utilize the cloud optimally, i.e., as cost-efficiently as possible, resource utilization must be known or analyzed to prevent an over- or under-commissioning of resources. As resource utilization may vary over time, a continuous monitoring of cloud resources is essential [12].

However, the flexibility provided by cloud computing introduces new challenges to the management and monitoring of the IT environment. Most current monitoring systems were originally not designed to monitor cloud environments with highly volatile virtual machines that may come and go in a matter of minutes, but rather for traditional hardware and slowly changing environments [4,27]. When business needs change, which leads to changes in the cloud environment, the use of a different monitoring system that better suits these business needs might be favorable. This challenge is amplified by the current state of IT departments, which are oftentimes understaffed and lack resources for new technologies [2,26]. Hence, they have problems keeping up with state-of-the-art technologies and deploying them in the company. Replacing a monitoring system is a time-consuming, error-prone, and therefore, expensive task that cannot be handled by every IT department. In general, each monitoring system comes with its own monitoring agent (in the following, only called agent), a software process collecting monitoring data inputs and reporting them to a monitoring system [17]. Therefore, replacing a monitoring system leads to the replacement of all of its agents. Usually, the agents are written in different programming languages, which requires learning new syntax and semantics. Furthermore, features of the agents, e.g., aggregation or filtering, may differ from agent to agent. Lastly, the newly created agents need to be deployed onto the virtual machines, which oftentimes needs to be conducted manually.

We address these challenges by introducing generic agent templates to create an abstraction level for the modeling of agents. Agents are modeled only once in an abstract and generic way and domain experts provide the transformation logic required to transform these agents into executable agents for specific monitoring systems. Therefore, the complexity and time needed to replace a monitoring system are heavily reduced, which also reduces monetary losses. Furthermore, we introduce an extended lifecycle for agents to support the management and automatic deployment of agents to enable scalability, which is essential in large-scaled cloud environments.

The remainder of this paper is structured as follows: Sect. 2 introduces a motivating scenario and its requirements for cloud monitoring. In Sect. 3, we present the generic agent templates and the extended agent life cycle. Section 4 discusses related work and lastly, Sect. 5 contains the conclusion of this paper as well as future work.

2   Motivating Scenario

Modern cloud computing leads to a plethora of data points that need to be monitored. For illustration purposes, in this motivating scenario, we only consider the monitoring of the CPU of a virtual machine running a Linux server. In general, each IT infrastructure monitoring (ITIM) system provides preconfigured agents for the most common tasks. Ligus [17] defines an agent as *"a software process that continuously records data inputs and reports them to a monitoring system"*. The agent collects the CPU information from the host system and sends it to a monitoring server where the data are stored and can be accessed by the system administrator. In addition to the CPU measurement, metadata such as the host name and timestamp are sent as well. Oftentimes it is sensible to perform certain actions on the monitoring data before sending it to the server which requires changes to the configuration of the agent. Those actions fall into one of two categories: processing—actions performed on each data sample, i.e., each collected CPU load measurement—and aggregation—actions performed on a set of data samples. In our example, processing could be used to transform the timestamp (usually in UNIX time format) into a human-readable format. Aggregation could be used to aggregate multiple measurements for a specified time interval, e.g., 60 s, and compute the mean, max, or min value and only send this result to reduce network traffic.

This way, the system administrator can create arbitrary agent configurations that suit the business needs of the company. Based on these configurations, agents are deployed on each virtual machine. Now, business needs or the cloud environment may change that require an adaptation of the monitoring system. For example, Linux servers may be replaced by Windows servers. However, not every ITIM system supports Windows as well as Linux. Therefore, either an additional ITIM system must be added that supports Windows, which increases management complexity since two systems need to be managed, or the original system must be replaced by a system that supports both operating systems.

In both cases, the system administrator has to recreate the agent configurations for the new ITIM system. Furthermore, as mentioned in the previous section, agents differ in their programming language and functionalities, which the system administrator has to learn first. Lastly, the agents must be deployed again. This often occurs manually, which leads to scalability issues when considering large-scale cloud environments. Furthermore, changes at runtime to adapt to a dynamic environment and automatic deployment are not supported by most monitoring systems. All this results in a time-consuming process whereby the actual task, i.e., CPU monitoring, stays exactly the same.

Out of the above, we deduce following requirements:

R1: **Genericity:** Instead of modeling solution-specific agents, a generic way for modeling agents to abstract from implementation details is required. This way, system administrators only need to model an agent once. Domain experts provide the transformation logic required to transform generic agents into solution-specific agents. This way, system administrator can save time and money and focus on translating business needs into generic agents instead of learning technical details of monitoring systems.

R2: **Expressiveness:** The expressiveness of generic agents needs to support all common tasks of an agent, such as processing and aggregation of monitoring data. This guarantees that many monitoring systems can be supported by generic agents.

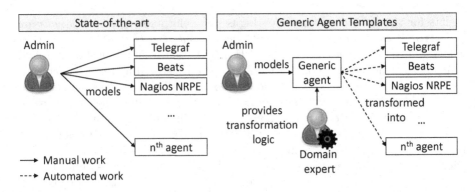

Fig. 1. Left: Admin has to model n agents; Right: Admin only has to model one agent

R3: **Automation:** With rising complexity in cloud environments, manual tasks must be minimized. Automation requires management in regard to the modeling and deployment of agents to unburden system administrators and enable them to focus on actual problems in the IT environment.

3   Generic Agent Templates

We introduce generic agent templates to fulfill the above-mentioned requirements $R1-R3$. System administrators model agents in accordance to the business needs and decide which resources need to be monitored, and how often and what kind of actions are performed on the monitoring data (cf. our motivating scenario). As shown in Fig. 1 (left), currently, a system administrator has to model n different agents to support n monitoring systems, even if the underlying tasks, e.g., *collect CPU load, transform timestamp into human-readable format, calculate mean over one minute, send to database*, stay the same. This process is done manually and in each company separately.

By introducing generic agent templates (Fig. 1 (right), the system administrator only has to model this underlying task once in a generic way. However, those generic agents are non-executable and need to be transformed into a solution-specific agent that is used within the specific company. Because generic agents conform to a predefined schema, domain experts with detailed technical knowledge about the several solution-specific agents can provide the transformation logic to transform generic agents into executable agents for the desired monitoring system. The transformation logic can be shared across all companies and, therefore, only needs to be provided once per supported agent.

To further support system administrators, we introduce an extended lifecycle for agents, as shown in Fig. 2. On the left, the current, inadequate lifecycle is shown consisting of two manual tasks *Modeling* and *Manual Deployment*, both performed by system administrators. Each time an agent is modeled or changed, the deployment onto each virtual machine must be repeated. To cope with the issues of this approach, we introduce an extended life cycle as shown in Fig. 2 (right), which comprises of the following four phases:

- *Phase 1 - Modeling:* The demand for genericity and usability requires changes in the modeling phase. We introduce *generic agent templates* to create an abstraction layer that is generic and supports an easy modeling of agents that can be used for multiple monitoring systems.
- *Phase 2 - Transformation:* Since the modeled agents are generic and non-executable, the generic agents are transformed into executable, solution-specific agents. To not further burden the system administrator, this task is automated since domain experts provide the transformation logic.
- *Phase 3 - Automatic Deployment:* Instead of deploying agents manually, we automate the deployment to tackle scalability issues in large-scale cloud environments using standards-compliant technologies.
- *Phase 4 - Adaptation:* We present an agent management to support adaptation of agents to the dynamically changing environment at runtime.

Phase 1 - Modeling: In general, agents are integrated into specific monitoring systems and cannot be used for different monitoring systems. There are a few exceptions, e.g., Nagios' NRPE agent, which, due to its wide distribution, is supported by many different monitoring systems. However, there are no agents that can be used in all monitoring systems. Also, each agent differs in its syntax, used programming language, and functionality. To enable genericity, we introduce the novel concept of *generic agent templates* to achieve an abstraction layer for the modeling of agents.

The basis of a generic agent template is the agent pipeline containing the four components *Input, Processor, Aggregator*, and *Output* nodes as shown in Fig. 3. This pipeline is based on the plugin-based architecture of agents, e.g., *Telegraf* [15] (TICK-Stack) or *Beats* [9] (ELK-Stack) and enables modeling a flexible and extendable agent due to its modular structure. With those components, a system administrator can model generic agents (exemplary configuration in Fig. 4) that can be transformed by the *Agent Mapper* to solution-specific agents. In the following, we describe the nodes and pipeline concept in detail.

- **Input Node** defines what metric the agent is collecting (e.g., *CPU load*). Besides meta data, such as *ID* and *name* of the node, for each input node, the sampling frequency can be set individually, e.g., *CPU load* is collected every second whereas *RAM load* is collected every ten seconds. Data from the Input Node can be sent to *Processor* and *Aggregator* nodes for further processing, or directly to an *Output Node*. A single agent can contain multiple Input Nodes.
- **Processor Node** is an optional node which contains functions that can be executed on single data samples. Examples are transformations of data (e.g., UNIX date to human readable date) or filtering (e.g., $if\ value < 200MB$). At the end, a data sample that passes a Processor Node may receive a tag

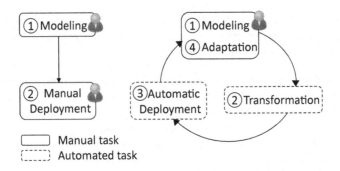

Fig. 2. Left: Current agent lifecycle; Right: New, extended agent lifecycle

that may be used for further routing inside the agent pipeline. Based on those tags, data are sent to Aggregator Nodes or Output Nodes.

- **Aggregator Node** is an optional node and acts similar to a Processor Node. The difference is that functions are not performed on single data samples, but on a set of data samples. Therefore, a window is defined, in which the Aggregator Node calculates statistics, e.g., calculate the mean CPU and RAM load over one minute. Again, tags can be added for further routing and data can be sent to Processor Nodes or Output Nodes.
- **Output Node** represents the endpoint of a pipeline and defines where data are sent to from an agent's perspective (e.g., to a database). Multiple Output Nodes can exist and can be chosen based on tags added to the data.

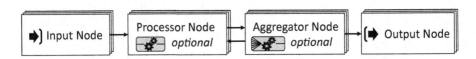

Fig. 3. Agent pipeline

The result of the modeling process is a generic agent in form of a JSON document. We define a schema[1] for the node definitions using JSON schema. A small excerpt of this schema for the input node is shown in Listing 1.1. The *id* is used as an identifier within an agent template. The *type* denotes the type of input node referring to premodeled input types like CPU input or RAM input. The *config* contains configurable parameters of the input node, such as sampling frequency. Finally, *next* describes the next node within the pipeline. Based on this schema, we implemented a graphical, web-based modeling tool[2] to ease the

---

[1] https://github.com/mormulms/agent-centric-monitoring/blob/master/generic-agent/mona-template-editor/src/assets/schema.json.

[2] https://github.com/mormulms/agent-centric-monitoring/blob/master/generic-agent/mona-template-editor.

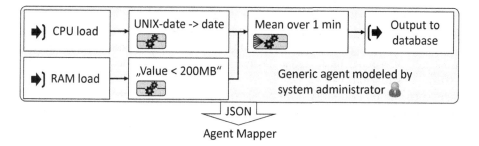

Fig. 4. Exemplary generic agent configuration sent to Agent Mapper

```
"GenericInput": {
    "type": "object",
    "properties": {
      "id": {
        "$ref": "#/definitions/InputId"
      },
      "type": {
        "$ref": "#/definitions/NodeType"
      },
      "config": {
        "$ref": "#/definitions/InputConfig"
      },
      "next": {
        "$ref": "#/definitions/NextArray"
      }
    },
    ...
}
```

Listing 1.1. Excerpt of the input node definition

modeling and alteration of agents without the need to understand all the specifics of a monitoring system.

Phase 2 - Transformation: The result of a modeled agent pipeline is a generic agent template. The *Agent Mapper* is the component that receives this generic agent template and transforms it into an executable agent for a specific monitoring system. Of course, for each supported monitoring system, the transformation logic must be implemented. Therefore, the complexity and variance of different monitoring systems is not diminished but rather shifted from the end user to a few domain experts who implement the transformation logic for specific monitoring agents. Those implementations must be shared publicly to gain an advantage. So far, in our current prototype, we only support transformations to Telegraf. To support further monitoring systems and agents in the future, we created an extendable, modular architecture for the Agent Mapper. Agents, such as Telegraf, already have the functionalities for processing and aggregating monitoring data and, therefore, are simple to transform to. However, agents that do not support one or more of those functionalities require a more complicated transformation. The most basic agent always has a means to collect data and send it to the destination, i.e., the monitoring server. However, if a modeled generic agent contains functionality such as aggregation and the system administrator wants to transform this generic agent into a solution-specific agent that does not support this functionality, there are two options: (i) the system administrator receives a warning that the transformation is not possible and needs to select a different solution-specific agent, which the generic agent should be transformed into. The second option is the use of a Complex Event Processing (CEP) engine or similar engines. The agent sends its collected monitoring data to this engine. Then, the processing and/or aggregation nodes can be translated into CEP queries to perform the needed functionalities and forward the monitoring data back to the agent which further forwards it to the monitoring server. This greatly increases the impact of generic agents since a transformation to many existing monitoring agents is possible.

Phase 3 - Automatic Deployment: Especially in large-scaled scenarios with a large number of VMs, any manual task becomes cumbersome, error-prone, and simply does not scale. Therefore, an automatic deployment is essential [25,29]. There are many existing tools and platforms for automated deployment, such as *Docker* [6] or *Vagrant* [11]. A few monitoring systems like *Splunk* [23] support this feature as well. However, a standardized deployment framework is desirable, since technologies tend to disappear over time.

An established deployment standard is the Topology and Orchestration Specification for Cloud Applications (*TOSCA*) [19] of the Organization for the Advancement of Structured Information Standards (*OASIS*). TOSCA enables a two-step software deployment approach. First, a topology template is created, modeling the application, platform, and infrastructure components. Second, this topology is used to deploy the components in the corresponding infrastructure.

One implementation of the TOSCA standard is OpenTOSCA [28]. OpenTOSCA provides an eco system consisting of the TOSCA topology modeler Winery [7], the OpenTOSCA container that handles the actual software deployment based on the topology, and a self-service portal called Vinothek [3]. TOSCA and its implementation OpenTOSCA can be used in our approach for automated deployment of monitoring agents. Since TOSCA is a standard, it provides a high degree of applicability and is future-secure.

We modeled a topology template using OpenTOSCA to deploy a monitoring system and a template to deploy agents. If a new virtual machine is started, an agent is deployed on it to guarantee a monitoring from the beginning.

Phase 4 - Adaption: According to Gartner [20], flexibility against a changing IT architecture is a key objective when investing in new monitoring systems. The modeling of agents is influenced by the current business needs and the status of the current cloud environment. However, both of these variables may change over time. In this case, starting a new agent life cycle is excessive when only

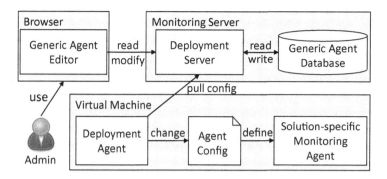

Fig. 5. Architecture for automatic adaptations at runtime

minor changes to the agent are required. Instead, the user should be able to access the previously modeled generic agents and change them to fit the current needs. The changes are then propagated to all agents originating from this generic agent. To support this process, we implemented a prototype as shown in Fig. 5. The user models a new generic agent or changes an existing one via the web-based modeling tool, which is connected to the *Deployment Server* on the *Monitoring Server*. The deployment server is responsible for the management of the generic agents, which are stored in the *Generic Agent Database*. On the virtual machines, a *Deployment Agent* periodically requests the configuration (the transformed generic agent) for the agent and resolves differences between the retrieved configuration and the currently used one. In case of changed configurations or failures, the deployment agent automatically restarts the agent.

4   Related Work

The transformation of abstract models to a concrete executable implementation is a commonly used means in order to abstract from technical details. Consequently, domain users are able to create models on a high level of abstraction without requiring technical knowledge about their realization. The use of a single abstract model leads to genericity and reduces the threat of vendor lock-in, since it is not dependent on specific technologies. In the following, approaches are described that aim at a similar approach for transforming abstract models into concrete ones.

Falkenthal et al. [10] aim towards a generic approach to transform abstract pattern languages to concrete solution implementations. Since their approach is generic, they do not focus on a specific domain but rather discuss how such a transformation can be conducted in general. In this paper, we apply this approach to the concrete domain of monitoring by introducing a mapping of generic monitoring templates to concrete, executable implementations.

Eilam et al. [8] introduce an approach for model-based provisioning and management of applications. Through transformations, application topologies are mapped onto different levels of abstraction in order to finally create executable implementations that can be deployed. However, Eilam et al. require a premodeling of concrete implementation artifacts, which is not necessary in our approach.

Furthermore, Eilam et al. introduce a combination of a model-based and workflow-based approach for the automated provisioning of the transformed applications. This approach creates a provisioning model based on workflow technology that can be used for automated deployment. In this paper, we also introduce an approach how the templates can be automatically deployed after transformation, which uses existing software deployment technologies instead of this heavy-weight workflow-based approach.

Similar approaches regarding the agent templates are introduced by Künzle et al. [16] and Cohn et al. [5] that use artifact-centric approaches. In these approaches, so-called business artifacts are created, i.e., abstract representations of software components with the goal of hiding technical details. These artifacts can be mapped onto concrete executable implementations, as shown by Sun et al. [24]. However, the business artifacts of Künzle et al. and Cohn et al. are vaguely described, i.e., only on a conceptual level, an example application is missing. In this paper, we introduce a concrete scenario our concepts can be applied to.

Reimann [21] introduces generic patterns for simulation workflows that are also mapped onto concrete executable implementations, in his approach onto workflow fragments provided in the Business Process Execution Language (BPEL) [30]. However, Reimann focuses exclusively on workflows. In contrast, we focus on agents and model their capabilities through tailored templates.

In our previous work [14], we introduced so-called Situation Templates, which represent abstract descriptions of situations to be recognized without the necessity to provide technical implementation details. These situation templates can be mapped onto various formats, for example, complex event processing queries. In this paper, we adapt this concept to the domain of cloud monitoring to make it suitable for modeling of agents.

5   Conclusion

The monitoring of complex cloud environments can lead to several challenges. Selecting an unsuitable monitoring system or evolving business needs may lead to a required replacement of the monitoring system. However, monitoring agents are spread across the IT environment and, oftentimes, can be tightly integrated into the monitoring system. Therefore, the replacement of the agents is a time-consuming task and the modeling of new agents requires technical expertise. For this, we introduce generic agent templates to unburden system administrators by creating an abstraction to the modeling of agents. System administrators model generic agents once and domain experts provide the transformation logic required to transform the generic agents to solution-specific agents. This way, generic agents can be transformed into several solution-specific agents without further additional work. Expressiveness is provided by the agent pipeline consisting of Input, Processor, Aggregator, and Output Nodes. Of each node, multiple instances can be created so that system administrators can model arbitrary agent configurations. Furthermore, if solution-specific agents do not support some of those functionalities, e.g., processing, a transformation to CEP queries is also possible to even support agents with missing functionalities. Lastly, an extended life cycle supports management, automatic deployment, and adaptation of agents at runtime using state-of-the-art and standardized technologies. Transformations to solution-specific agents are automated via the Agent Mapper. The automatic deployment is enabled using the deployment standard TOSCA. Adaptations to generic agents at runtime are automatically propagated to all according agents.

In future work, we plan to validate our approach on multiple monitoring systems and extend the usage to the IoT domain by using the open-source IoT platform MBP [13,22]. Furthermore, we plan to apply the concept of generic templates to other parts of the monitoring system as well to further support a possible replacement and reduce the risk of vendor lock-ins. Similar to agent configurations, alerting rules are defined by system administrators to inform them about problems in the monitored IT environment. Therefore, analogous to generic agent templates, generic alerting rules may present similar benefits.

References

1. Gartner forecasts worldwide public cloud revenue to grow 17.3 percent in 2019 (2018). https://www.gartner.com/en/newsroom/press-releases/2018-09-12-gartner-forecasts-worldwide-public-cloud-revenue-to-grow-17-percent-in-2019
2. Bowen, D.A.: Challenges archivists encounter adopting cloud storage for digital preservation. In: Proceedings of the International Conference on Information and Knowledge Engineering (IKE), pp. 27–33 (2018)
3. Breitenbücher, U., Binz, T., Kopp, O., Leymann, F.: Vinothek - a self-service portal for TOSCA. In: Herzberg, N., Kunze, M. (eds.) Proceedings of the 6th Central-European Workshop on Services and their Composition (ZEUS 2014). CEUR Workshop Proceedings, vol. 1140, pp. 69–72. CEUR-WS.org (2014)
4. Calero, J.M.A., Aguado, J.G.: Comparative analysis of architectures for monitoring cloud computing infrastructures. Future Gen. Comput. Syst. **47**, 16–30 (2015)
5. Cohn, D., et al.: Business artifacts: a data centric approach to modeling business operations and processes. Bull. IEEE Comput. Soc. Tech. Committee Data Eng. (2009)
6. Docker Inc: Docker (2019). https://www.docker.com/
7. Eclipse: Winery (2019). https://eclipse.github.io/winery/
8. Eilam, T., et al.: Pattern-based composite application deployment. In: IM, pp. 217–224. IEEE (2011)
9. Elasticsearch, B.V.: Beats (2019). https://www.elastic.co.de/products/beats
10. Falkenthal, M., Barzen, J., Breitenbücher, U., Fehling, C., Leymann, F.: From pattern languages to solution implementations. In: Proceedings of the Sixth International Conference on Pervasive Patterns and Applications (PATTERNS), pp. 12–21 (2014)
11. HashiCorp: Vagrant (2019). https://www.vagrantup.com/
12. Hauser, C.B., Wesner, S.: Reviewing cloud monitoring: towards cloud resource profiling. In: 2018 IEEE 11th International Conference on Cloud Computing (CLOUD), pp. 678–685 (2018)

13. Hirmer, P., Breitenbücher, U., da Silva, A.C.F., Képes, K., Mitschang, B., Wieland, M.: Automating the provisioning and configuration of devices in the internet of things. Complex Syst. Inform. Modeling Q. **9**, 28–43 (2016)

14. Hirmer, P., et al.: Situation recognition and handling based on executing situation templates and situation-aware workflows. Computing 1–19 (2016)

15. InfluxData: Telegraf (2019). https://www.influxdata.com/time-series-platform/telegraf/

16. Künzle, V., et al.: PHILharmonicFlows: towards a framework for object-aware process management. J. Softw. Maintenance Evol.: Res. Practice (2011)

17. Ligus, S.: Effective Monitoring and Alerting, p. 7. O'Reilly Media, Inc., Sebastopol (2012)

18. Mell, P., Grance, T., et al.: The NIST definition of cloud computing (2011)

19. OASIS Open: Topology and Orchestration Specification for Cloud Applications Version 1.0 (2019). http://docs.oasis-open.org/tosca/TOSCA/v1.0/os/TOSCA-v1.0-os.html

20. Prasad, P., Bhalla, V.: Use This 4-Step Approach to Architect Your IT Monitoring Strategy (2018). https://www.gartner.com/en/documents/3882275/use-this-4-step-approach-to-architect-your-it-monitoring

21. Reimann, P., Schwarz, H., Mitschang, B.: A pattern approach to conquer the data complexity in simulation workflow design. In: Meersman, R., Panetto, H., Dillon, T., Missikoff, M., Liu, L., Pastor, O., Cuzzocrea, A., Sellis, T. (eds.) OTM 2014. LNCS, vol. 8841, pp. 21–38. Springer, Heidelberg (2014). https://doi.org/10.1007/978-3-662-45563-0_2

22. Franco da Silva, A.C., Hirmer, P., Schneider, J., Ulusal, S., Tavares Frigo, M.: MBP - not just an IoT platform. In: Proceedings of the 18th Annual IEEE International Conference on Pervasive Computing and Communications Demonstrations (2020)

23. Splunk Inc: Splunk (2019). https://www.splunk.com/

24. Sun, Y., et al.: Modeling data for business processes. In: Proceedings of the 30th IEEE International Conference on Data Engineering (ICDE), Chicago, USA (2014)

25. Taherizadeh, S., Jones, A.C., Taylor, I., Zhao, Z., Stankovski, V.: Monitoring self-adaptive applications within edge computing frameworks: a state-of-the-art review. J. Syst. Softw. **136**, 19–38 (2018)

26. Technology, R.H.: All things are digital in business, but finding digital talent is a tall order (2017). http://rh-us.mediaroom.com/2017-11-01-All-Things-Are-Digital-In-Business-But-Finding-Digital-Talent-Is-A-Tall-Order

27. Trihinas, D., Pallis, G., Dikaiakos, M.: Monitoring elastically adaptive multi-cloud services. IEEE Trans. Cloud Comput. (1) (2015)

28. University of Stuttgart: Opentosca (2019). https://www.opentosca.org/

29. Ward, J.S., Barker, A.: Observing the clouds: a survey and taxonomy of cloud monitoring. J. Cloud Comput. **3**(1), 1–30 (2014). https://doi.org/10.1186/s13677-014-0024-2

30. Weerawarana, S., Curbera, F., Leymann, F., Storey, T., Ferguson, D.F.: Web Services Platform Architecture: SOAP, WSDL, WS-Policy, WS-Addressing, WS-BPEL. WS-Reliable Messaging and More. Prentice Hall PTR, Upper Saddle River (2005)

Challenges of Data Management in Industry 4.0: A Single Case Study of the Material Retrieval Process

Antonello Amadori[1], Marcel Altendeitering[2(✉)], and Boris Otto[1,2]

[1] TU Dortmund University, August-Schmidt-Strasse 4, 44227 Dortmund, Germany
{antonello.amadori,boris.otto}@tu-dortmund.de
[2] Fraunhofer ISST, Emil-Figge-Strasse 91, 44227 Dortmund, Germany
{marcel.altendeitering,boris.otto}@isst.fraunhofer.de

Abstract. The trend towards industry 4.0 amplifies existing data management challenges and requires suitable data governance and data quality measures. Although these topics have been previously discussed in literature, companies are still struggling to cope with the resulting challenges and fully exploit the benefits of industry 4.0. In this paper, we conducted a single case study in an automotive company. We exemplary used the material retrieval process in automotive manufacturing to uncover what challenges there are hindering the utilization of industry 4.0. We were able to identify six major challenges in the domains of data quality and data governance.

Keywords: Industry 4.0 · Data quality · Data governance · Single case study · Expert interviews

1 Introduction

In a globalized and digital world companies are striving to flexibly respond to the rapidly changing demands of the market. As a result, the digital world is increasingly networking with industry to make production processes smarter [10,13]. These developments opened up new technological potentials, with disruptive consequences for the economy and society [3]. In the international debate this new technological trend is referred to as 'the second machine age', 'third industrial revolution' and in the German-speaking area as the 'fourth industrial revolution' or 'industry 4.0' [19].

Industry 4.0 covers the optimization of the entire production and supply chain by connecting production facilities through the Internet of Things (IoT) [19]. As a result, 'Cyber Physical Systems' (CPS) are created and used in industrial production to enable machines, storage systems and equipment to be networked globally and across facilities [13].

Particularly affected by these developments are large car manufacturers, who operate a complex global production network in order to generate cost advantages and ensure high utilization of the production sites [19]. In this production

© Springer Nature Switzerland AG 2020
W. Abramowicz and G. Klein (Eds.): BIS 2020, LNBIP 389, pp. 379–390, 2020.
https://doi.org/10.1007/978-3-030-53337-3_28

network, a seamless flow of information across different locations is vital and as important as the flow of physical materials [3,13].

Thus data management is of utmost importance for automotive manufacturers that want to leverage the business potentials of industry 4.0. However, the difficulty of using this data in accordance with the requirements and exploiting the strategical capabilities of industry 4.0 is often based on the historical buildup of the companies as well as the complex value creation chain [16,21]. Since data governance and data quality are common issues we decided to further investigate these problems. We therefore formulated the following two research questions:

- What are the data governance challenges in the material retrieval process?
- What are the data quality challenges in the material retrieval process?

For answering our research questions we conducted a single-case study on the material retrieval process in automotive manufacturing and utilized Template Coding (TC) for data analysis, using open-coding from Grounded Theory Methodology (GTM) as an example. We derived six challenges in the domains of data quality and data governance that provide an introduction to wider research and guide managerial activities addressing the identified issues. The remainder of the paper is structured as follows. We start by presenting the current state of research (Sect. 2). We continue by describing the methodological approach we followed for collecting and analyzing data and answering our research questions (Sect. 3). Afterwards, we present the results of our qualitative analysis (Sect. 4) and conclude our study by discussing the results, derive concrete recommendations and highlight paths for future work (Sect. 5).

2   Background

As the fourth industrial revolution is gaining attention so is the research around industry 4.0 increasing. Most works in the field are focusing on the technical aspects, while some others are studying the economical and organizational implications [16]. There are two studies [12,21] providing an systematic review of current literature in the field. However, current research often takes on a broad managerial perspective but lacks an in-depth investigation of challenges on the process level [12].

Kiel et al. conducted a multiple case study on the challenges and benefits of industry 4.0 with 46 managers from the machine, electrical and automotive engineering industries [16]. They found that companies often experience a low quality in data transfers and the lack of a suitable data and corporate structure. They summarized these issues under the terms 'Technical Integration' and 'Organizational Transformation'. Horwáth and Szabó focused their research on the differences between multi-national enterprises (MNEs) and small-and-medium sized enterprises (SMEs) [12]. For MNEs they made similar findings to [16] and concluded that they are particularly challenged by technological and process integration factors (e.g. lack of standards, lack of unified communication protocols, etc.). Hermann et al. derived four industry 4.0 design principles, which are addressing typical challenges: 'Technical Assistance', 'Interconnection', 'Decentralized Decisions' and 'Information Transparency' [11]. The aspects 'Interconnection' and 'Information Transparency' are summarizing data governance and data quality issues respectively.

By getting an overview of the existing research we found that data governance and data quality issues are always mentioned under similar terms. Solutions to these data management aspects have been discussed for a while under the broader concept of 'IT governance' [29]. They are separated in organizational (e.g. IT strategy), architectural/technical (e.g. data marts and warehouses) and cultural (e.g. decision making, promotion of use) measures [15,25,29,30]. However, although there are solutions available companies are still struggling to realize the benefits of industry 4.0 [16,21].

3 Methodology

3.1 Data Collection

In order to inform our research questions and gain an understanding of data governance and data quality challenges a thick-description of the field is required [7]. We chose to pursue a qualitative research approach and conduct a single case study as other studies in this field have done (e.g. [12,16]). According to [6,22] a case study is suitable for gaining detailed insights into new phenomena and helps to answer questions about 'why' and 'how' something is happening.

A single case study was sufficient for getting a deep understanding for two reasons. Firstly, the business process under investigation is very specific and important to an automotive company. It is therefore difficult to get access to such processes and obtain the necessary data. We thought that getting a deep understanding of the challenges in one case was more valuable than gaining a general understanding on a higher level in several cases [9,33]. Secondly, the automotive company under investigation is one of the leading automotive manufacturers worldwide and we argue that our findings are therefore representative for the whole industry.

For the selection of participants, we aimed to get insights from different steps in the business process as each step requires different data and has different requirements towards data handling. We furthermore wanted our interviews to include several factories in order to discover potential differences among the factories and support the generality of our findings. This selection yielded in seven interviews with nine participants (A - I) and covered five different factories (North, West, South, East, North-East). We found that this number of participants was sufficient for getting a complete overview of the business process under investigation and answering our research questions. The first contact was made by email, which was followed by a telephone call discussing the nature of the interview and preparing the interviewee for the main interview.

Since our goal was to derive the interviewees' personal experiences in working with data and the participants have different experience levels, we decided to follow a semi-structured interview approach [24]. The questions asked cover three topics: (1) the interviewees' role in the process and what information he/she is dealing with; (2) the individual perspective and experiences in working with data; and (3) the interviewees' future perspective on working with data. We conducted the interviews in German language during May and June 2019. Each interview lasted between 60 and 90 min and was audio-recorded. We transcribed the audio recording and wrote up an individual memo containing impressions of the interviewee within 24 hours after the interview as recommended by Yin [33].

3.2 The Material Retrieval Process

For our study we chose to focus on the process of material retrieval at the assembly line. The process describes the use of so called 'delivery call-off' systems for providing material to the place of installation. We selected this process as it is typical for the automotive industry and exists in different factories in similar ways. It thus supports the representativeness of our study and limits the researchers bias. Furthermore, the process is of high importance to the manufacturer, produces a lot of data and involves different machines and information systems, making it a good example for an industry 4.0 use case.

During assembly, the vehicle is completed in sequence according to customer requirements. The required parts are provided by the logistics department. The process begins with the material call-off. The retrieval is implemented on the system side by system MA, whereby the demand-driven and consumption-controlled retrieval is handled in two different subsystems (see Fig. 1). The consumption-oriented call-off will be triggered by an employee at the assembly line or by an automated event such as triggering a sensor. Demand retrieval is automated by the BO system, which forecasts the anticipated demand for a specific demand location. The system uses the information of the cycle time, the replacement time and the location of the vehicle. The location of the vehicle is determined using the detection points of the MES system. The information which parts have to be installed results from the respective pitch at which a certain part is requested or needed. After receiving the material by the MA system, MA transfers the demand of requested parts to the DEPOT system. The DEPOT system contains information about which parts are available in which warehouse. This information is recorded when the parts are delivered to the interim storage facilities in DEPOT. MA also receives the information of the time for replenishment from DEPOT since the information about the storage location is stored there and the distances between the respective demand and storage locations can be taken into consideration. In addition, DEPOT gives feedback to MA on what and how many parts of material are available. Finally, the ROUTE system calculates the optimal route for the driver of the 'material train' as needed and forwards the information to a driver. The material is then delivered to the required location.

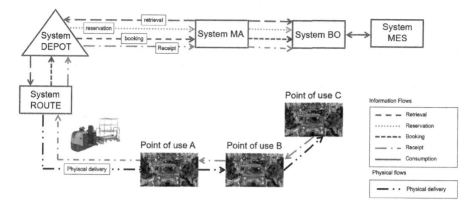

Fig. 1. Overview of the material retrieval process

3.3   Data Analysis

The topics of data governance and data quality are well-established in literature and have been previously identified and discussed in the context of industry 4.0. We therefore decided to conduct a theoretical informed data analysis and apply TC using open coding, as practiced in GTM, as a guiding example [17,28]. TC distinguishes from open-coding in the way that the codes are drawn a priori from research and literature. We found that there was little use in following an open-coding approach in this field and rather took on the interpretive nature template coding was offering us [1]. This way helped us to include existing findings in our study and avoid getting findings that are too close to our data and produce narrow theories [6]. One of the problems of template coding is to decide on what template to use for analysis [1]. For this, we returned to existing studies in the fields of data governance and data quality and selected categories of previously identified challenges as our codes. Table 1 provides a summary of the template codes we selected and the corresponding literature.

Table 1. Template codes

Data governance	Sources
Data life cycle	[15, 29]
Data definitions	[15, 32]
Education of users	[18, 29]
Locus of decision making	[29, 32]
Data Quality	**Sources**
Accuracy	[5, 31]
Accessibility	[14, 20, 31]
Completeness	[5, 20]
Redundancy	[14, 20]
Relevancy	[5, 31]

In order to assure the quality and integrity of the selected template codes we undertook an initial coding using the first two interviews as suggested by [17]. Codes we did not need were dropped and new themes were added. The verified set of codes was then applied to the remaining dataset. During the analytic process we assigned the template codes to the participants' responses and derived the associations among the codes until internally consistent themes emerged. This process involved three-passes over the data as suggested by [23]. Each of the passes was conducted independently of previous passes by the first two authors in order to limit the researchers bias and ensure qualitative rigor. Minor inconsistencies in the coding were solved in subsequent discussions.

Based on the Gioia-Methodology [8] we developed a data structure visualizing our findings in a data structure consisting of the two research questions, nine template codes and six identified challenges (see Fig. 2).

4   Challenges in the Material Retrieval Process

4.1   RQ1: Data Governance

Challenge 1.1: Non-unified Data Definitions. One of the most important problems we noted in all interviews is the lack of uniform data definitions. This means that data users do often not understand the provenance of data and it is unclear who the right contact person for an information is. Interviewee D gave a concrete example: *"I cannot tell you in which way system DEPOT is connected to system ROUTE"*. Interviewee B said something similar and noted: *"I cannot tell you what data originates from what system"*.

A possible reason for the lack of a standardized information model is the complex and heterogeneous system landscape. Everyone noted that the complexity is causing problems and the system landscape needs to be simplified for future success. For example, a change in one system often causes a large number of changes in associated systems. Interviewee I said the following: *"So, the systems [...] are very, very, very complex and changes need to be done very, very, very fast. Thereby you sometimes forget to set a few values or activate some functionality"*.

Challenge 1.2: Lack of a Single Source of Truth. Closely associated with the lack of unified information models is the need for a single source of truth. All participants noted that they sometimes spend time searching for the right information and transforming data. In response to the question how the current way of searching for information could be improved the participants suggested a central information system or interface for data access. Interviewees F and G described that they regularly need their personal contacts to get the right information. This is particularly difficult for new employees. Interviewee F said: *"It was very difficult to get information when I started working here. Now I have a network of contacts and know that there is someone in the logistics department who can help me find the information I need"*.

Research Questions   Template Codes   Identified Challenges

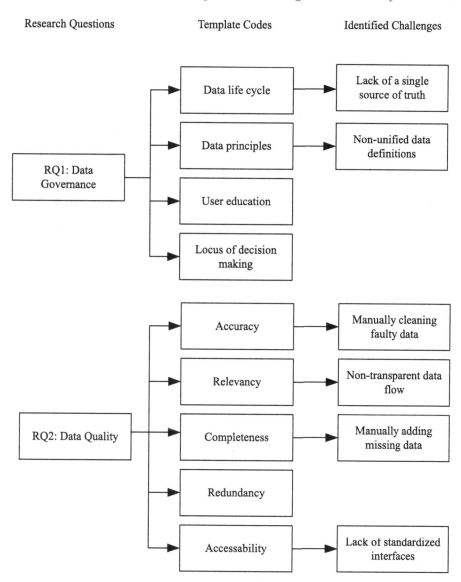

Fig. 2. Derived data structure

We observed that the lack of a single source of truth is a problem for some factories but not so much for others. Factory West is the only factory that recently upgraded to a novel enterprise information system. Interviewees B and C, who work at factory West, did not mention problems finding data in their own systems but rather highlighted problems accessing data of other factories or other companies.

4.2   RQ2: Data Quality

Challenge 2.1: Lack of Standardized Interfaces. All interviewees noted that the average level of information quality is quite well as long as the required data is coming from one system. However, once it becomes necessary to access multiple systems the experienced data quality plummets. Interviewee B gave a good description of the situation: *"We are mostly fighting with data from associated systems, which are sending data into our system. This causes most of our work and is very challenging, I would even say frustrating"*.

This problem causes manual data integration tasks. Exemplary, she described the following situation: *"We are sometime using dummy-values to enable data transfers to system MA when otherwise the system would not be able to process our data"*. The participants also noted that they would prefer less and more standardized interfaces, which are easier to operate and enable a seamless data transfer. Interviewee E formulated the following ideal scenario: *"So currently there are many interfaces that are always differently specified. If we could realize a central interface that is suitable for many systems, it will be beneficial for the whole process"*.

Challenge 2.2: Non-transparent Data Flow. Similar to challenge 2.1 is the problem of a non-transparent data flow. Interviewee B highlighted that there is no technical transparency in data handling across multiple systems. Employees often do not understand what data they are receiving and this data is incompatible with the data in another information system. Interviewee B provided a possible explanation for this situation and indicated that the willingness to take responsibility for good data quality in the overall process is quite low. She stated that data transfers are often designed not taking requirements of other systems into account and problems are simply forwarded to the next department or interface.

Challenge 2.3: Manually Cleaning Faulty Data. Participants noted that the data they are working on is often inconsistent or inaccurate and require manual data cleaning. This is either due to errors in data integration or manual errors in data processing. Since the manual correction of data causes a lot of work, the participants often associated good data quality with no need for manual work and a seamless process execution. Interviewee E exemplary defined good data quality: *"[...] as an integrated process. Meaning when the business process runs smoothly then my data must be of good quality"*. More specifically, an insufficient level of data accuracy could lead to wrong parts or materials being delivered, losing track of materials or having materials in a wrong order, which causes troubles in Just-in-Time processes. In worst case, this could halt production and 'data task forces' are established to manually and forward data and materials.

Challenge 2.4: Manually Adding Missing Values. In addition to manual data cleaning, the participants also noted missing data as a problem. We found that missing data was either due to errors in data-acquisition or intentionally left empty, because the data is only needed in a later process-step. For data-acquisition problems interviewee H noted: *"The materials management department has no idea what the field 'point of use' means and has to look like. So we often get values that are wrong"*. Interviewee C provided an example of data intentionally left blank: *"There is material information that needs to be manually recorded. And if this work isn't done we can't ship our materials and there is nothing I can do about it"*. Both of these problems are causing manual work in gathering and adding missing values. For the future interviewee E noted that she envisions a system that automatically checks data for its completeness and takes appropriate actions if necessary.

5   Conclusion

Building on existing data governance and data quality research we were able to answer our research questions and derived six challenges relevant to our case. Given the nine template codes we found in literature we can infer that *user education*, *locus of decision making* and *redundancy* were not seen as problems in material retrieval process, but should be considered in future research. Due to the focus on a single case our study is of limited generality. Nevertheless, we argue that our findings can be generalized for other automotive companies facing similar problem instances [9] for two reasons. First, our research covers multiple factories as embedded units of the single case we investigated [6,33]. Second, the material retrieval process is common in automotive manufacturing and exists in several companies in similar ways.

5.1   Discussion

Managerial Implications. The identified *lack of standardized interfaces* and the issue of *non-transparent data flows* are based on the gradual growth of the company. Many interfaces are incompatible and intermediary-systems were created for achieving compatibility. To resolve these problems holistic data and system architectures are currently developed. Towards this end, we recommend the establishment of a centralized and shared data space, which offers standardized data access points as suggested by our interviewees. A similar approach is the creation of a central data catalog system that stores and manages the metadata of valuable data objects. The latter approach is currently discussed in the company under investigation.

Potential technical solutions must be accompanied by appropriate governance measures [29]. Such governance measures can act as a means for the *lack of a single source of truth* and *non-unified data definitions*. We thus recommend that companies should develop common data definitions as part of a general information model and assign a responsible single source of truth to each data object.

Tallon et al. provide some guidance on selecting an appropriate governance model [29]. An establishment of these practices can support enterprises in establishing architectural standards, which help to achieve the required level of information agility and sustain a firm's competitive advantage [2].

Addressing the problems of *manually cleaning faulty data* and *manually adding missing data* is particularly challenging for automotive manufacturers due to specific data formats and complex system landscapes. We thus recommend companies to establish dedicated roles responsible for data quality. These can be supported by automated data quality checks as envisioned by interviewee E. In literature, Karkouch et al. propose a data quality middleware that actively manages the quality of heterogeneous data sources in IoT settings [14].

Scientific Implications. In the area of data governance our interviewees put an emphasis on the need for unified data definitions and a single source of truth for information. Constantinides et al. made similar findings and argued that standardization in data handling can support the creation of common information governance practices [4]. Without these, "information is going to grow in a negative way" (p. 160). A problem that interviewee A is familiar with, she said: *"Too much data is just causing problems"*. We therefore argue that research should address how data governance principles can be established in light of heterogeneous data sources and how existing barriers can be overcome.

To address the data quality issues we identified a seamless data integration across systems on the inter and intra-organizational levels is of utmost importance. Rai et al. empirical validated this need and found that a cross-functional application integration alongside a high level of data accuracy has a positive influence on a firms' performance [26]. While it is difficult to establish data collaborations within an organization, this is even more challenging on the inter-organizational level. Companies often fear losing control over their data and face potential privacy risks [30]. We argue that further research is required on the potentials of novel digital infrastructures (e.g. blockchains) for data collaborations in industry 4.0 settings [4]. Additionally, given the trend towards big data the veracity (i.e. correctness and accuracy) of data will become even more important and difficult to assure [27]. Further research is required on automated data quality checks and the tracability and provenance of data.

5.2   Limitations and Future Work

Although we applied a high level of rigor, our research has a few limitations. Most importantly, our study is focused on one business process in one industry and therefore of limited generality. In the future we want to extend our research to a multiple case study and include other processes and automotive companies. More precisely, we want to further investigate the identified challenges and how these could be addressed. We believe the manufacturing of the car body could be a valuable process for future research as it comprises a lot of sensor data and interfaces to other facilities.

In addition to our qualitative study it would be beneficial to conduct a quantitative case study on the subject. Such a study would provide a broad view on data-related challenges in industry 4.0 settings in contrast to the in-depth analysis of qualitative studies. Solving data-related challenges in production as well as in general is critical for automotive companies in face of current trends such as big data or autonomous logistics systems.

References

1. Blair, E.: A reflexive exploration of two qualitative data coding techniques. J. Methods Measur. Soc. Sci. **6**(1), 14–29 (2015). https://doi.org/10.2458/v6i1.18772, https://journals.uair.arizona.edu/index.php/jmmss/article/download/18772/18421
2. Boh, W.F., Yellin, D.: Using enterprise architecture standards in managing information technology. J. Manag. Inf. Syst. **23**(3), 163–207 (2006). https://doi.org/10.2753/MIS0742-1222230307
3. Brynjolfsson, E., McAfee, A.: The Second Machine Age: Work, Progress, and Prosperity in a Time of Brilliant Technologies. WW Norton & Company (2014)
4. Constantinides, P., Henfridsson, O., Parker, G.G.: Introduction–platforms and infrastructures in the digital age. Inf. Syst. Res. **29**(2), 381–400 (2018). https://doi.org/10.1287/isre.2018.0794
5. DAMA UK: The six primary dimensions for data quality assessment: Defining data quality dimensions
6. Eisenhardt, K.M.: Building theories from case study research. Acad. Manag. Rev. **14**(4), 532–550 (1989)
7. Gibbs, G.R.: Analyzing Qualitative Data, vol. 6. Sage (2018)
8. Gioia, D.A., Corley, K.G., Hamilton, A.L.: Seeking qualitative rigor in inductive research: notes on the Gioia methodology. Organ. Res. Methods **16**(1), 15–31 (2013)
9. Gustafsson, J.: Single case studies vs. multiple case studies: a comparative study (2017)
10. Hartmann, M., Halecker, B.: Management of innovation in the industrial internet of things. In: 26th ISPIM Conference Proceedings, pp. 1–17 (2015)
11. Hermann, M., Pentek, T., Otto, B.: Design principles for industrie 4.0 scenarios. In: 49th Hawaii International Conference on System Sciences (HICSS), pp. 3928–3937. IEEE (2016). https://doi.org/10.1109/HICSS.2016.488
12. Horváth, D., Szabó, R.Z.: Driving forces and barriers of industry 4.0: do multinational and small and medium-sized companies have equal opportunities? Technol. Forecasting Soc. Change **146**, 119–132 (2019). https://doi.org/10.1016/j.techfore.2019.05.021, http://www.sciencedirect.com/science/article/pii/S0040162518315737
13. Kagermann, H., Helbig, J., Hellinger, A., Wahlster, W.: Recommendations for implementing the strategic initiative INDUSTRIE 4.0: Securing the future of German manufacturing industry; final report of the Industrie 4.0 Working Group. Forschungsunion (2013)
14. Karkouch, A., Mousannif, H., Al Moatassime, H., Noel, T.: Data quality in internet of things: a state-of-the-art survey. J. Netw. Comput. Appl. **73**, 57–81 (2016). https://doi.org/10.1016/j.jnca.2016.08.002
15. Khatri, V., Brown, C.V.: Designing data governance. Commun. ACM **53**(1), 148–152 (2010)

16. Kiel, D., Müller, J.M., Arnold, C., Voigt, K.I.: Sustainable industrial value creation: benefits and challenges of industry 4.0. Int. J. Innov. Manag. **21**(8), 1–34 (2017). https://doi.org/10.1016/j.jnca.2016.08.002

17. King, N.: Template Analysis. Sage Publications Ltd (1998)

18. Kooper, M.N., Maes, R., Lindgreen, E.R.: On the governance of information: introducing a new concept of governance to support the management of information. Int. J. Inf. Manag. **31**(3), 195–200 (2011)

19. Lasi, H., Fettke, P., Kemper, H.G., Feld, T., Hoffmann, M.: Industry 4.0. Bus. Inf. Syst. Eng. **6**(4), 239–242 (2014)

20. Lennerholt, C., van Laere, J., Söderström, E.: Data access and data quality challenges of self-service business intelligence. In: 27th European Conference on Information Systems (ECIS), pp. 1–13 (2019)

21. Lu, Y.: Industry 4.0: A survey on technologies, applications and open research issues. J. Ind. Inf. Integr. **6**, 1–10 (2017)

22. Madnick, S.E., Wang, R.Y., Lee, Y.W., Zhu, H.: Overview and framework for data and information quality research. J. Data Inf. Qual. **1**(1), 1–22 (2009). https://doi.org/10.1145/1515693.1516680

23. Marshall, C., Rossman, G.B.: Designing Qualitative Research. Sage Publications (2014)

24. Meuser, M., Nagel, U.: The expert interview and changes in knowledge production. In: Bogner, A., Littig, B., Menz, W. (eds.) Interviewing Experts, pp. 17–42. Springer, Heidelberg (2009). https://doi.org/10.1057/9780230244276_2

25. Otto, B.: Organizing data governance: findings from the telecommunications industry and consequences for large service providers. Commun. Assoc. Inf. Syst. **29**, 45–66 (2011)

26. Rai, A., Patnayakuni, R., Seth, N.: Firm performance impacts of digitally enabled supply chain integration capabilities. MIS Q. **30**(2), 225–246 (2006). http://dl.acm.org/citation.cfm?id=2017307.2017310

27. Shankaranarayanan, G., Blake, R.: From content to context: the evolution and growth of data quality research. J. Data Inf. Qual. **8**(2), 1–28 (2017). https://doi.org/10.1145/2996198

28. Strauss, A., Corbin, J.M.: Grounded Theory in Practice. Sage (1997)

29. Tallon, P.P., Ramirez, R.V., Short, J.E.: The information artifact in IT governance: toward a theory of information governance. J. Manag. Inf. Syst. **30**(3), 141–178 (2013)

30. van den Broek, T., van Veenstra, A.: Modes of governance in inter-organizational data collaborations. In: 23rd European Conference on Information Systems (ECIS), pp. 1–12 (2015). https://doi.org/10.18151/7217509

31. Wang, R.Y., Strong, D.M.: Beyond accuracy: what data quality means to data consumers. J. Manag. Inf. Syst. **12**(4), 5–33 (1996)

32. Weber, K., Otto, B., Österle, H.: One size does not fit all–a contingency approach to data governance. J. Data Inf. Qual. **1**(1), 1–27 (2009). https://doi.org/10.1145/1515693.1515696

33. Yin, R.K.: The case study as a serious research strategy. Knowledge **3**(1), 97–114 (1981). https://doi.org/10.1177/107554708100300106

Design of an Architecture of a Production Planning and Control System (PPC) for Additive Manufacturing (AM)

Wjatscheslav Baumung$^{(\boxtimes)}$

Reutlingen University, Alteburgstraße 150, 72762 Reutlingen, Germany
Wjatscheslav.Baumung@reutlingen-university.de

Abstract. Additive Manufacturing is increasingly used in the industrial sector as a result of continuous development. In the Production Planning and Control (PPC) system, AM enables an agile response in the area of detailed and process planning, especially for a large number of plants. For this purpose, a concept for a PPC system for AM is presented, which takes into account the requirements for integration into the operational enterprise software system. The technical applicability will be demonstrated by individual implemented sections. The presented solution approach promises a more efficient utilization of the plants and a more elastic use.

Keywords: Production planning and control · PPC · ERP · Information systems · Additive Manufacturing · 3D-printing · Enterprise architecture

1 Introduction

Additive Manufacturing (AM) is a manufacturing process in which objects are produced from 3D model data by applying individual layers on top of each other [12]. Using the slicing process, the manufacturing object is first broken down into a set of individual 2D layers, followed by the generation of a machine code for each of the layers, from which an overall machine code is calculated for the entire object. This manufacturing process is in contrast to traditional subtractive and formative manufacturing methods. Compared to traditional manufacturing methods, AM does not require specific component-dependent tools [7,13,25]. This makes AM technology a universal production technology in the sense of Industry 4.0 [10,17], which can be used without being tied to individual production steps. In contrast, traditional manufacturing is characterized by the use of large quantities of task-specific equipment. This results in the requirement for production planning and control (PPC) to synchronize the individual plants and their operating sequences to all tasks necessary for the manufacture of specific products in order to avoid downtimes [23].

© Springer Nature Switzerland AG 2020
W. Abramowicz and G. Klein (Eds.): BIS 2020, LNBIP 389, pp. 391–402, 2020.
https://doi.org/10.1007/978-3-030-53337-3_29

Therefore, the design of the information system must take into account that the use of AM will lead to a paradigm shift compared to traditional PPC systems. Instead of coordinating a large number of individual task-specific production plants, it is necessary to coordinate a large number of production plants with the same tasks in order to achieve a high overall output. This requirement leads to the first research question (RQ1): *How should the architecture of a PPC for AM be designed to achieve efficient use of production facilities with the same area of responsibility?* Changes and short-term events are an inseparable part of every production system, such as changes in the order, personnel availability, machine availability and components. The fast and direct response to the changes that occur determines the competitiveness of a production system [24]. The need to consider changes in the planning and control of production leads to the second research question (RQ2): *Which methods should be used for considering events and changes occurring typically in production to achieve the production targets when using AM?* Increasing digitalization creates new opportunities for companies and their employees to perform well-known tasks. While flexible working hours are the norm for office workers, they are an exception for production workers. Due to AM's new production capabilities, these may lead to flexible working hours for production workers, which is the aim of the third research question (RQ3): *How should the process flows for planning AM manufacturing orders be designed to allow flexible working hours for employees?*

2   Related Work

As mentioned in the introduction, AM is a promising manufacturing technology which is also being investigated by the scientific community. Unfortunately, the scientific focus so far has mainly been on the manufacturing process and the materials themselves. In addition to the current research in the field of technology, due to the requirements of industrial application, research in the field of information systems is essential to develop a planning tool that has long been available for traditional manufacturing processes. In the context of AM production, Kucukkoc et al. are investigating the planning of several additive manufacturing systems with different capacity specifications, in which a genetic algorithm for the calculation of parallel additive manufacturing orders based on release and due dates was developed as a solution approach [15]. Kucukkoc et al. continue to investigate the planning of AM manufacturing systems in order to achieve maximum resource utilization while maintaining delivery times. For the problem of complex nesting problems, a solution is proposed in which parts from different sales orders are grouped together and distributed to the various associated manufacturing plants [8,16]. Li et al. examines the order acceptance and scheduling (OAS) for AM manufacturing plants in order to be able to make correct statements about price and delivery date in case of customer inquiries [19]. Complementary this article focuses on the integration of AM and the necessary control and planning processes into the internal system landscape of companies. Therefore modules of the operative business software are first examined in order to consider them in the architecture.

2.1   ERP Modules Related to the Handling of AM

The ERP system stores, organises, manipulates and analyses all company-relevant data related to business processes and their results and makes it available to employees at various levels [22]. The Sales and Distribution (SD) module processes and creates orders for this purpose, including data on the delivery commitment depending on the available production capacity [11]. All personnel issues are covered by the HR module, such as qualifications and qualification measures and attendance processes [18]. Due to the strong operational networking to the service providers and the departments among each other, all relevant information about the products is managed and distributed in Product Lifecycle Management (PLM) and Product Data Management (PDM). PLM manages all internal and external development and value-added processes along the entire product life cycle and makes this information available to employees [6]. The PDM is part of PLM, which is responsible for the data management and storage of the product lifecycles, including CAD data, models and other documents, and for which the versioning of product data can also be used [6]. The concept provides that all data is managed in the context of a product. The usual PPC system, further referred as traditional PPC, deals with the handling of production order related data, such as bills of material, calculation of material requirements, scheduling, creation and release of sequences of production orders at production lines. The utilization of the production facilities is also monitored and reports are prepared for quality management.

2.2   Derivation of the Requirements to the PPC for AM

In order to determine the requirement for PPC in additive production, it is appropriate to first review the nature of the traditional production process. In contrast to AM technology, traditional manufacturing processes have a significantly lower degree of digitization, which is reflected in a large number of manual steps [9]. Due to their long existence, however, the manufacturing processes are well known and can therefore be well planned in companies. Therefore, in the following we will briefly discuss the manufacturing process and the planning of subtractive manufacturing (SM), which has a representative share of traditional manufacturing processes, in order to illustrate the changes resulting from the use of AM. In SM, objects are manufactured by material removal. SM requires specific tools and fixtures that vary according to the objects to be produced. The use of object-dependent specific tools requires that assembly and dismantling times are taken into account when scheduling production orders [14]. The sequence of objects to be produced must be considered so that a large number of specific tools can be reused for the next part in production 1 [21]. On the one hand, this large number of necessary work steps requires long and complex planning and, on the other hand, the production of the individual objects is associated with high costs, as the assembly and disassembly times are no longer in proportion to the actual processing time. This results in rigid planning, which requires a great deal of effort to be able to react to changes. The scheduling and scheduling of manufacturing orders is the task of the PPC systems [20]. Precise planning with high utilization of the production machines has a considerable influence on the company's production costs [20]. The basis for an exact planning are the usable data sets, in the form of time measurements for the necessary work steps and the respective throughput times. A problem for the time estimation are the work steps which are carried out by the employees. Either historical values are used or estimates of the necessary working time for the work step are made in the first step and later corrected on the basis of the historical values. This form of planning is particularly critical for the production of individual parts, since there are no directly usable historical values and therefore the entire planning process is based on a large number of estimations, which results in a high risk of the planning deviating. In contrast, AM does not require a special tools, which means that assembly and disassembly times are eliminated. Consequently, on the production cost side, there is no difference between serial and individual parts. The required work steps, disassembly and release, which are carried out by personnel, are not component-specific and therefore have no influence on production costs. In summary, AM enables the continuous production of individual parts. These advantages can also be transferred to the production of serial parts, which enables dynamic planning of production orders. One aspect that is not required under SM, but is necessary for AM planning, is the ability to combine different parts from different products and jobs in one print job in order to achieve optimum utilization of the individual production facilities.

3   Proposed Architectural Design for the AM PPC System

The following describes an architecture that combines a complete automation of different types of AM production plants, in which all processes are networked through the exchange of information and are thus computer-controlled. The necessary manual work steps are digitally recorded and scheduled to perform sequential processing of the tasks in the AM production plants. Once the manual work instructions have been issued, the production systems are released for production again. The production start of a system is automatically at the scheduled time. The desired result is the shortest possible downtime of the production systems. The interaction and the areas of responsibility of the individual plant sections are shown in the following subchapters.

3.1   The Overall Architecture

For the development of a production plan for the AM production facilities, a multi-layer model must be used due to the many interconnected processes. The various parameters from the HR, MM, SD, PLM and traditional PPC modules must be considered. Due to the interdependence of the tasks involved, changes affect not only current processes but also subsequent processes. Depending on the specifics of the change and the nature of the change itself, this can lead to a complete rescheduling of a production period. The described production architecture uses a number of process-influencing input parameters, such as: 1) changes in customer and production orders from the SD module, 2) the availability of employees, from the HR module, and machines as well as CAD models from the PLM module. The following section describes the system architecture of the PPC for use in networked AM systems, the respective processes for the various input parameters are described and a detailed description of the various types of changes from the ERP modules is contained in Sect. 3.3 to 3.5.

The first prerequisite for scheduling is the determination of available production resources, including employees, materials and production equipment. Following, the required production resources, such as the CAD model and customer orders are compared as input parameters. To obtain information about the required production resources, a machine code is generated for each model of each type of production plant. Using the machine code, parameters can be read out by simulating the production process to determine how much production resources are required and the working time at the respective production plant. This data is linked to the CAD model and stored in the AM-PPC's Manufacturing data model, rather than in the PDM. This results from the fact that production plants of the same type can have different parameters. A later exchange of a model, e.g. due to a revision, would lead to a complete new entry of the required production resources for the model. The requirements for manufacturing order batches are calculated from the BOM (PPC) and sales orders (SD). For this purpose, the available production area of a production plant is used as much as possible, taking into account the maximum usable time period. The models that can be combined with each other are performed by a nesting or packaging algorithm. A precise sequence of the necessary steps for a time-oriented combination of objects has already been described in an article [8]. A restriction on the free combinability of different models is the special section on grouping. The grouping describes a set of parts to be completed simultaneously, as required for the assembly of the individual components. The parts can be bundled together and distributed to several printers during subsequent completion to ensure they follow each other during sequential removal. For these dispatched orders, the routings with the necessary work steps are scheduled. Sequential processing at the plants is maintained, whereby the completion time of a production plant together with the working time for the respective task forms the completion time of the next production plant. The start-up of an individual production plant then begins automatically via the manufacturing execution system (MES). Unexpected events, such as the occurrence of errors in an individual production plant, are also reported to the MES, which then triggers the appropriate actions, such as a technician's request message and the event of a change in machine availability. The respective products or components are then received as output parameters. A distinction as to whether the output parameter is a product or a component is made by the input parameter customer or production order, since further processing steps are required for a production order and therefore further planning must be carried out in the non-AM PPC and ERP system. In a production order, any combination of products and components is carried out on the basis of the available production resources. The architecture is shown schematically in the Fig. 1.

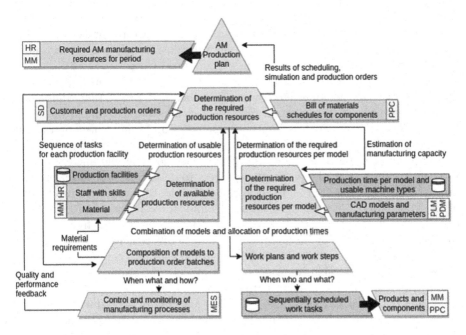

Fig. 1. Architecture for a AM PPC with dynamic planning and networked components

3.2   Scheduling of Production Batches

Backward scheduling is used for scheduling production orders. This type of scheduling is advantageous at this point due to the flexibility for changes that can be taken into account in the production orders before the start of the production process. If an order is changed in quantity or a new one is added, these changes can be taken into account in the period up to the planned start of printing by rescheduling the models to be printed or by adding new models in order to be able to use the available production period. This approach also allows for adaptability to changing production conditions. When allocating production orders, not only the current utilization of the individual machines is taken into account, but also the workpieces sent by the machine and in production. If errors are detected by the machine operator during the production process or after completion, the input stream of new workpieces can be blocked at the respective production line. The rejected faulty parts can be distributed to the machines and produced within the time window in which changes can be reacted to. In this way, a capacity bottleneck is avoided and compliance with the set deadlines is striven for (Fig. 2).

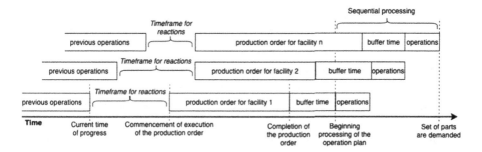

Fig. 2. Backward scheduled production orders with illustration of the response time for changes

3.3   Handling the Data from the SD Modulel

If the order situation changes, the capacity utilization of the individual machines is corrected in the form of individual production orders. This change is made due to cancellations, changes in order quantity or the acceptance of new orders. If the order quantity is reduced, all planned orders can be rescheduled and parts redistributed so that individual machines can be shut down without unnecessary operating costs. If an order increases beyond its own maximum production quantity, additional production capacity from other manufacturers can be consolidated directly, since apart from the digital object and the production parameters, no other parts or tools are needed that would require transport logistics or adjustment work.

3.4   Handling Data from the HR and MM Modules

The input parameters of the availability of the material, the production facilities and the employees represent a central module for scheduling. First of all, it must be checked when which employees with certain qualifications are available in specific periods. This ensures that the architecture always follows the personnel specifications without specifying them. Available production times are thus known and an initial filtering of the 3D models to be produced can be carried out on the basis of the available production facilities. Starting from the production facilities the materials and parameters such as type, manufacturer and colour have to be filtered further. The different model combinations are then calculated for the production orders, while maintaining and checking the available production times and the calculated material consumption for the respective combinations. When changes occur in a staff attendance or qualified work performance, a new planning of the production orders for the respective work phase has to be carried out. Missing work can either be shifted to a next work phase or lead to a standstill of the respective production plant. When a production line confirms the completion of an order, the system checks whether the confirmation took place in the planned period and how much material was actually consumed. If there is a deviation from the planned period, the system triggers a change in machine availability. If this occurs frequently, a different simulation of the production orders must be selected for the forecast of the execution times.

3.5   Handling the Data from the PLM Module

Components and products may change due to errors in design or production parameters, improvements or model maintenance, or complete replacement, ultimately resulting in changes to the associated CAD models. A CAD model updated in the PLM module can be directly considered for all production orders that have not yet been started by the event that has occurred. This allows a complete rescheduling for all production orders with the affected model to be carried out directly. However, it is important to consider how to proceed with the components already belonging to a product. Either all remaining components that belong to the product must be produced and then converted to the new series, or the system must convert directly to the new product series and dispose of the produced parts.

4   Implementation of the Architecture

The technical feasibility of the designed architecture and its sections was tested by individual implemented sections of the AM PPC, which are integrated into an operative business software. As business software the ERP system of Odoo [2] with the modules Sales, HR, Manufacturing and Inventory was used, as it is available as open source and offers the possibility to integrate self-developed modules. The PDM module is an experimental feature as a distributed version control system (DVCS) was used. The use of a DVCS enables the management of new and modified CAD data in which the life cycles of the products are managed in repositories and used in the AM PPC module for the order and release processes. The release process can be triggered by committing a new CAD model to the development repository of a product with the associated manufacturing data and can only be transferred to the product version repository after it has been checked by AM engineer, manufacturing and quality management for a specific product version. However, it should be noted that most CAD modelling software does not offer the possibility of merging multiple changes in one CAD file, as there is no tool to display the individual changes. Using this type of modeling software, a central version control system (CVCS) can be used, which provides the ability to set file locks to prevent multiple editing and overwriting changes. In contrast to the usual interactive CAD modelling software, the OpenSCAD [3] software offers modelling by script files, whereby individual changes can be displayed and merged. Due to the described release process and the use of repositories for different product versions, a DVCS was used for the implementation as PDM. For this purpose git with a self-hosted gitlab[1] instance was used. The AM-PPC module itself was written as a web-based application using the Python programming language and the Django framework. In the following, the implemented process flow logic for the creation of production orders is explained and can be seen in Fig. 3, divided into the areas for all objects and on demand.

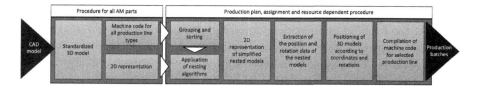

Fig. 3. Implemented process flow for the creation of production orders

The implementation requires a standardized 3D model of the CAD model as the first input parameter from the PDM module. The model is then adjusted and checked for errors, which would lead to complications later in the process. To extract the AM PPC relevant manufacturing data from the models, a machine code with the manufacturing parameters for each machine type is generated for each model. This extracted data, such as production time, material consumption and object volume, is stored in the AM-PPC module. The grouping and sorting is done according to defined priorities and completion times of the models. The grouped models are then aggregated into further constellations for the production lines based on the sum of their production times and a percentage buffer time that is added until the available production time is reached. The buffer time is an estimate, since the effective print time of all models of an order differs from the sum of the print times of the individual models due to the additional positioning procedures for which no material is extruded. The records obtained in this way were then checked using an interleaving algorithm to see whether the objects could be combined. The objects that did not fit on the production area were exchanged with other objects that have a similar production time but differ in dimensions or volume. In order to maximize the utilization of the individual production lines, different nesting methods can be used. The nesting algorithm has been separated as a separate service that can be accessed via an API. The nesting methods differ in the efficiency, but also in the required computing capacity. Nesting is done by shadowing the 3D manufacturing object on the Z-plane. The shadow casting itself provides the most precise 2D representation of the individual models and thus also the possibility for maximum utilization of the building space. At the same time, it also requires the most computing power for the calculation of the nesting. A maximum simplification of the object as a 2D representation is represented by the maximum dimensions themselves. Although the rectangles formed from this simplification allow nesting with the lowest computing capacity requirement, they also exhibit the most inefficient. Therefore, when selecting the nesting algorithm, the available IT resources, the number of objects to be nested and the available computing time, the selection of a suitable simplification of the models must also be considered, as reported in [8]. The merging of the different 3D models was done in OpenSCAD and then exported as a single stl file and sliced. This tedious approach had to be used because there is currently no way to automatically manipulate multiple stl files using the command-line interface (CLI) of the tested slicers Cura [5] and Slic3r [4] for positioning. This is a limitation, as only models with the same manufacturing parameters can be used for the creation of production orders. After nesting is complete, the rotation and positions for each body are read out and assigned to the objects. Based on these data, the 3D objects are then ported, from which a machine code for the respective production plant is generated.

5 Conclusion

This paper proposes an architecture for a PPC system for a networked AM plant. The architecture is supplemented by a description of the functional requirements for the individual modules of the PPC. The architecture is complemented by a description of the functional requirements of the individual modules and their integration logic for the execution of the process flow, which can be used for subsequent system implementation and testing. The PPC architecture described provides a suitable approach for controlling a large number of AM production plants sharing the same tasks. The implementation and simulation-based testing of the proposed architecture has to answer the question whether the design goals set have been achieved: namely, whether the proposed PPC contributes to the reduction of manufacturing costs due to the increased utilization of the AM plants and an early detection of over- and under-capacities due to the planning runs. In terms of distributed manufacturing, a DVCS was used during implementation to map an incremental change history for distributed collaboration between designer and manufacturer. The advantages and problems of using DVCS as PLM were also discussed. In addition, the proposed control logic allows variable working hours of the producing employees. The planning is not fixed for the employees (as is the case in the traditional production model), instead the production processes are aligned with the working times of the maintenance personnel and allow changes (in the employee's work schedule) and interruptions. During the test runs, a limitation of the currently usable slicers was noticed in the handling of stl files using the CLI, which means that an automated machine code cannot be produced jointly for models with different manufacturing parameters. This limitation should be explored, as by grouping the models with different production parameters, larger groups can be used to generate the production orders, which would lead to higher efficiency. Finally, the presented approach allows to plan and control the production of components and products with AM for a specific location or independently from a specific location in a networked production facility.

References

1. The first single application for the entire DevOps lifecycle. https://about.gitlab.com/. library Catalog: about.gitlab.com
2. Open-Source-ERP und -CRM—Odoo. www.odoo.com. library Catalog
3. OpenSCAD-The Programmers Solid 3D CAD Modeller. www.openscad.org. library Catalog
4. Slic3r - Open source 3D printing toolbox. https://slic3r.org/

5. Ultimaker/CuraEngine. Ultimaker, March 2020
6. Abramovici, M.: Future trends in product lifecycle management (PLM). In: Krause, F.L. (ed.) The Future of Product Development, pp. 665–674. Springer, Heidelberg (2007). https://doi.org/10.1007/978-3-540-69820-3_64
7. Attaran, M.: The rise of 3-D printing: the advantages of additive manufacturing over traditional manufacturing. Bus. Horiz. **60**(5), 677–688 (2017). https://doi.org/10.1016/j.bushor.2017.05.011
8. Baumung, W., Fomin, V.V.: Optimization model to extend existing production planning and control systems for the use of additive manufacturing technologies in the industrial production. Procedia Manuf. **24**, 222–228 (2018). https://doi.org/10.1016/j.promfg.2018.06.035
9. Birtchnell, T., Urry, J.: A New Industrial Future? 3D Printing and the Reconfiguring of Production, Distribution, and Consumption. No. 2 in Antinomies: Innovations in the Humanities, Social Sciences and Creative Arts, 1st edn. Routledge, Taylor & Francis Group, London (2016)
10. Frank, A.G., Dalenogare, L.S., Ayala, N.F.: Industry 4.0 technologies: implementation patterns in manufacturing companies. Int. J. Prod. Econ. **210**, 15–26 (2019). https://doi.org/10.1016/j.ijpe.2019.01.004
11. Gardiner, S.C., Hanna, J.B., LaTour, M.S.: ERP and the reengineering of industrial marketing processes. Ind. Mark. Manag. **31**(4), 357–365 (2002). https://doi.org/10.1016/S0019-8501(01)00167-5
12. Gibson, I., Rosen, D., Stucker, B.: Additive Manufacturing Technologies. Springer, New York (2015). https://doi.org/10.1007/978-1-4939-2113-3
13. Gu, D.D., Meiners, W., Wissenbach, K., Poprawe, R.: Laser additive manufacturing of metallic components: materials, processes and mechanisms. Int. Mater. Rev. **57**(3), 133–164 (2012). https://doi.org/10.1179/1743280411Y.0000000014
14. Herrmann, F., Manitz, M.: Ein hierarchisches Planungskonzept zur operativen Produktionsplanung und -steuerung. In: Claus, T., Herrmann, F., Manitz, M. (eds.) Produktionsplanung und –steuerung, pp. 7–22. Springer, Heidelberg (2015). https://doi.org/10.1007/978-3-662-43542-7_2
15. Kucukkoc, I., Li, Q., He, N., Zhang, D.: Scheduling of multiple additive manufacturing and 3D printing machines to minimise maximum lateness, February 2018
16. Kucukkoc, I., Li, Q., Zhang, D.: Increasing the utilisation of additive manufacturing and 3D printing machines considering order delivery times, February 2016
17. Lasi, H., Fettke, P., Kemper, H.-G., Feld, T., Hoffmann, M.: Industry 4.0. Bus. Inf. Syst. Eng. **6**(4), 239–242 (2014). https://doi.org/10.1007/s12599-014-0334-4
18. Lengnick-Hall, C.A., Lengnick-Hall, M.L.: HR, ERP, and knowledge for competitive advantage. Hum. Resour. Manag. **45**(2), 179–194 (2006). https://doi.org/10.1002/hrm.20103
19. Li, Q., Kucukkoc, I., He, N., Zhang, D., Wang, S.: Order acceptance and scheduling in metal additive manufacturing: an optimal foraging approach, February 2018
20. McKay, K.N., Wiers, V.C.S.: Planning, scheduling and dispatching tasks in production control. Cognit. Technol. Work **5**(2), 82–93 (2003). https://doi.org/10.1007/s10111-002-0117-4
21. Salonitis, K., Ball, P.: Energy efficient manufacturing from machine tools to manufacturing systems. Procedia CIRP **7**, 634–639 (2013). https://doi.org/10.1016/j.procir.2013.06.045
22. Subba Rao, S.: Enterprise resource planning: business needs and technologies. Ind. Manag. Data Syst. **100**(2), 81–88 (2000). https://doi.org/10.1108/02635570010286078

23. Sydow, J., Möllering, G.: Produktion in Netzwerken: make, buy & cooperate. Vahlens Handbücher, Verlag Franz Vahlen, München, 3, aktualisierte und überarbeitete auflage edn. (2015). oCLC: 920809509
24. Szelke, E., Monostori, L.: Reactive scheduling in real time production control. In: Brandimarte, P., Villa, A. (eds.) Modeling Manufacturing Systems, pp. 65–113. Springer, Heidelberg (1999). https://doi.org/10.1007/978-3-662-03853-6_5
25. Zhang, Y., Bernard, A., Gupta, R.K., Harik, R.: Evaluating the design for additive manufacturing: a process planning perspective. Procedia CIRP **21**, 144–150 (2014). https://doi.org/10.1016/j.procir.2014.03.179

A Model Management Platform for Industry 4.0 – Enabling Management of Machine Learning Models in Manufacturing Environments

Christian Weber[1]([⊠])(ID), Pascal Hirmer[2](ID), and Peter Reimann[1,2]

[1] Graduate School of Advanced Manufacturing Engineering, University of Stuttgart, Stuttgart, Germany
{christian.weber,peter.reimann}@gsame.uni-stuttgart.de
[2] Institute for Parallel and Distributed Systems, University of Stuttgart, Stuttgart, Germany
pascal.hirmer@ipvs.uni-stuttgart.de

Abstract. Industry 4.0 use cases such as predictive maintenance and product quality control make it necessary to create, use and maintain a multitude of different machine learning models. In this setting, model management systems help to organize models. However, concepts for model management systems currently focus on data scientists, but do not support non-expert users such as domain experts and business analysts. Thus, it is difficult for them to reuse existing models for their use cases. In this paper, we address these challenges and present an architecture, a metadata schema and a corresponding model management platform.

Keywords: Model management · Machine learning · Metadata management · Industry 4.0

1 Introduction

Through the emergence of cyber-physical systems and initiatives, such as Industry 4.0 [8] and the Industrial Internet [13], manufacturing environments are equipped with a variety of sensors that produce large amounts of data. Analyzing these data enhances various use cases in manufacturing. These use cases include predictive maintenance, product quality control, root cause analysis of quality problems, and the optimization of manufacturing processes [6,16]. In the analysis process, data scientists experiment with different algorithms, frameworks and tools for machine learning (ML). The result of such data analyses are several promising candidates for the final machine learning model. By estimating the generalization error and conducting A/B tests, the most suitable model is selected and used for inference or prediction [3,9].

In the process of creating and maintaining a ML solution, a high amount of diverse models is created even for a single use case. The reason for this is

© Springer Nature Switzerland AG 2020
W. Abramowicz and G. Klein (Eds.): BIS 2020, LNBIP 389, pp. 403–417, 2020.
https://doi.org/10.1007/978-3-030-53337-3_30

the variety and multitude of manufacturing processes, machines and product variants that require individual models. This is further reinforced by the fact that companies implement many different uses cases that multiply the amount of models being created. Due to the high dynamics of the manufacturing processes, such models tend to become stale very quickly, e.g., because of concept drifts in data [1,4,17]. For example, a machine learning model that predicts the wear of a tool based on multiple sensors may lose quality because patterns in data change, e.g., through changes in air pressure or damage to a sensor. Thus, stale models need to be continuously replaced by new models. Since a large number of new models are created every day and others become stale at the same time, the management of these models turns into a great challenge. Hence, for all stakeholders, it becomes a tedious task to keep an overview of all models in the manufacturing environment. This prevents effective reuse of machine learning models and may result in costly reengineering tasks. So, data scientists often recreate models for each new use case, although they could possibly reuse an existing model that has been created for a similar use case.

In order to cope with these issues, we introduce a Model Management Platform (MMP) for I4.0 that enables to store, index, and retrieve machine learning models in a manufacturing context. Our platform enables central access to all involved models and provides domain experts with a clear overview of different versions of models and their current status, i.e., whether they are currently planned, experimental, in use, maintained or retired. A rich interface to query context data facilitates the effective discovery of models for similar use cases and ML tasks. Furthermore, the platform provides reporting functionalities for domain experts to keep the overview on all models within the company.

Related work does not cover these issues to a full extend, i.e., they do not consider the manufacturing or business context in the lifecycle of a machine learning model. Furthermore, to the best of our knowledge, current solutions do not provide status tracking to machine learning models.

The remainder of this paper is structured as follows: In Sect. 2, we provide an exemplary model management use case which guided the design of our platform and also serves as base for our later case study. In Sect. 3, we describe foundations and discuss related work in the scope of model management and related platforms. Section 4 describes our main contribution: the Model Management Platform (MMP), including its metadata schema and architecture. Section 5 discusses the implementation of the platform and presents the features of the platform by a case study. Finally, Sect. 6 concludes this paper and lists possible future work.

2   Exemplary Use Case and Requirements

We provide an exemplary use case in which a platform for model management becomes crucial. The use case was derived by conducting interviews with our industrial partners from the manufacturing sector.

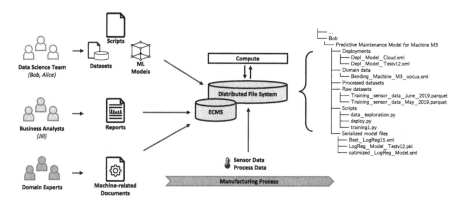

Fig. 1. Example showing the current IT infrastructure of the model management use case of the steel plate manufacturer. Each data scientist owns a ML project folder on the distributed file system with his or her files. Business Analysts and Domain Experts manage their files in an Enterprise Content Management System (ECMS).

2.1   Use Case Environment

We consider that a medium-sized manufacturer produces tailor-made steel plates for customers all over the world. In order to serve the market in the best possible way, the steel plate manufacturer owns several plants in different countries. The products are of good quality, but in the past there have often been problems in meeting delivery times. Completing orders took longer than planned because machines suddenly failed due to wear on machine parts. Therefore, based on the experience of production employees, the manufacturer introduced fixed maintenance intervals to replace machine parts preventively. Nevertheless, there were still occasional machine downtimes, as machine parts were worn out faster than planned. Furthermore, the maintenance personnel noticed that some machine components showed only slight signs of wear and were replaced much too early. In order to better determine the maintenance intervals individually for each machine component, the steel plate manufacturer introduced a ML solution which can make predictions based on historical wear data for the machine components. At the moment, the data of machines from different locations are collected centrally. This includes time-series data, such as temperature, humidity, and vibration data, as well as high-resolution image data from cameras.

2.2   Use Case for Model Management

We imagine the data scientists Bob and Alice, who are located in different departments. Together with other data scientists, they form a decentralized data science team. In addition, Jill is a business analyst whose task is to keep an overview on ML projects in the company. Bob, Alice, and Jill share a common IT infrastructure (see Fig. 1). A distributed file system is part of this IT infrastructure to store datasets, scripts, and model files of ML projects.

Bob wants to create a machine learning model that predicts downtimes for a specific machine type. After receiving data about the machine, he performs several machine learning experiments by writing a Python script and trying out several algorithms for logistic regression. Each time the script is run, it creates a new version of the model. Some of the model versions provide good evaluation results and meet the target thresholds of evaluation metrics. Therefore, he marks them as candidates for the final model. Through A/B testing, he selects the best performing model and deploys it to a ML serving system. The deployed model may however become stale at some point of time, because the data from machines is non-stationary and concept drifts occur. Bob retrains and redeploys the model from time to time to keep it up-to-date, which results in a new model version. Each time he trains or retrains a model, he stores the raw data, his script, and model files on the distributed file system, thereby using his individual folder structure and file names as shown in Fig. 1. Exemplary names for model files include *LogReg_Model_Testv12*, *optimized_LogReg_Model*, *Depl_Model_Testv12*.

Later, Alice begins to develop a logistic regression model for another machine that includes the same sensor types as Bob's machine. Unfortunately, she knows nothing about Bob's experiments. After performing a short search on the distributed file system, she cannot find any related models. She is confused by the names of the model files and wonders to which machines which model belongs. Therefore, she decides to develop the model from scratch and spends a lot of time on it. By chance she meets Bob, who tells her that he has developed a model for a similar type of machine. Bob shows Alice where he saved his model files. After examining his files and doing A/B tests on the models, Alice finds that Bob's final model is more accurate than any of the models she created on her own. She uses Bob's model and puts it into production. She copies Bob's files into her folder on the distributed file system.

Some time later, Jill needs to create a report for the chief digital officer. The report should contain an overview of currently used models, the involved departments and business processes. She talks to Alice and Bob to acquire the necessary information. With this information, Jill creates a report using a spreadsheet. Jill, however, complains about the slow reporting process. This is because Bob and Alice must manually examine the model serving system and cannot do so immediately due to other tasks.

2.3   Required Functions of a Model Management Platform

Overall, the steel plate manufacturer's analytics team has several problems establishing an efficient way of model management. For these reasons, the manufacturer's data science department wants to establish a model management platform to cope with the depicted issues. In the following, we discuss the issues of the use case and identify required functions a model management should offer.

Although the analytics team stores data sets, scripts and model files in a customized folder structure on a distributed file system, it is difficult to keep track on which model is the most recent one and which files relate to it. This can even lead to re-deploying model versions which have even been retired. Here, a function to *versionize models*, according datasets, and scripts would be beneficial.

In addition, information about assets, such as machines and other manufacturing equipment for which the models are built, is hidden in an ECMS where domain experts store their documents. Therefore, models are isolated from these documents and it is hard to discover related models used for similar assets and for corresponding ML tasks. These management problems let data scientists (re-)develop models, even if they exist already. Thus, a function to add *manufacturing and business context* to models is required. Afterwards, the platform can use the added context to provide a better discovery of models and to provide views for effective model reporting. For example, users may search for models that are appropriate for a certain business process or machine. In addition they may also want to have an overview on models that are used within different departments of the company.

Further difficulties exist to distinguish between models. That is, experimental versions of the model are stored along with model candidates and the final version in the same folder. Beyond managing the files themselves, it is also unclear which model is currently in use or stale. This requires a function to distinguish these different *states of ML models*.

In sum, required model management functions include a) *model versioning*, b) *providing context*, and c) *tracking the status* of models across their lifecycle.

3   Related Work and Challenges

Our work is based on the process model for managing machine learning models by Weber et al. [15]. This process is depicted in Fig. 2. It shows all the lifecycle steps a machine learning model passes through and that must be supported by model management. In the following, we discuss the steps in the process model and show how the derived model management functions a), b) and c) (see Sect. 2.3) can provide support for each process step. Afterwards, we discuss related work on model management platforms and possible tool support.

In the first step, new use cases serve as initiators to plan models that can help, e.g., to improve manufacturing processes. Planning includes, e.g., specifying the involved data sources, describing specifics of the use case, defining acceptance criteria, or documenting expected results. Different stakeholders are required that bring in domain-specific or technical expertise. In this step, function b) providing context to models would help to identify and re-use existing models for similar use cases. After that, if the model passes a feasibility check, the model is created and evaluated by data scientists and programmers (step 2). In order to document results of experiments and to store different model files, function a) model versioning is required. In case the evaluation of the model shows promising results, the model can be deployed into its target environment (step 3). Otherwise, e.g., because the data quality is poor, the model needs to be replanned or even retired (step 6). After deployment, the model is used, e.g., to make continuous predictions (step 4a). In parallel, it is constantly monitored

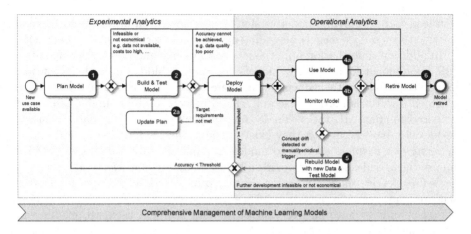

Fig. 2. Process of the model lifecycle [15] that is covered by the platform. Blue lines: experimental loop, green lines: update loop, purple line: upgrade loop. (Color figure online)

to recognize concept drifts (step 4b). In case of a concept drift, the model needs to be rebuilt with new data and tested once again before it can be redeployed (step 5). Here, function a) is also needed because this step is similar to step 2 in that different versions of a model may be created. Once a model is not useful or not needed anymore, it is retired (step 6). Function c) tracking the status of models is required throughout the whole model lifecycle to check the status of a model in each step, e.g., if it is already in production or just a candidate for future application. This prevents users from selecting models that are outdated or do not conform to their requirements.

In order to evaluate solutions for implementing the functions proposed in Sect. 2.3, we also conducted an evaluation of different tools for managing models in machine learning projects. For function a) model versioning, data scientists can rely on versioning tools that provide a repository such as Git LFS [2] or more specialized tools like DVC [11] for bigger model files. Additionally, for storing metadata about how the model files were produced, data scientists can use experiment management tools. Examples for such tools are ModelDB [14] and MLflow [18], which is available both in the Microsoft Azure platform and as a standalone application. These tools store settings and metrics about training runs, and they track the history of experiments along with their evaluation results. This makes it possible for data scientists to reproduce experiments and share their results with others. Furthermore, the tools can be used for saving metadata about experiments, as well as materialization and versioning of model files produced in step 2) and 5) of the ML lifecycle. MLflow provides a model registry which allows to add states to a model version, *staging* and *production*. Multiple model versions belong to a model that has a unique name. However, we argue that the status information should be kept by the model entity as well because status information about other lifecycle steps is required. A platform that addresses the whole lifecycle would need additional states for each lifecycle step, e.g. planned (step 1), experimental (step 2), staging (step 3), production (step 4a, 4b), stale (step 4a, 4b), maintenance (step 5) and retired (step 6). Thus, c) tracking the status of models across their lifecycle is partially supported. Since these tools are intended to support data scientists rather than domain experts and business analysts, requirement b) providing context information to models is also an open issue that we want to support with the platform.

Based on the required functions resulting from the model management use case as well as our discussion of related work and tool support, we identified open challenges that must be addressed by the platform. These challenges can be summarized as follows:

① **Tracking the status of models across their lifecycle** –
According to the process model for the model lifecycle, a model passes several steps. In each step, metadata is generated that can be used to further describe the model. However, current tools and related work do not focus on tracking metadata across all lifecycle steps to describe the whole lifecycle. The following states require support by a model registry component and an appropriate metadata schema: planned (step 1), experimental (step 2), staging (step 3), production (step 4a, 4b), stale (step 4a, 4b), maintenance (step 5) and retired (step 6).

② **Adding context to models for discovery/reporting** –
Currently, there is no related work on adding context to machine learning models to enable functions for search and discovery that are tailored to domain experts. In order to discover the most suitable model for a given use case, it is essential that users are provided with context information. However, this information is kept in files that are often located in Content Management Systems. These files may contain semantic information about manufacturing assets that conforms to standards for Industry 4.0 or the Industrial Internet. Due to broader use of these standards in the future, they need to be supported by different platforms such as model management platforms. To cope with these formats in manufacturing, a platform requires concepts to extract, store, and link information about manufacturing assets to a machine learning model. This would allow data scientists to search for a model version that matches a specific machine or sensor type. Moreover, through adding business context, business analysts can be supported with model reporting functionalities which reduces manual efforts to collect the required information from different stakeholders.

4   Model Management Platform for Industry 4.0

Our model management platform (MMP) builds on the process model of the model lifecycle by supporting users in each step. In this section, we present a metadata schema, as well as the MMP's architecture and its components. We follow up with details about the implementation of the components.

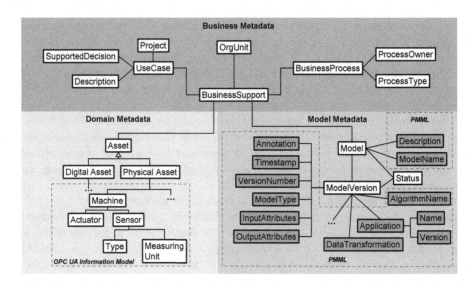

Fig. 3. Extract from the MMP metamodel, showing the combination of the model, business, and domain metadata package.

4.1 Metadata Schema

For storing models with according metadata for the different steps of the lifecycle process, we provide a metadata schema as depicted in Fig. 3.

The model metadata are provided via a model metadata package. In order to provide structure to massive amounts of ML models, the lifecycle of machine learning models is represented by the *model entity*. A model entity is comprised of several *model versions* that are produced within steps 2) and 5) of the ML lifecycle process. It contains a model name, a description and the status of a model. Thus, the metamodel supports challenge ①.

A model version contains descriptive attributes, e.g., the version number, model type, annotation, a creation timestamp, and the input and output parameters. Furthermore, a model version contains lineage metadata that describe the process to generate the model version, e.g., metadata about experiments – the algorithm used, data transformations used for preprocessing the input attributes and quality metrics. These metadata can be parsed from model files by a model metadata extractor component (attributes are color-coded in grey).

In contrast, the domain metadata package describes the manufacturing context. Different *assets*, e.g., *machines*, stations and manufacturing equipment that provide input data to models can be linked to the model's lifecycle. In order to support linking models to as much manufacturing entities as possible, we rely on the concept of Gröger et al. [5] which stores links between models and manufacturing entities in a data warehouse. However, this concept does not provide means to link models to semantical information contained in files that are not part of the datawarehouse, e.g. XML-based OPC UA files that describe machines. We hence support parsing these files, extracting their content, storing it, and linking it to a model. Currently, we support parsing semantical descriptions of machines. By providing the machine's *actuator*, as well as *sensors* with their *type* and *measuring unit*, we can add context to models, e.g., to search for models for a certain machine and/or sensor type. Thus, we extend the concept of Gröger et al. by providing further details on machine entities.

We further define a business metadata package to support reporting for models. It supports linking *business processes, organizational units*, and *use cases* of *projects* to the model lifecycle entity. A use case contains the supported decision for which the model should provide support and a description. Examples for decisions are e.g. root cause analysis for production failures or the prediction of maintenance schedules. The description then explains more domain specific details about the problem, e.g., about quality issues, downtimes, the production process, production materials, and machines. This enables to provide visualizations such as dashboards and overviews on models used within different departments of a company.

By combining the business, domain and model metadata package via a *BusinessSupport* entity, the metamodel supports adding context to models and facilitates model discovery and reporting according to challenge ②.

4.2   Architecture

Our architecture (see Fig. 4) consists of a frontend component, a backend component, and data storages for metadata and model files. The frontend component offers a graphical user interface to manipulate and query the metadata and model store. In the backend component, corresponding functions are provided to manage models and according metadata based on user input. The detailed components are presented in the following. Thereby, we point out how these components offer the required functions discussed in Sect. 2.3 and especially how they cope with the challenges identified in Sect. 3.

Frontend Components. The frontend provides functions for model management and model reporting. It communicates with the backend components through a REST API. Model management functionalities include model upload, model comparison, model deployment and model discovery. Model reporting comprises a model landscape visualization which provides a holistic overview of models used in different organizational units and business processes for business analysts. Furthermore, a model dashboard provides multiple KPIs, e.g., about the amount of current models in use and added models over the past months.

Model Metadata Store. The model metadata store contains metadata that are generated in each step of the lifecycle process. These are organized according to the metadata model we provided in Sect. 4.1. The model metadata store enables functions for model discovery, model comparison, and model reporting

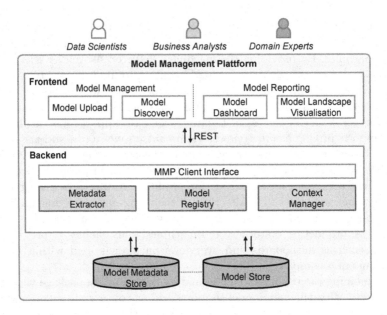

Fig. 4. MMP Architecture showing the backend and frontend components connected via REST interfaces

to store and retrieve information about models. For example, the stored metadata allows users to perform semantic queries to discover related models for a particular use case. By materializing metadata in the model metadata store, the platform can provide faster access than parsing model files at run-time for each user request.

Model Store. The model store is used to store machine learning models that should be materialized for future use. It is closely tied to the model metadata store, because metadata of models need to be linked to materialized model files. Additionally to different serialization formats for ML models, the model store contains information models of machines.

Model Metadata Extractor. The metadata extractor is a functional component that extracts metadata from model files by parsing their content. The model metadata extractor's functionality is of utmost importance to handle the indexing of all the model versions which are created in steps 2) and 5) of the model lifecycle process. By using the metadata, users can search for models with, e.g., specific input attributes or model types (e.g. Regression, Tree Model, Neural Network). The extracted metadata are inserted into the model metadata store and a link is set to the corresponding model file in the model store. For experimental models that should not be materialized, the user can decide that the extractor should only store the available metadata.

Due to the plenty amount of available tools and libraries for machine learning, data scientists also use different formats to persist their model files. A commonly used format is the *Predictive Modeling Markup Language (PMML)*. It is an open standard for an interchange format that is supported by a variety of tools, frameworks, and ML serving systems. As depicted in Fig. 3, grey attributes represent attributes that can be extracted from PMML files. PMML supports a variety of metadata and is therefore a suitable model format for storing models. The format is XML-based and allows to store the model and its metadata in one file. Hence, it can be parsed using common XML parsing libraries. However, there also a lot of other formats, e.g., *scikit-learn* models use *pickle*[1] as serialization format, *RDS*[2] is used for *R*, and *HDF5*[3] for *Tensorflow*. Often, these formats are binary formats that are just readable by the according libraries. For these model formats, the metadata extractor uses the *jpmml*[4] library to convert them into the PMML format. Thus, the metadata extractor also supports users who want to use their custom model formats. The models are kept in their converted and original model file format for users that want to deploy them quickly to different ML serving systems.

Model Registry. The model registry keeps track on model files and the actual status of a model. Thus, the component addresses challenge ①. The status is set for both, model version and model entity. The status of the model entity depends on the collective status information of its underlying model versions. For example, if a model contains multiple model versions with status "experimental" and another model version which is in production, the status of the model entity is set to "production". The same applies when the actual model version becomes stale, then the overall status of the model is "stale". Users can set the status information manually or by inserting specific REST-calls in their training scripts to collect it automatically. For example, if the user rebuilds a model, a setModelStatus-Call can be inserted right before the training routine to set it to maintenance. After the training routine and the deployment, another call sets the status to "production" again. Old model versions can then be set to retired.

Context Manager. The context manager is a component to realize function b) of the use case and to provide a solution to challenge ②. It provides interlinking between models, manufacturing entities, and business entities. In order to support model discovery for domain-related queries, we link machine learning models to assets in the manufacturing environment, e.g., machines. For that purpose, we rely on information models [12] that semantically describe machines. For example, the information model can describe device information, process

---

[1] Pickle: https://docs.python.org/3/library/pickle.html.
[2] RDS: https://www.rdocumentation.org/packages/base/versions/3.6.1/topics/read RDS.
[3] HDF5: https://www.hdfgroup.org/solutions/hdf5/.
[4] jpmml: https://github.com/jpmml.

variables or machine capabilities. Information models are part of the Open Platform Communications Unified Architecture (OPC UA) [10]. We consider OPC UA information models as suitable because they can be modeled according to companion specifications that are uniform across different machine manufacturers. Therefore, a machine learning model for a certain type of machine can be re-used across machines of the same type from different machine manufacturers. The linking of machine learning models to information models enhances the functions for model discovery. In order to also provide functions for model reporting to business users, it is possible to attach an enterprise architecture repository that contains information about business processes and organizational units. By interlinking these with the model metadata, we can provide powerful visualizations for business analysts such as the model landscape view (see Fig. 5). Such views are common in the scientific field of enterprise architectures and provide a holistic overview [7]. The model Landscape view shows business processes on top, departments on the left and corresponding models highlighted in green color. The view provides two benefits. First, business analysts can get an overview on all the models within the company. Second, they can also identify spots with missing models in the model landscape.

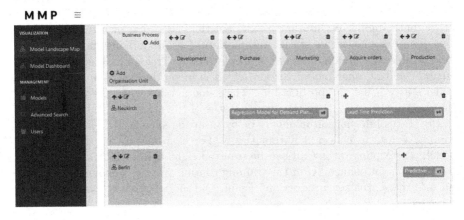

Fig. 5. The MMP Frontend – showing the model landscape of machine learning models in the manufacturing environment. Models (green) are listed with a name and a version number. A grid organizes the models via business processes on top and according organizational units on the left. (Color figure online)

5   Case Study

In this section, we assess how the platform addresses the main challenges ① **Tracking the status of models across their lifecycle**, and ② **Adding context to models for discovery/reporting**. For the assessment, we use the steel plate manufacturer's use case for model management.

The steelplate manufacturer develops a new type of machine that must be supported by a machine learning model for predictive maintenance. In the plan model step, Bob creates a new project on the MMP and adds a label and description to it. For documenting purposes, he also adds the business process "Machine Maintenance" and the department "Machine Maintenance Services" that should be supported by the model. After that, domain experts provide a dataset with historical records and an OPC UA information model of the machine to Bob. With the supported data, he conducts multiple machine learning experiments in the build & test model step. He screens the resulting models and selects the best performing model from a list of candidates. After uploading the model to the platform, the metadata extractor stores the model in the model store and then extracts its metadata and saves it to the model metadata store. Bob assigns the status "experimental" to the models created. In order to support model discovery for future experiments, he also uploads the OPC UA information model to the platform. The information model is parsed and its semantic information is linked to the machine learning model and stored in the model metadata store. He deploys the model into production and sets its status to "production". From time to time, he updates it on detected concept drifts (model status: "stale"). Models that are currently updated are set to status "maintenance" through entries in the training pipeline that interface with the MMP. When a new model version is deployed, the old version is set to status "retired" and the new version to status "production".

Some time later, Alice creates a new project in the MMP for the model she wants to develop. The status "planned" is assigned to the model. She does not know about Bob, but she has access to the model management platform. Therefore, she searches for similar models in the model store that satisfy her use case requirements. For instance, she searches for models that belong to a certain machine and sensor type. She discovers Bob's model as a possible candidate. With this information, she can also get more context, e.g., about the machine in which the sensor is built in and she also retrieves information about Bob, who built the model. She finds out that the model was developed for the same type of machine and that it already satisfies her requirements. Therefore, she copies it into the newly created project. Through the MMP's support, Alice saves time by finding and re-using existing models. The MMP stores a link between the copied version and the original model version of Bob to keep track of the history for later analysis.

Again, some time later, Jill wants to create the report about all models that are used for predictive maintenance in the company. With the MMP, she now can generate reports faster and provide them to the chief digital officer. She searches the repository for models with the status "production" and the term "predictive maintenance" in the use case description. By using the visualization function of the platform, she can display all the models with corresponding business processes and departments, e.g., in a model landscape view (see Fig. 5). This relieves Alice and Bob from the burden of creating such visualizations manually.

Right now, the status experimental can be automatically tracked via REST calls to the MMP which are inserted into the training scripts by data scientists. Candidate models need to be explicitly flagged on the platform, because it is a manual selection process. The same applies to the final model selected. When data scientists create a script to put the model into production, REST calls in this script put the model into status production. Typically such scripts contain a training routine that is also used to retrain a model when a model becomes stale. When the script is executed, the current model version is set to maintenance via a REST call. Stale models that are replaced by new ones are set to retired automatically. If a model is not used anymore data scientists set it to retired.

To sum up, the proposed platform and its function address both challenges introduced in Sect. 3: ① Tracking the status of models across their lifecycle, and ② adding context for model reporting and discovery. This constitutes an advancement to the state of the art, because existing model management platforms to do not offer such dedicated means. More precisely, our platform goes beyond existing platforms in that it provides status information across the whole lifecycle of machine learning models. Other platforms provide this information just for the steps *build and test model* and *use model* according to the process model we provide in Sect. 3. Furthermore, our platform supports adding context to machine learning models to enable functions for search and discovery that are tailored to domain experts and business analysts. Thus, it is the first platform that provides such functionalities for model management.

6   Conclusions and Future Work

Providing structure and context to machine learning models is a core requirement in ML projects. Machine learning models that are well organized facilitate the effective re-use of artifacts and collaboration between data scientists and business analysts. In this paper, we focus on the challenge of adding context to models to enable model discovery and reporting as well as tracking the status of models across their lifecycle. To address these challenges, we propose a model management platform. The model metadata is then extended to include additional context, namely, manufacturing and business metadata for Industry 4.0 use cases. The platform integrates with existing ML environments for training and model serving systems through standardized interfaces.

Our future work will focus on the collection of metadata from operational pipelines and lineage tracking across different steps in the life cycle of machine learning models. In addition, we are going to support more industrial standards for adding context and to develop more views to enhance user experience.

References

1. Breck, E., Cai, S., Nielsen, E., Salib, M., Sculley, D.: The ML test score: a rubric for ML production readiness and technical debt reduction. In: 2018 IEEE International Conference on Big Data (Big Data), pp. 1123–1132. IEEE (2017)

2. Carlson, B.M., Schneider, L., Schuberth, S., et al.: Git large file storage. https://git-lfs.github.com/
3. Ding, J., Tarokh, V., Yang, Y.: Model selection techniques: an overview. IEEE Sig. Process. Mag. **35**(6), 16–34 (2018)
4. Gama, J., Žliobaitė, I., Bifet, A., Pechenizkiy, M., Bouchachia, A.: A survey on concept drift adaptation. ACM Comput. Surv. **46**(4), 44:1–44:37 (2014)
5. Gröger, C., Schwarz, H., Mitschang, B.: The deep data warehouse: link-based integration and enrichment of warehouse data and unstructured content. In: 2014 IEEE 18th International Enterprise Distributed Object Computing Conference, pp. 210–217. IEEE (2014)
6. Gröger, C., Schwarz, H., Mitschang, B.: The manufacturing knowledge repository - consolidating knowledge to enable holistic process knowledge management in manufacturing. In: ICEIS (2014)
7. Kirchner, A., Scheurer, S., Weber, C., Wiechmann, A.: Architektur eines Cockpits zur interaktiven Analyse von Enterprise Architectures auf Basis von Viewpoints. In: Porada, L. (ed.) Informatiktage 2014, pp. 139–142. GI-Edition, Gesellschaft für Informatik, Bonn (2014)
8. Knafla, F., Loewen, U., et al.: Implementation strategy Industrie 4.0: report on the results of the Industrie 4.0 platform. http://www.zvei.org/Publikationen/Implementation-Strategy-Industrie-40-ENG.pdf
9. Kumar, A., McCann, R., Naughton, J., Patel, J.M.: Model selection management systems: the next frontier of advanced analytics. ACM SIGMOD Rec. **44**(4), 17–22 (2015)
10. Mahnke, W., Leitner, S.H., Damm, M.: OPC Unified Architecture, 1st edn. Springer, Berlin (2009). https://doi.org/10.1007/978-3-540-68899-0
11. N.N.: Data version control. https://dvc.org/
12. OPC Unified Architecture: Part 5 : Information Model (2017)
13. Sisinni, E., Saifullah, A., Han, S., Jennehag, U., Gidlund, M.: Industrial internet of things: challenges, opportunities, and directions. IEEE Trans. Ind. Inf. **14**(11), 4724–4734 (2018)
14. Vartak, M., Madden, S.: MODELDB: opportunities and challenges in managing machine learning models. IEEE Data Eng. Bull. **41**, 16–25 (2018)
15. Weber, C., Hirmer, P., Reimann, P., Schwarz., H.: A new process model for the comprehensive management of machine learning models. In: Proceedings of the 21st International Conference on Enterprise Information Systems , ICEIS, vol. 1, pp. 415–422. INSTICC, SciTePress (2019)
16. Weber, C., Wieland, M., Reimann, P.: Konzepte zur Datenverarbeitung in Referenzarchitekturen für Industrie 4.0: Konsequenzen bei der Umsetzung einer IT-Architektur. Datenbank-Spektrum **18**(1), 39–50 (2018)
17. Wuest, T., Weimer, D., Irgens, C., Thoben, K.D.: Machine learning in manufacturing: advantages, challenges, and applications. Prod. Manuf. Res. **4**(1), 23–45 (2016)
18. Zaharia, M., Chen, A., et al.: Accelerating the machine learning lifecycle with MLflow. IEEE Data Eng. Bull. **41**(4), 39–45 (2018)

BIS 2019

Refining Rule Bases for Intelligent Systems: Managing Redundancy and Circularity

Abir Boujelben[(✉)] and Ikram Amous

MIRACL Laboratory, University of Sfax, Sfax, Tunisia
abeer.bujleben@gmail.com, ikram.amous@isecs.rnu.tn

Abstract. Intelligent systems are technologically advanced machines that perceive and respond to the world around them. They can take many forms: facial recognition programs, personalized shopping suggestions, healthcare tools, etc. Research in intelligent systems faces numerous challenges, many of which relate to automatic reasoning. Intelligent systems' knowledge bases are founded on facts and rules. Rules updates are essential to ensure that the system adapts to its environment evolution. In this paper, we aim to facilitate the automation of rule bases management by eliminating redundancies and handling circularity. This research work is part of the proposition of an approach for automating the management of rule bases. Our method is based on dependency relationships that may exist between the rules. The experimentation results show that our proposition succeeded in eliminating redundancies and detecting a great number of cycles.

Keywords: Rule base · Verification · Optimization · Semantic web technologies · Rules dependency

1 Introduction and Preliminaries

Intelligent systems (IS) exist all around us in digital televisions, traffic lights, health care tools, web services among a great number of other possibilities. An IS has the capacity to gather and analyze data and communicate with other systems. Other criteria for IS include the capacity for remote monitoring and management, the capacity to learn from experience and to adapt according to current data. The periodic, and in some cases manual, updates are complex and may cause raising inevitable inconsistencies in rule bases governing such systems. This problem may disturb the system's performance and its service quality. On the other hand, due to the increasing sizes of rule bases, the automation of their management tasks (adding, editing and deleting rules, in addition to inconsistency elimination) has become a necessity. This crucial need is still tackled. For example, [13] addresses the problem of automating rule bases verification. [8,12,17] try to manage context-aware systems and their evolutionary context rules.

© Springer Nature Switzerland AG 2020
W. Abramowicz and G. Klein (Eds.): BIS 2020, LNBIP 389, pp. 421–433, 2020.
https://doi.org/10.1007/978-3-030-53337-3_31

Besides, there is a variety of available tools that offer help to experts. For example, there are tools that allow a comprehensive representation of the rules, others try to explain the progress of the some knowledge inference [1, 16]. But the reasoning process within a huge rule base is still toilsome to follow. Recently, the successful adoption of Semantic Web technologies by IS in many areas of application has led to new challenges for solving the problem of rule bases management. In this research framework, eliminating and handling anomalies that may exist in rule bases are still tackled [18]. To address this problem, we have studied types of redundancies and cases of cycles proposed in the literature. We have applied them on toy-examples of rule bases. These allowed us to discover cases of untreated anomalies. Thus, in this paper, we list the cases of redundancies and cycles, and we propose solutions for their treatments. We relied on the dependency relationships that may exist between the rules, which we studied in a previous work [3].

Semantic Web Technologies: Semantic Web technologies have been stated by the World Wide Web consortium (W3C). They are designed to describe and relate data on the web and inside enterprises. Ontologies are considered as one of the pillars of the semantic Web offering powerful research and inference capabilities. The OWL language [14] is a basis for defining ontologies. It defines the fact base using four types of entities: (i) classes, (ii) object properties connecting two classes, (iii) data properties representing attributes for the defined classes, and (iv) individuals which are the classes' instances. OWL may not suffice for all applications as its modelling constructs are not always adequate, and there are statements it cannot express [7,11]. To encounter such a problem, W3C stated the Semantic Web Rule Language (SWRL). It combines ontologies (expressed in OWL-DL) and rules (expressed in RuleML). It allows encoding rule bases associated to ontologies. SWRL rules contain an antecedent part, and a consequent part. The antecedent and the consequent consist of positive conjunctions of atoms. Each atom is an expression of the form: P(arg1, arg2,...,argn) where P is a predicate symbol (class, property...) and arg1, arg2,...,argn are arguments (individuals, data values, variables, classes, ...). Some atoms types are displayed in Table 1. Hereafter, we present two examples of rules.

Table 1. Examples of SWRL atoms types

Atom type	Example
Class	Person(x)
Object property	hasParent(x,y)
Data property	hasAge(x,18)
Built-in	lessThan(n,25)

Example 1. $Rule - x$ and $Rule - y$ are two examples of SWRL rules.
$Rule - x : Person(p), hasAge(p, age), greaterThanOrEqual(p, 18) \longrightarrow Adult(p)$
$Rule - y : Adult(p) \longrightarrow canGetDrivingLicense(p, true)$
The first rule indicates that if a person is 18 or more, then he is considered as adult. The second rule states that if a person is adult, then he can have a driving license.

Dependency Relationships: The atom in $Rule - x$'s consequent and the first atom of $Rule - y$'s antecedent are referencing the same class *Adult*. So, $Rule - x$ produces knowledge that can be used by $Rule - y$. We say that $Rule - y$ depends on $Rule - x$. In this way, there are dependency relationships between elements of a rule base. These relationships reflect knowledge about the state of the base, including the anomalies that it may contain.

Anomalies in Rule Bases - Redundancy and Circularity: There are several types of anomalies that may exist in a rule base. In this manuscript, we treat redundancy and circularity.

Redundancy: It means that there are rules that accomplish the same role with the same inputs. Redundant rules may cause useless inferences. They even clutter up the rule base management process. Eliminating redundancy is important for a rule base optimisation. There are three types of redundancy: equivalence, subsumption and transitivity.

Equivalence: In literature, two rules are said to be equivalent if they are composed of the same atoms (Example 2, *R01* and *R02* are equivalent)

Example 2
$R01 : Person(x), hasBrother(y, z), hasParent(x, y) \longrightarrow hasUncle(x, z)$
$R02 : Person(x), hasParent(x, y), hasBrother(y, z) \longrightarrow hasUncle(x, z)$

Subsumption: There is a subsumption between two rules if they have the same consequent, while the antecedent of one rule subsumes the antecedent of the other (Example 3, R06 subsumes R05).

Example 3.
$R05 : Person(x), hasBrother(y, z), hasParent(x, y) \longrightarrow hasUncle(x, z)$
$R06 : Person(x), Person(y), hasBrother(y, z), hasParent(x, y) \longrightarrow has Uncle(x, z)$

Transitivity: Transitivity exists if a rule can be deduced from two or more other rules. In Example 4, *R11* can be deduced from *R09* and *R10*. We say that *R11* is a redundant rule.

Example 4. $R9 : Person(p), hasAge(p, a), greaterThan(a, 17) \longrightarrow Adult(p)$
$R10 : Adult(p) \longrightarrow hasInsuranceID(p, true)$
$R11 : Person(h), hasAge(h, a), greaterThan(a, 17) \longrightarrow hasInsuranceID(h, true)$

Circularity: It means that there is a set of rules that can run endlessly during an inference process. For example, if Ra, Rb and Rc are three rules that draw the path $\langle Ra \rightarrow Rb \rightarrow Rc \rightarrow Ra \rangle$ during the inference process, we say that they are circular rules. Cycles may cause the system suspension during the inference process. It may go in a blocking state or an infinite loop state. It is important to check circular rules to keep a rule base consistency.

Manuscript Organization: In this research work, we propose a method to optimize and verify rule bases by treating redundancies and cycles. In a previous work [3], we proposed a method to build a rules dependency graph. In this manuscript, we based our proposal on a dependency graph $DepG$ extracted for a rule base Ω. We define, detect and handle redundancy and circularity using $DepG$.

The rest of this paper is organized as follows: Sect. 2 presents a literature review. Sect. 3 introduces our proposal and details its modules performance. The results of its evaluation are provided and discussed in Sect. 4. Section 5 presents a conclusion and our future work.

2   Literature Review

Rule base verification means to verify whether the base is flawed or not from the structural correctness point of view. This requires checking several anomalies absence, such as redundancy and circularity. This task is difficult to be achieved in practice because of the large number of rules interacting during inference process. In this paper we are interested in eliminating redundancy and dealing with circularity. In literature, there are several works that have been interested in solving these types of anomalies. Dani et al. [6] proposes an approach for managing changes in a rule base. The elimination of anomalies is based on computing the rules execution frequency when applied to some examples. Different rules producing the same results are called redundant. Yunchuan et al. [21] proposes an algebraic method to eliminate the redundancies, the conflicts and the cycles of a rule base. To solve redundancies and the circularity, it is based on calculating the Closure[1] of the base and comparing the literals. Benaissa et al. [2] reformulates the ontology and its associated rule set into a constraint satisfaction problem. Then it uses filtering techniques to reduce the anomaly search space. This work proposes and discusses two approaches (a complete approach and an incomplete one) for the consistent manipulation of changes. Cheng et al. [4] introduces a method for resolving errors in some rules-based knowledge. It uses similarity matrices between antecedents and similarity matrices between the consequent parts of the rules. This technique has also been integrated in other works that have addressed the same problem such as Yan et al. [20] which suggests detecting conflicts through antecedent analysis and consequent analysis. Cota et al. [5] proposes to detect the inconsistencies during the analysis of the dependencies between the rules, during the extraction of the order of execution

---

[1] The closure of a rule base R: Closure(R)=$\{R \cup \{all\ the\ rules\ implied\ by\ R\}\}$.

and during the analysis of use case graphs. The analysis of these different graphs is manual, which means that the proposal can be applied only to small bases.

Synthesis: The major insufficiency that we noticed in almost all the studied works is the fact that in each algorithm it is necessary every time to go through all the rules and to redo the same treatments. Besides, the studied works stated that there are three types of redundancies: equivalence, subsumption and transitivity. While eliminating equivalent and subsumption rules they did not consider hierarchical relationships between classes and between properties. Besides, some of the proposed algorithms are exponentially complex (e.g. [2]). Moreover, dependency relationships between the rules play an important role in their execution. These relationships may reflect important knowledge about the state of the rule base. They can be represented by a matrix (as in [10] and [9]) or by an ontology (as in [3]). These two representation techniques make it possible to lighten rule bases maintenance process.

3   Proposed Method

Rule bases may include some anomalies, namely redundancies and cycles. The process of rule bases verification must be able to facilitate their elimination while guaranteeing the non-loss of the knowledge they embody. In this manuscript, we propose a graph based method which includes two steps: (i) redundancy elimination and (ii) cycle handling. It takes as input the rule base and its dependency graph, and returns a new version of the rule base and its associated new dependency graph (Fig. 1). We present a definitive order for the elimination of anomalies. We start by eliminating the redundant rules to reduce the number of rules processed in cycle management and reduce the number of processed cycles.

In this section, we begin by listing the different anomalies. Then we present their order of elimination. Finally, we clarify the progress of each step. Our work is based on the rule dependency graph associated with the base Ω [3]. This one reflects lots of knowledge on the rule base. We take advantage of the knowledge on the relations between the elements of the base. So, firstly we propose to group rules on the basis of incoming and outgoing dependency relationships. The rules having the same incoming edges and the same outgoing edges belong to the same

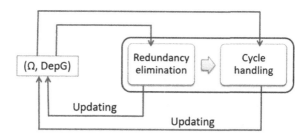

Fig. 1. Proposed method for eliminating redundancies and handling cycles

group. They bear semantically on the same classes and the same properties in their antecedents and their consequents. As we will explain in a further section, this will reduce the number of rule pairs to deal with when treating redundancy and circularity.

Hereafter, we present some notations that we will use in the following subsections.

Notations: For the rest of the paper, we consider:

- Ω a rule base, and $DepG = <\Omega, E>$ its associated rules dependency graph.
- E is the set of edges in $DepG$. Rx and Ry are two rules from Ω.
- $In(Rx)$: the set of incoming edges to a rule Rx from Ω.
- $Out(Rx)$: the set of outgoing edges from a rule Rx from Ω.
- *Matching atoms:* $a_i = pred_i(arg_1..arg_n)$ and $a_j = pred_j(arg_1..arg_m)$ are two matching atoms if:
  - predicates $pred_i$ and $pred_j$ are referencing equal or equivalent entities e_i and e_j(e_i and e_j are two classes or two properties)
  - or there is a hierarchical relationship between e_i and e_j (e_i is a sub-class or a sub-property of e_j).
  Matching antecedents (resp. consequents): two antecedent (resp. consequents) are matching, if all their atoms are matching.
- *Gi:* ($i \in [1.. \mid \Omega \mid]$) a group of rules that are semantically similar. Their antecedents relate to the same classes and their consequents cover also the same classes. Each Gi is defined such as: $\forall$ *Rk and Rl from Gi*, $In(Rk) = In(Rl)$ *and* $Out(Rk) = Out(Rl)$.

3.1   Redundancy Elimination

Redundancy occurs when there are unnecessary rules. It does not only increases the rule base size, but it can also cause additional and pointless inferences. There are three types of redundancies: equivalence, subsumption and transitivity. In the following we present each of them, we define it using the graph of dependencies and we present the solution(s) to eliminate it.

Equivalence. Two rules are equivalent means that they accomplish the same action or infer the same knowledge using the same conditions. The difference might be the atom order and/or the usage of different variables and/or the usage of equivalent or sub classes.

Example 5

$R01 : Person(x), hasBrother(y, z), hasParent(x, y) \longrightarrow hasUncle(x, z)$
$R02 : Person(x), hasParent(x, y), hasBrother(y, z) \longrightarrow hasUncle(x, z)$
$R03 : Person(x), hasBrother(y, z), hasFather(x, y) \longrightarrow hasUncle(x, z)$
$R04 : Person(x), hasBrother(y, z), hasParent(x, y) \longrightarrow hasRelative(x, z)$

It is obvious that equivalent rules belong to the same group Gi. So, instead of examining the equivalence relation between each pair of rules in Ω, it will be enough to examine each couple of rules within the same group Gi. Thus, for each pair of rules Rx, Ry belonging to the same group, this task is done by analyzing the match between *Antecedent(Rx)* and *Antecedent(Ry)* if Rx and Ry have the same atoms in their consequents, and between *Consequent(Rx)* and *Consequent(Ry)* if Rx and Ry have the same atoms in their antecedents. We distinguish three equivalence cases.

Case 1: Two rules have the same atoms in their antecedents and the same atoms in their consequent (Example 5: R01 and R02). In such situation, just delete one of them.

Case 2: Two rules have the same consequent, and their antecedents match due to an hierarchical relationship between two atoms (Example 5: *R01* and *R03*, *hasFather* property is a sub-property of *hasParent*). Here, the rule to be deleted is the one with the *less general* entity (Example 5: *R03* is the rule to be deleted because *hasFather* is less general than *hasParent*).

Case 3: Two rules have the same antecedent, and their consequents match because of an hierarchical relationship between two atoms (Example 5: *R01* and *R04*, *hasUncle* property is a sub-property of *hasRelative*). The rule to delete is the one with the *most general* entity (Example 5: *R04* is the rule to be deleted because *hasUncle* is more general than *hasRelative*).

Definition 1. *Two rules Rx, Ry are equivalent if:*

- $\{Rx,\ Ry\} \in Gi$
- *Rx and Ry have matching antecedents and matching consequents.*

Subsumption. Subsumption rules share the same consequent while the antecedent of the first subsumes the antecedent of the second. In such a situation, the first rule has to be eliminated.

Example 6
R05 : $Person(x), hasBrother(y, z), hasParent(x, y) \longrightarrow hasUncle(x, z)$
R06 : $Person(x), Person(y), hasBrother(y, z), hasParent(x, y) \longrightarrow has\ Uncle(x, z)$
R07 : $hasFather(x, y), hasBrother(y, z) \longrightarrow hasUncle(x, z)$
R08 : $hasMother(x, y), hasBrother(y, z) \longrightarrow hasUncle(x, z)$

We distinguish two subsumption cases.

Case 1 - Simple Subsumption: Two rules have the same consequent and the antecedent of the first one is included in the second (Example 6: *R06* subsumes *R05*, *R06* is the rule to delete).

Case 2 - Semantic Subsuption: There is a general rule and some other rules representing special cases, but all of them infer the same result (Example 6: *R07* and *R08* are two special cases of *R05* (and special cases of *R06* too)).

To detect subsumption redundancies, we propose to start by analyzing incoming edges and outgoing edges of rule groups. Subsumption may exist between rules from the same group[2] or between rules from a group Gi and a group Gj such as:

- $Ri \in Gi$
- $Rj \in Gi$
- $In(Gi) \subset In(Gj)$
- $Out(Gi) = Out(Gj)$

Transitivity. A transitivity exists if a rule can be deduced from two or more other rules. In such a case, the transitive rule must be removed. Based on the dependency graph DepG, we define a transitive rule Rt by the fact that it can be replaced by a path $CH =< Ra, .., Ri, .., Rn >$. This implies that Rt's antecedent uses all the data used in the antecedents of all the rules Ri in path CH, and that it has the same consequent as Rn (the last rule in CH). After that, CH and Rt must be applicated on fictive individuals to distinguish true transitive rules from false ones.

The detection of a transitive rule Rt is formulated using Algorithm 1.

Algorithm 1. Transitive rule detection

Require: $\Omega : rulebase$
Require: $DepG =< \Omega, E >$: dependency graph associated to Ω
Ensure: *decision*
  1: **VAR**
  2: CH : path
  3: a, i, n : Integer
  4: Ra, Ri, Rn, Rt : rule
  5: $S1$: a set of fictive individuals
  6: $result1, result2$: set of knowledge
  7: **BEGIN**
  8: **if** $In(Rt) = \cup \{In(Ri)$ such as $Ri \in CH$ and $i \in [a..n]\}$ **then**
  9:    **if** $Out(Rt) = Out(Rn)$ **then**
 10:       result1 $\leftarrow$ apply Rt on $S1$
 11:       result2 $\leftarrow$ apply CH on $S1$
 12:       **if** result1 $\equiv$ result2 **then**
 13:          decision $\leftarrow$ Rt is a transitive rule
 14:       **else**
 15:          decision $\leftarrow$ Rt is NOT a transitive rule
 16:       **end if**
 17:    **end if**
 18: **end if**
 19: **END**

---

[2] Example 6: *R05* and *R06* are two rules from the same group Gi.

3.2 Circularity Management

Some rules are called circular rules if their execution chaining forms a cycle [15,19]. So, using a dependency graph, the dependency relationships connecting circular rules form a cycle. The experimentation results proved that some of the detected cycles are false ones. Example 7 shows a true cycle, while Example 8 shows a false one. We propose to apply the chain on fictive individuals. We propose that a true cycle can be detected using the following two checks:

- checking incoming and outgoing edges for each group Gi
  - $CH =< Ra, .., Ri, .., Rn, Ra >$ is a path in DepG
- checking using fictive individuals
  - $CH =< Ra, .., Ri, ..Rn, Ra >$ is a true cycle if, when applied on a set of individuals, the results inferred by Rn are the same knowledge already used by Ra.

Example 7. <R12,R13,R14> is a true cycle

$R12 : hasChild(x, y), Woman(y) \longrightarrow hasDaughter(x, y)$
$R13 : hasChild(x, y) \longrightarrow hasParent(y, x)$
$R14 : hasParent(x, y), Woman(x) \longrightarrow hasDaughter(y, x)$

<R12, R13, R14, R12> is a cycle extracted from DepG. Alice and Sophie are two individuals on whom we will apply the rules R12, R13 and R14. Assumed that Alice and Sophie are two women, and Sophie is Alice's mother.

$R12 : hasChild(Sophie, Alice), Woman(Alice) \longrightarrow hasDaughter(Sophie, Alice)$
$R13 : hasChild(Sophie, Alice)) \longrightarrow hasParent(Alice, Sophie)$
$R14 : hasParent(Alice, Sophie), Woman(Alice) \longrightarrow hasDaughter(Sophie, Alice)$

It is obvious here that hasDaughter(Sophie,Alice) is a sub-property of hasChild(Sophie,Alice). So, R12 will be run again with the same data. This proves that <R12,R13,R14,R12> is a true cycle.

Example 8. <R15,R13,R16,R15> is not a true cycle

$R15 : hasParent(x, y), hasSibling(x, z) \longrightarrow hasParent(z, y)$
$R13 : hasParent(x, y), Woman(x) \longrightarrow hasDaughter(y, x)$
$R16 : hasNiece(x, y), hasDaughter(z, y) \longrightarrow hasSibling(x, z)$

$< R15, R13, R16, R15 >$ is a cycle extracted from DepG. Sophie, Alice, Amelia and Ben are four individuals on whom we will apply the path $< R12, R13, R14, R12 >$. It is assumed that Alice and Amelia are Sophie's daughters, and that Ben is Sophie's brother.

$R15$: $hasParent(Alice, Sophie), hasSibling(Alice, Amelia)$ $\longrightarrow$ $hasParent$ $(Amelia, Sophie)$

$R13$: $hasParent(Amelia, Sophie), Woman(Amelia)$ $\longrightarrow$ $hasDaughter$ $(Sophie, Amelia)$

$R16$: $hasNiece(Ben, Amelia), hasDaughter(Sophie, Amelia)$ $\longrightarrow$ has $Sibling(Ben, Sophie)$

It is noteworthy that $R15$ will be performed again but this time it will use another combination of individuals. Thus we can deduce that the path $< R12, R13, R14, R12 >$ is a false cycle.

Our method saves true cycles to avoid them during the inference process. This also allows the experts to consult them in order to eliminate the rules they consider not useful or defective. Our method starts by searching and saving the *k-cycles* (cycles formed of k edges) by incrementing k. It starts by looking for *3-cycles* then *4-cycles*, etc. The progress of the cycle handling is presented by Algorithm 2. True-cycles are saved in a set named $Tcyc$, while the false-cycles are saved in a set named $Fcyc$. The latter makes it possible not to consider the false cycles already extracted during cycles searching. We note that the two sets $Tcyc$ and $Fcyc$ are updated each time the rule base Ω is edited.

Algorithm 2. Cycle handling

Require: Ω : rule base
Require: $DepG = < \Omega, E >$: dependency graph associated to Ω
Ensure: $Tcyc, Fcyc$: set of paths
 1: **VAR**
 2: k : Integer
 3: cyc : path
 4: **BEGIN**
 5: $k \leftarrow 3$
 6: **repeat**
 7:     **while** there are unsaved k-cycles **do**
 8:         cyc $\leftarrow$ find an unsaved k-cycle
 9:         **if** cyc is a true-cycle **then**
10:             $Tcyc \leftarrow Tcyc \cup \{cyc\}$
11:         **else**
12:             $Fcyc \leftarrow Fcyc \cup \{cyc\}$
13:         **end if**
14:     **end while**
15:     $k \leftarrow k + 1$
16: **until** there are no unsaved cycles
17: **END**

---

4   Experimentations and Results

In the previous section, we proposed a method to eliminate redundancies and handling cycles on the basis of dependency graphs. Our algorithm complexity is $O(n^2)$.

To evaluate our proposal, we applied our method to three rule bases associated to OWL ontologies: (i) an ontology to match security policies (ii) another one for providing appropriate meal components for diabetic patients, and (iii) a third one for road traffic management. The rule bases were designed by experts in the field who are not among the authors. To check the redundancy and cycle management, we made random changes to the rule bases. Then we applied our method. Finally, we compared the results to the initial correct bases. We calculated the number of redundant rules and the number of undetected cycles in each case study.

Table 2. Initial state of data sets

	Security policy matching	Diabetic food ontology	Traffic ontology
# Rules (initial)	225	24	77
# Redundant rules	+22	+8	+17
# Cycles	68	36	54

Table 3. Experimentation results

	Security policy matching	Diabetic food ontology	Traffic ontology
# Redundant rules	2	0	0
# Saved true-cycles	43	12	34

In Table 2 we present the datasets and in Table 3 we show the experimentation results. Table 3 shows that our method has deleted all redundant rules in the second and the third rule bases. We checked the deleted rules and the new rule base state, we found no rules deleted but redundant ones. Table 3 shows also that our method has left two redundant rules in the first rule base. This is due to the fact that the Semantic Policy matching rule base includes a great number of built-in atoms which some of them may have similar roles. We think that a semantic analysis of the atoms' predicated should eliminate such a deficit. On the other hand, the detected true-cycles may be consulted by the experts for remediation. They may be also avoided by reasoners during the inference process.

5   Conclusions

Rule bases supporting IS performance are continuously modified to be up to date with their environment new state. These changes may cause inevitable inconsistencies such as redundancy and circularity. On one hand, redundant rules may cause useless and pointless inferences and treatments. On the other hand, cycles may lead to deadlocks during the reasoning process. In this paper, we proposed a method eliminate redundant rules and deal with cycles. We defined redundancy and circularity using dependency graphs and we presented solutions for their management. In our work, we have cited cases of redundancies that have not been considered in literature. The obtained results are promising. In our future work, we are planning to propose a method that can manage all types anomalies to better verify and optimize rule bases.

References

1. Bak, J., Nowak, M., Jedrzejek, C.: Graph-based editor for SWRL rule bases. In: RuleML (2) (2013)
2. Benaissa, M., Lebbah, Y.: A constraint programming based approach to detect ontology inconsistencies. Int. Arab J. Inf. Technol. **8**(1), 1–8 (2011)
3. Boujelben, A., Amous, I.: A new method for rules dependency extraction. In: 22nd International Conference on Knowledge Based and Intelligent Information Enginecring Systems. Elsevier (2018)
4. Cheng, M.Y., Huang, C.J.: A novel approach for treating uncertain rule-based knowledge conflicts. J. Inf. Sci. Eng. **25**(2), 649–663 (2009)
5. Cota, É., Ribeiro, L., Bezerra, J.S., Costa, A., da Silva, R.E., Cota, G.: Using formal methods for content validation of medical procedure documents. Int. J. Med. Inform. **104**, 10–25 (2017)
6. Dani, M.N., Faruquie, T.A., Karanam, H.P., Subramaniam, L.V., Venkatachaliah, G.: Rule set management, 15 April 2014. US Patent 8,700,542
7. Dautov, R., Veloudis, S., Paraskakis, I., Distefano, S.: Policy management and enforcement using OWL and SWRL for the internet of things. In: Puliafito, A., Bruneo, D., Distefano, S., Longo, F. (eds.) ADHOC-NOW 2017. LNCS, vol. 10517, pp. 342–355. Springer, Cham (2017). https://doi.org/10.1007/978-3-319-67910-5_28
8. Davtalab, M., Malek, M.R.: A spatially aware policy conflict resolution for information services. J. Ambient Intell. Smart Environ. **10**(1), 71–81 (2018)
9. Dolinina, O., Shvarts, A.: Algorithms for increasing of the effectiveness of the making decisions by intelligent fuzzy systems. J. Electr. Eng. **3**, 30–35 (2015)
10. Hassanpour, S., O'Connor, M.J., Das, A.K.: Exploration of SWRL rule bases through visualization, paraphrasing, and categorization of rules. In: Governatori, G., Hall, J., Paschke, A. (eds.) RuleML 2009. LNCS, vol. 5858, pp. 246–261. Springer, Heidelberg (2009). https://doi.org/10.1007/978-3-642-04985-9_23
11. Höffner, K., Walter, S., Marx, E., Usbeck, R., Lehmann, J., Ngonga Ngomo, A.C.: Survey on challenges of question answering in the semantic web. Semant. Web **8**(6), 895–920 (2017)
12. Khattak, A.M., Khan, W.A., Pervez, Z., Iqbal, F., Lee, S.: Towards a self adaptive system for social wellness. Sensors **16**(4), 531 (2016)
13. Ksystra, K., Stefaneas, P.: Formal analysis and verification support for reactive rule-based web agents. Int. J. Web Inf. Syst. **12**(4), 418–447 (2016)
14. McGuinness, D.L., Van Harmelen, F., et al.: Owl web ontology language overview. W3C Recomm. **10**(10), 2004 (2004)

15. Nguyen, T.A., Perkins, W.A., Laffey, T.J., Pecora, D.: Checking an expert systems knowledge base for consistency and completeness. In: IJCAI, vol. 85, pp. 375–378 (1985)
16. O'Connor, M.J., Das, A.: The SWRLTab: an extensible environment for working with SWRL rules in Protégé-OWL. In: RuleML (2006)
17. Salfinger, A., Retschitzegger, W., Schwinger, W.: Staying aware in an evolving world specifying and tracking evolving situations. In: 2014 IEEE International Inter-Disciplinary Conference on Cognitive Methods in Situation Awareness and Decision Support (CogSIMA), pp. 195–201. IEEE (2014)
18. Simiński, R., Nowak-Brzezińska, A., Simiński, M.: Experimental implementation of web-based knowledge base verification module. In: Nguyen, N.T., Pimenidis, E., Khan, Z., Trawiński, B. (eds.) ICCCI 2018. LNCS (LNAI), vol. 11056, pp. 268–278. Springer, Cham (2018). https://doi.org/10.1007/978-3-319-98446-9_25
19. Suchenia, A., Potempa, T., Ligęza, A., Jobczyk, K., Kluza, K.: Selected approaches towards taxonomy of business process anomalies. In: Pełech-Pilichowski, T., Mach-Król, M., Olszak, C.M. (eds.) Advances in Business ICT: New Ideas from Ongoing Research. SCI, vol. 658, pp. 65–85. Springer, Cham (2017). https://doi.org/10.1007/978-3-319-47208-9_5
20. Sun, Y., Wu, T.Y., Li, X., Guizani, M.: A rule verification system for smart buildings. IEEE Trans. Emerg. Top. Comput. 5(3), 367–379 (2017)
21. Yunchuan, S.: Managing rules in semantic web: redundancy elimination and consistency check. JDCTA: Int. J. Digit. Content Technol. Appl. 5(2), 191–200 (2011)

Audio-Visual Emotion Recognition System for Variable Length Spatio-Temporal Samples Using Deep Transfer-Learning

Antonio Cano Montes$^{(\boxtimes)}$ (iD) and Luis A. Hernández Gómez (iD)

Signal, System and Radiocommunications Department, Grupo de Aplicaciones del Procesado de Señal (GAPS), Universidad Politécnica de Madrid, 28040 Madrid, Spain
antonio.cano.montes@hotmail.com
http://www.upm.es/observatorio/vi/index.jsp?pageac=grupo.jsp&idGrupo=284

Abstract. Automatic Emotion recognition is renowned for being a difficult task, even for human intelligence. Due to the importance of having enough data in classification problems, we introduce a framework developed with the purpose of generating labeled audio to create our own database. In this paper we present a new model for audio-video emotion recognition using Transfer Learning (TL). The idea is to combine a pre-trained high level feature extractor Convolutional Neural Network (CNN) and a Bidirectional Recurrent Neural Network (BRNN) model to address the issue of variable sequence length inputs. Throughout the design process we discuss the main problems related to the high complexity of the task due to its inherent subjective nature and, on the other hand, the important results obtained by testing the model on different databases, outperforming the state-of-the-art algorithms in the SAVEE [3] database. Furthermore, we use the mentioned application to perform precision classification (per user) into low resources real scenarios with promising results.

Keywords: Emotion recognition · Multimodal deep learning · Transfer learning · Variable sequence length · Model fusion · Convolutional neural network · Bidirectional recurrent neural network

1 Introduction

Emotions constitute an essential part of the human being. Human emotional behavior offers a variable and rich range of different information which depends on many factors such as cultural, genetic, chemical, personality, environmental, etc. Recognizing emotion from an inside out perspective is not trivial, even between humans, which make automatic emotion recognition a problem hard and complex to solve regarding its pronounced subjective nature.

The research community has been working to understand emotions from both psychological and computational perspective. Recently, this topic has become

© Springer Nature Switzerland AG 2020
W. Abramowicz and G. Klein (Eds.): BIS 2020, LNBIP 389, pp. 434–446, 2020.
https://doi.org/10.1007/978-3-030-53337-3_32

more important as a result of its fundamental role in the effectiveness of real human computer interactions [1], the development of Artificial Intelligence, medicine and personalize advertisement, among others.

Humans show emotions by communicating with each other, but, furthermore, we use conscious and unconscious physical expressions. Therefore, the exchanged information is largely encapsulated in this natural, multimodal format [2]. Automatic emotion recognition is challenging because of the existing emotional gap between emotions and audio-visual features. One of the main problems to face comes from the complexity of facial and oral expressions along with variable spatio-temporal dimension and the lack of a semantic context in these characteristics. Different neural network's techniques have been applied to model spatio-temporal learning context within dynamic samples.

In this paper we approach the problem of Automatic Emotion Recognition by applying our algorithms to well-known databases. We focus on the development of a model capable of working in real-life scenarios with complex and insufficient data. For that purpose, we try to take advantage of the strong feature-learning ability of Deep Neural Networks and the development of a *"precision"* model. In that way, we explore the use of Transfer Learning to extract high-level features with less resources, from audio-visual emotion data focusing on the basic emotions recognition (happiness, sadness, anger and neutral state).

2   Related Work

2.1   Datasets

According to [2], we can divide emotion datasets into acted and natural in regards to the type of data recording. The first group corresponds to controlled environments from a record perspective where, generally, actors play emotions emphasizing tone and facial expression, semantic context or both. Second group corresponds to data manually labeled from scratch using material available on the internet, films, etc. or datasets recorded in real environments with or without scripts.

Regarding our problem, we need to differentiate those databases which are recorded under self-perspective and those which have audio-visual data well labeled. It is not trivial to found previous specific condition adjusted to basic emotions (including neutral state) on the available databases. For those reasons, to measure the effectiveness of our proposed audio-video deep architecture without semantic context, we select from well-known datasets, those which correspond to acted and controlled scenarios to reduce the variability and fit with our demands:

- Surrey Audio-Visual Expressed Emotion (**SAVEE** [3]): consist of 15 English sentences for basic emotions being one file per sentence and 30 sentences for 'neutral'. Utterances are recorded in a supervised way and they are extreme acted for each emotion. Four actors play the same sentences, being each one of variable duration.

- Interactive Emotional Dyadic Motion Capture (**IEMOCAP** [4]): consists of about 12 h of audiovisual data from 10 different actors. All recordings have a structure of dialogs between a man and a woman, either scripted or improvised on a given topic, in English. It is composed by a total of 4448 utterances (Angry: 600, Happy: 1205, Neutral: 1078, Sad: 635 utterances).
- Berlin Database of Emotional Speech (**EmoDB** [5]): contains about 500 German utterances spoken by 10 different actors and 10 different scripts.

2.2   Features Vector Extraction

At that point, we briefly describe the related work regarding automatic emotion recognition in speech, facial and its combination, in that order. We focus on traditional views and deep learning techniques, highlighting best approaches and ideas founded in the field.

Previous to the increasing development of deep learning, **speech** emotion recognition systems relied on feature engineering to extract low level features which could be categorized into prosody features (e.g. pitch, energy), voice quality features and spectral features (e.g. Mel-Frequency Cepstral Coefficient (MFCC)) [6].

S. Lalitha in [9] used pitch and prosody hand-crafted features extracted from the raw audio and with the use of Support Vector Machine (SVM) classifier, they obtain a 81% of accuracy on 7 classes on EmoDB. Despite those results, recent studies and publications show that low-level features could not be sufficiently for discriminate human emotions. Instead, researchers are exploring automatic feature extraction from audio, which is increasing the attention among the computer vision community. Stefano Pini [7] aims to extract high level features directly from audio applying 1-D convolutional operations on a single stream combined with multimodal techniques. P. Harár [8] uses a end-to-end convolutional network to model audio segments from the whole audio envelope. Authors are also developing temporal models for audio modality. One of them is Wan Ding [10] who mixes Low level Descriptors (LLDs) with Recurrent Neural Networks (RNN).

More connected to this paper are [12] and [11], they both merge frequency images extracted from audio data into a RGB image as input of a CNN network. In the case of [12], they merge LLD features after shallow networks and feature vectors are extracted from feeding MFCC images into a CNN and RNN in audio stream. In [11], Shiqing Zhang combines MFCC and dynamic features into a RGB image as a pre-trained AlexNet [13] input for audio stream.

Regarding video images, emotion recognition is traditionally related to **facial expression** [15]. We can divide publications following the criteria of static [10] and dynamic inputs, which are interesting for this present work. CNNs have exhibited promising performance on feature learning from images but recently we denote an increasing use of RNN to quantify visual motion. Other authors, like Samira Ebrahimi [14], are combining CNN with RNN to model temporal facial expression in video with favorable results. At present, authors like [11] are exploring 3D-Convolutional Networks (C3D) which computes feature maps from both spatial and temporal dimensions.

2.3   Multimodality Fusion

Multimodality Fusion refers to methods which use audio and visual modalities together with different fusion strategies: score-level fusion (*late fusion*), feature-level fusion (*early fusion*), decision-level fusion and model-level fusion.

Regarding the first group, we found in [11] promising results which aims to uses such technique. Decision-level fusion [16] tends to combine several unimodal emotion recognition results through an algebraic combination, but it cannot model well mutual correlation between modalities. Feature-level fusion [16] is controversial and some authors [11] affirm that this strategy cannot model complicated relationship from both data types.

Model-level fusion joints feature representation of both audio visual modalities into one feature vector. Y. Kim [17] work out in this strategy concluding that a simple shallow network cannot merge enough information to fit a classifier, instead some deep network should use. The same conclusion is adopted by [11] which to solve it, they implement Deep Belief Networks (DBN) as fusion method. Publications [10,11,17,18] use models architectures based on two-stream convolution network, with a final classifier after fully connected layers such as SVM or DBN, obtaining better performance than single (isolated) data modalities.

2.4   Transfer Learning (TL)

The concept of TL is so simple to understand, it refers to the process of using model knowledge for a task (i.e. network*weigths*), for perform another task or as the starting point of a model in a second different (or equal) task. This technique is becoming more and more popular because its improvements on many deep learning activities with limited data [19]. Related to emotion recognition, recent publications show favorable models achievements in multi-modal emotions recognition. [18] uses *VGG* [21] and *Resnet* [22] feature extractor at face frame level and *C3D* [23] for temporal samples. Another example of successful implementation comes from [11] which uses a fine-tuned version of *AlexNet* [13] as audio RGB feature extractor and *C3D-Sports-1M* [24] models to video face samples which the aims of learning high-level patters from the raw (preprocessing) data.

As a general analysis, numerous researches are tending to work with pretrained models (based on ImageNet [20]) as feature extractor and use them as the input through other architecture. All of the results in previous works that uses TL, show higher accuracy when applying automatic feature extraction instead of handcrafted features. Furthermore, it is possible to see that those models which keep temporal information provide better results than all the rest. So far, model-level fusion seems to perform better than others fusion techniques. However model robustness for the final fussed model is still an issue and multimodal data fusion remains an important challenge in emotion recognition systems. Table 1 presents a summary of previous research on emotion databases that we will use to compare our model performance.

3   Proposed Deep Learning Multimodal Framework

In this section we present our proposal which is formed by an Android application to collect data and the deep learning architecture composed by four modules: **input data preparation and preprocessing, static part, temporal (dynamic) part and output** (with different fusion strategies).

The proposed multimodal framework is presented in Fig. 1 which can be thought as common for both modalities (audio and video) but also for hybrid models with some prior or late changes for multimodal data.

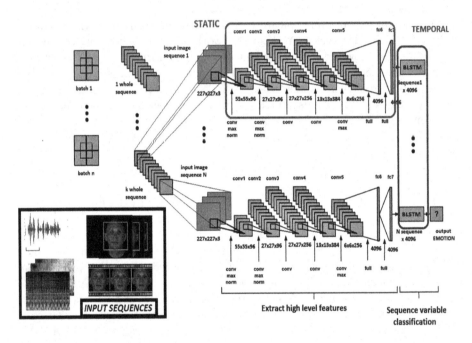

Fig. 1. Overall structure of the proposed variable sequence spatio-temporal model for audio, video and multimodal inputs. In left-bottom the multimodal input sequences are showed: the first image is the spectral MFCC representation of an audio sentence, and the second image represents the corresponding sequence of facial video frames.

3.1   Android Application to Create Our Spanish Database

Due to the lack of real emotion datasets available in **Spanish**, we decide to create our own dataset (**OwnDB**) for audio modality. For that purpose, we developed an Android application which function as a database recorder and as a model evaluator, like we will see later on. The application facilitates data collection and give us enough variability respecting number of volunteers (different pitchs, ways of speaking, etc.) and record condition (microphone, noise). The workflow is simple: a user logging into the application, after that one audio sample is recorded and in real-time user's receive emotion model's prediction. Finally, users need to give a feedback to the system of which emotions they pretend to record and the network is re-trained for each user (isolated). For our research purpose, it was very important to have under control who record each utterance so we implemented a login system to guarantee speaker independence, with an eye to future real applications. The database is continuously growing with random volunteers (speakers) recording themselves, in almost all cases, until they get perfect recognition. So far, the database has 14 users with 1158 utterances in total.

3.2   Data Preparation and Preprocessing

In this section we describe how to generate the inputs of audio and video sequences to our pre-trained network. First, we split samples into a variable number of overlapping segments per audio-video file:

1) *Audio preparing:* it is known that 2-D spectrogram represent low-level acoustic information related to the speaker's emotion expression (such as energy, pitch, formants). We transform 1-D raw audio signal into an array 3D where we extract three channels of log Mel-spectrogram segments including velocity and acceleration components. We split each amplitude normalized audio utterance into 64 frames, named as **context window**, with a shift of 30 frames. Internally, it is applied Short-Time Fourier Transform (STFT) with 2048 points with a window size of 25 ms and 10 ms of overlapping. It is important to remark that we use an energy-based Voice Activity Detector to remove silents.

This STFT array is used to compute the Mel spectrogram with 64 filters. This result becomes into a matrix of **64(Mel-filter banks)** $\times$ **64(frames)**. We consider this form as a image and we stack it as the first channel of a 3 channels RGB image. To capture temporal information of Mel-Spectrogram, the second dimension comes from computing the velocity of previous and future windows known as delta (Δ). Finally, the last channel of the image is given by computing the acceleration of the Mel-Spectrogram as in the second dimension, called delta-delta (Δ^2). As a consequence, each segment has 64 frames length and its temporal duration is $10 \cdot (64 - 1) + 25 \, \text{ms} = 655 \, \text{ms}$ like in [11]. As a final step, the RGB image resultant is transformed (with bilinear interpolation technique) into a size suitable for the pre-trained model, **227** $\times$ **227** $\times$ **3**.

2) *Face cropped from whole video:* the same sequence splitting procedure is applied directly to the raw video, that is considered as a static images sequences. So a context window of 64 frames (665 ms), like in audio, is applied straight to the raw video resulting in N variable (same in audio) frames per second segments. To ensure a fixed number of frames per window to all the video formats (25,60... fps), 16 frames were randomly selected from the whole set of frames in a window. Due to the strong relevance of the facial information in emotion recognition, for each frame of the video segment, we run a pre-trained face detection, Multi-Task Cascaded Convolutional Network (MTCNN) [25]. It treats the problem

as cascade of sub-tasks and uses trained CNNs for each task. More specifically three tasks are defined: coarse background cutting, face candidate detection and fine bounding box localization at facial landmarks detection. By presuming each video contains only one person's face, we take highest confidence detected face from different candidates for the next step of feature extraction, we obtain a perfect detection for each one of the frames of all the videos. The bounding box are elongated by a empirical rate and the images are resized (interpolated) into the pre-trained size $227 \times 227 \times 3$.

3.3 Static Module: Pre-trained AlexNet [13]

We label this part as "static" because temporal information is not used and each image, both from audio and video, is treated as independent from the rest.

Because of the promising results in task related to emotion recognition in images and small database size, we decide to use the power of a pre-trained model. Inspired by the positive performance showed by [11] in different databases, we built a fine-tuned version of AlexNet trained on ImageNet [20]. As the reader can identify from the static part on Fig. 1, AlexNet is composed by 5 convolutional layers and 3 fully connected layers where, in our case, the last one is omitted and connected to the RNN block. Our structure could be thought as a trainable high feature extractor in which convolutional layers are deterministically frozen and the fully connected (**FC**) layers are trained to output all the original sequence stacked as a sequence of feature vectors to train the RNN.

To reduce error model and help the convergence on it, all the input data were preprocessing in the same way as the original AlexNet on ImageNet [20]. The reason to use AlexNet in our emotion classification problem is that we will train some layers from the network but the rest have already been used for object classification. Therefore, it seems that its low level filters could extract suitable information to emotion recognition requirements and then could be applied to our problematics.

As an output we get a feature vector $F^i(i_s; \theta^i)$ denoting the 4096-D input modality features (FC7) with network parameters θ^i for each image input.

3.4 Dynamic Module: Bidirectional Long Short-Term Memory (BLSTM) [26]

Bidirectional LSTM cells are based on the idea that the output O_t not only depending on previous elements, X_{t-1} but also depending on future elements X_t. They are just two LSTMs stacked on top of each other. The output is then computed based on the hidden state of both LSTMs. They present each training sequence forwards and backwards to both RNN which are connected to the same output.

In our architecture, BLSTM receive batch sequences of extracted features by CNN and they are treated as variable sequence, labeled as dynamic part. As an output of the BLSTM we obtain a feature vector $F^h(h_s; \alpha^h)$ for each hidden state.

Variable Sequence Length and Output. As a final step, we get the greedy output from the whole set of hidden states. For that purpose, we apply a mask from the maximum zero padding sequence to obtaing the real output regarding input length. Note that mathematical demonstration of this procedure is out of the scope of this implementation. We additionally mask the cost function to assure convergence only with real sequences per training batch:

$$L_i = \sum_{k=1}^{l} y \cdot log(softmax(W^h \cdot F^h(h_i \alpha_{real}{}^h), y_i)$$

$$L_i = mean(\frac{\sum_{i=1} L_i}{\sum_{j=1} ones_j (real_{length})}), ones \equiv SignMask$$

As a final step, real length hidden neurons are passed through a FC layer that output the four basic emotions probability vector.

3.5 Multimodal Fusion Modalities

As we introduced in Sect. 2.3, there are different multimodality fusion techniques. It is important to highlight that most of our effort was related to create robust enough networks for each modality isolated, so we did not enter very deep within fusion modalities. In that sense, we apply three different types of fusions [11]:

- Early Fusion (**EF**): both inputs modalities are concatenated into a RGB image. To maintain the concordance between audio and video, only one frame of video samples is selected, that's why we refer to this method as 'poor'. As a prior conclusion, we consider audio information is less representative for emotion recognition than facial images. Therefore, adding audio as an extra channel to facial images only adds noise and the model cannot learn the complicated relationship between features.
- Model-level Fusion (**MF**): we concatenate two stream feature vectors outputs, for each audio and video modality, into a huge feature vector to feed BLSTM network. Other authors probe that this technique is still not effective in modeling highly non-linear correlations between audio and video modalities. In our case, because our model becomes too complex and because of memory limitations, we couldn't make the network deeper enough and so we find some limitations in the model's performance.
- Late Fusion (**LF**): in this case we refer to fuse scores once the model are trained separately. Because video modality has an extra dimension we weight the fusion depending on audio sample: $S_{multimodal} = S_{Video} \cdot (1 - \frac{N_{frames}}{100}) + S_{Audio} \cdot \frac{N_{frames}}{100}), N = 16$

4 Experiments and Results

To assess the effectiveness of our **AlexNet BLSTM model pretrained on variable-length sequence data**, we evaluate the model over both isolated audio and video modalities and audio-visualdata using the three fusion types described in previous section. We use public databases detailed in Sect. 2.1, with the goal to compare the performance of the proposed architecture (and hyperparameters) that we will use in our own Spanish emotion Database, with other reference research.

Before presenting the results it is important to notice the difficulty of finding publications that use the same databases with the same purpose as this one; recognizing the four basic emotions. For that reason, together with our goal of developing a real application in the future, most of our tests have been focused in audio modality. It is vital to comment that, although all the networks are fitted with the same hyperparameters, the convergence was reached for all the models, achieving stable accuracy levels. Basically, we use Stochastic Gradient Descent (SGD) as optimizer with learning rate $\mu = 0.0005$ and 512 neurons for the BLSTM network. Our results are accompanied with their corresponding Confusion Matrix (CM) and accuracy is obtained from Unweighted Accuracy (UA). We randomly split each database into train-test (80%–20%) without speaker independence. To compare our final results we select the best public results on the same databases and basic emotions (neutral, happiness, sadness and anger).

For audio emotion recognition we take as reference SAVEE database, computing best hyperparameters for this dataset and then generalizing for the other databases and modalities.

Table 1. State-of-the-art results and audio results over the set of databases and our own database

AUDIO	IEMOCAP	SAVEE	EmoDB	OWNDB
BEST (%)	68.5 [27]	86.7 [28]	96.9 [8]	–
OURS(%)	65.6	91.6	82.5	75

Table 2. Our video results over SAVEE

Modality	SAVEE
VIDEO (%)	96.67

Table 3. Our multimodal results on SAVEE

Multimodal AUDIO-VIDEO	SAVEE
EF (%)	86.67
MF (%)	54
LF (%)	95.06

As it is shown in Table 1 for IEMOCAP and EmoDB databases, we get a slightly lower performance than the reference systems. However, for SAVEE we get higher accuracy therefore outperforming the state-of- the-art. For the evaluation on our Spanish Databse (OwnDB), as described in Sect. 3.1, we used 14 users with 1158 utterances in total. We trained four emotions classes (neutral, angry, sad and happy) using a homogenous distribution of 250 audios per category, splitting in train-test (80%/20%). As it can be shown in Table 1, results on

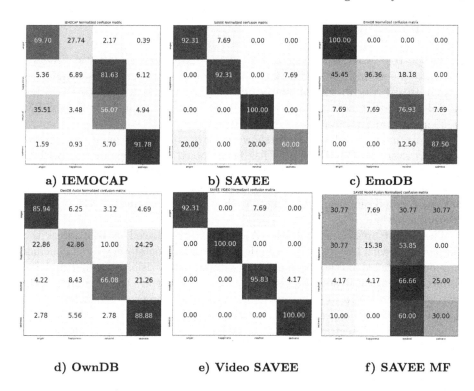

Fig. 2. Confusion matrices: a), b), c) and d) correspond to Audio modality, e) represent video results and f) shows model-fusion in SAVEE.

audio data achieved a 75% accuracy, similar to the values obtained for other similar datasets. Testing was performed in two ways: 1) normal test over the whole database achieving 75% of accuracy and 2) real-time test using the applications explained in Sect. 3.1, to make a *precision* task. Precision here comes from a personal unique train and test sts for each isolated speaker in the system (hence the need of unique log in), which empirically translate to a perfect classification after N (variable per user) samples per emotion. The user record and labels itself using the application in an unsupervised way and the samples collected are used to create the general model with all the users in our Spanish database.

When evaluation emotion recognition on video for SAVEE database (Table 2) we reach almost perfect test classification, 96.67%. Figure 2 illustrates the confusion matrixes for the different tests we performed. As it can be seen in the figure the states most difficult to differentiate are usually happy and angry followed by neutral and sad. The excellent results for video on SAVEE are due to the highly acted expressions of faces in this database (Table 3).

5   Conclusions

In this current work, we have presented an end-to-end Automatic Emotion Recognition system from audio-visual information. The systems use **variable-length input audio-visual sequences** and it is based on a **pre-trained model (TL)** that acts as high-level feature extractor that feeds a **BLSTM** the temporal information from original data.

Our evaluation results on audio show better or equal performance than state-of-the-art systems. But the model has been shown **robust** enough to perform **audio** emotion recognition in our Spanish scenario. Furthermore, we can affirm that it is capable of extract high-level relationships between audio-video information, in acted samples, according to each language and culture (among others external aspect).

The use of pre-trained models (**AlexNet**) has allowed us to address this complex task with less resources and deep networks. Therefore, we can conclude that our precision method could work without supervision in small and low resources scenarios. Results on video data also demonstrate the flexibility of the proposed architecture to be applied on multimodal scenarios. Regarding fusion methods, as stated in previous research as in [11], we have shown the need of more powerful architectures to manage the highly non-linear correlation between audio and video samples.

Acknowledgments. This work has been partially funded by the Spanish Ministry of Economy and Competitiveness and the European Union (FEDER) within the framework of the project DSSL: "Deep and Subspace Speech Learning (TEC2015-68172-C2-2-P)". We also thank ESAI S.L. Estudios y Soluciones Avanzadas de Ingeniería for their support and partially funding of this research towards an automatic emotion recognition system.

References

1. Brave, S., Nass, C.: Emotion in human-computer interaction. In: The Human-Computer Interaction Handbook: Fundamentals, Evolving Technologies and Emerging Applications (2002)
2. Sebe, N., Cohen, I., Huang, T.S.: Multimodal emotion recognition, **4** (2004)
3. Haq, S., Jackson, P.J.B., Edge, J.: Audio-visual feature selection and reduction for emotion classification. In: Proceedings of Internatioanl Conference on Auditory-Visual Speech Processing (AVSP 2008), Tangalooma, Australia, September 2008
4. Busso, C., et al.: IEMOCAP: interactive emotional dyadic motion capture database. Lang. Resour. Eval. **42**(4), 335–359 (2008). https://doi.org/10.1007/s10579-008-9076-6
5. Burkhardt, F., Paeschke, A., Rolfes, M., Sendlmeier, W., Weiss, B.: A database of German emotional speech, vol. 5, pp. 1517–1520 (2005)
6. Ayadi, M., Kamel, M.S., Karray, F.: Survey on speech emotion recognition: features, classification schemes, and databases. Pattern Recognit. **44**, 572–587 (2011)

7. Pini, S., Ahmed, O.B., Cornia, M., Baraldi, L., Cucchiara, R., Huet, B.: Modeling multimodal cues in a deep learning-based framework for emotion recognition in the wild. In: ICMI 2017, 19th ACM International Conference on Multimodal Interaction, ROYAUME-UNI, Glasgow, United Kingdom, Glasgow, 13–17th November 2017 (2017)

8. Harár, P., Burget, R., Dutta, M.K.: Speech emotion recognition with deep learning. In: 2017 4th International Conference on Signal Processing and Integrated Networks (SPIN), pp. 137–140, February 2017

9. Lalitha, S., Madhavan, A., Bhushan, B., Saketh, S.: Speech emotion recognition. In: 2014 International Conference on Advances in Electronics Computers and Communications, pp. 1–4, October 2014

10. Ding, W., et al.: Audio and face video emotion recognition in the wild using deep neural networks and small datasets. pp. 506–513 (2016)

11. Zhang, S., Zhang, S., Huang, T., Gao, W., Tian, Q.: Learning affective features with a hybrid deep model for audio-visual emotion recognition. IEEE Trans. Circuits Syst. Video Technol. **28**(10), 3030–3043 (2018)

12. Gu, Y., Chen, S., Marsic, I.: Deep multimodal learning for emotion recognition in spoken language. CoRR, abs/1802.08332 (2018)

13. Krizhevsky, A., Sutskever, I., Hinton, G.E.: Imagenet classification with deep convolutional neural networks. In: Pereira, F., Burges, C.J.C., Bottou, L., Weinberger, K.Q. (eds.) Advances in Neural Information Processing Systems, vol. 25, pp. 1097–1105. Curran Associates Inc. (2012)

14. Kahou, S.E., Michalski, V., Konda, K., Memisevic, R., Pal, C.: Recurrent neural networks for emotion recognition in video. In: Proceedings of the 2015 ACM on International Conference on Multimodal Interaction, ICMI 2015, pp. 467–474. ACM, New York (2015)

15. Tarnowski, P., Kołodziej, M., Majkowski, A., Rak, R.: Emotion recognition using facial expressions. Procedia Comput. Sci. **108**, 1175–1184 (2017)

16. Busso, C., et al.: Analysis of emotion recognition using facial expressions, speech and multimodal information. In: Proceedings of the 6th International Conference on Multimodal Interfaces, ICMI 2004, pp. 205–211. ACM, New York (2004)

17. Kim, Y., Lee, H., Provost, E.M.: Deep learning for robust feature generation in audiovisual emotion recognition. In: 2013 IEEE International Conference on Acoustics, Speech and Signal Processing, pp. 3687–3691, May 2013

18. Xi, O., et al.: Audio-visual emotion recognition using deep transfer learning and multiple temporal models. In: Proceedings of the 19th ACM International Conference on Multimodal Interaction, ICMI 2017, pp. 577–582. ACM, New York (2017)

19. Latif, S., Rana, R., Younis, S., Qadir, J., Epps, J.: Transfer learning for improving speech emotion classification accuracy, pp. 257–261 (2018)

20. Deng, J., Dong, W., Socher, R., Li, L.-J., Li, K., Fei-Fei, L.: ImageNet: a large-scale hierarchical image database. In: CVPR 2009 (2009)

21. Simonyan, K., Zisserman, A.: Very deep convolutional networks for large-scale image recognition. CoRR, abs/1409.1556 (2014)

22. He, K., Zhang, X., Ren, S., Sun, J.: Deep residual learning for image recognition. CoRR, abs/1512.03385 (2015)

23. Tran, D., Bourdev, L.D., Fergus, R., Torresani, L., Paluri, M.: C3D: generic features for video analysis. CoRR, abs/1412.0767 (2014)

24. Karpathy, A., Toderici, G., Shetty, S., Leung, T., Sukthankar, R., Fei-Fei, L.: Large-scale video classification with convolutional neural networks. In: CVPR (2014)

25. Zhang, K., Zhang, Z., Li, Z., Qiao, Y.: Joint face detection and alignment using multi-task cascaded convolutional networks. CoRR, abs/1604.02878 (2016)

26. Hochreiter, S., Schmidhuber, J.: Long short-term memory. Neural Comput. **9**(8), 1735–1780 (1997)
27. Yenigalla, P., Kumar, A., Tripathi, S., Singh, C., Kar, S., Vepa, J.: Speech emotion recognition using spectrogram and phoneme embedding, pp. 3688–3692 (2018)
28. Huang, Z., Xue, W., Mao, Q.: Speech emotion recognition with unsupervised feature learning. Front. Inf. Technol. Electron. Eng. **16**(5), 358–366 (2015). https://doi.org/10.1631/FITEE.1400323

Author Index

Printed in the United States
By Bookmasters